Municipal Wastewater Management in Developing Countries

Municipal Wastewater Management in Developing Countries

Principles and Engineering

Edited by

Zaini Ujang
University Technology of Malaysia

and

Mogens Henze
Technical University of Denmark

Published by   **IWA Publishing**
Unit 104–105, Export Building
1 Clove Crescent
London E14 2BA, UK
Telephone: +44 (0)20 7654 5500
Fax: +44 (0)20 7654 5555
Email: publications@iwap.co.uk
Web: www.iwaponline.com

First published 2006
© 2006 IWA Publishing

Chapter 4 published with agreement of Polyteknisk Forlag, Lyngby, Denmark

British Library Cataloguing in Publication Data
A CIP catalogue record for this book is available from the British Library

ISBN: 9781843390305 (hardback)
ISBN: 9781789065367 (paperback)
ISBN: 9781780402505 (eBook)

Contents

PART 2: TECHNOLOGY OPTIONS

Preface

This book addresses the state-of-the-art of wastewater management in developing countries, from the sustainable sanitation point of view. It covers conceptual, legal, management, financial, and engineering aspects. The target readers are postgraduate students, researchers, practitioners, consulting engineers and policy makers. This book project assembles professors and practitioners from both developed and developing nations. Four distinct characters of this book are as follows:

- State-of-the-art
- Integrated approach leading towards sustainable sanitation
- Examples on real case studies
- Examples on calculations related to the case studies

Zaini Ujang
University of Technology, Malaysia
Mogens Henze
Technical University of Denmark

Contributors

Tom Curtis Senior Lecturer, School of Civil Engineering and Geosciences, Newcastle University, UK. Email: Tom.curtis@ncl.ac.uk

Haniffa Abdul Hamid General Manager for Operation, Indah Water Konsortium, Malaysia. Email: haniffah@iwk.com.my

Mogens Henze Professor of Environmental Engineering & Head, Environment & Resources, Technical University of Denmark, Denmark. Email: moh@er.dtu.dk

Goen Ho Professor & Director, Institute of Environmental Science, Murdoch University, Perth, Australia. Email: g.ho@murdoch.edu.au

Robert Hughes Institute of Environmental Science, Murdoch University, Perth, Australia. Email: Robert.Hughes@murdoch.edu.au

Thorkild Hvitved-Jacobsen Professor, Department of Environmental Engineering, Aalborg University, Denmark. Email: i5thj@civil.auc.dk

Mohd. Ridhuan Ismail Deputy Director-General, Department of Sewerage Services, Ministry of Energy, Water and Communications, Malaysia. Email: ridhuan@ktak.gov.my

Blanca J. Jimenez Professor, Universidad Nacional Autonoma de Mexico, Mexico. Email: B.Jimenez@iingen.unam.mx

Wang Lin Professor, Water Pollution Control Research Centre, Harbin Institute of Technology, China. Email: l_wang@public.hr.hl.cn

Duncan Mara School of Civil Engineering, University of Leeds, UK. Email: d.d.mara@leeds.ac.uk

Kuruvilla Mathhew Institute of Environmental Science, Murdoch University, Perth, Australia. Email: mathew@essun1.murdoch.edu.au

Chongkrak Polpraset Professor & Dean, Urban Environmental Engineering Program, Asian Institute of Technology, Thailand. Email: chongrak@ait.ac.th

Eddy Soedjono Professor of Environmental Engineering, Institut Teknologi Sepuluh Nobember, Surabaya, Indonesia. Email: eddy_soedjono@its.edu.id

Zaini Ujang Professor of Environmental Bioengineering and Director, Institute of Environmental & Water Resource Management (IPASA), University of Technology, Malaysia. Email: zaini@utm.my

Jes Vollertsen Associate Professor, Department of Environmental Engineering, Aalborg University, Denmark. Email: jv@bio.aau.dk

1

Sustainable sanitation for developing countries

Zaini Ujang and Mogens Henze

1.1 INTRODUCTION

Ever since the Industrial Revolution in Europe, and rapid industrialisation in the United States and Japan, many innovations have been developed within water and wastewater management systems. The main purpose was public health and to protect the environment, in which water-borne diseases have been the major focus. The principal vehicle for the transmission and spread of a wide range of water-related communicable diseases is human excreta (Feachem *et al.* 1982). In 2000, the estimated mortality rate due to water sanitation hygiene-associated diseases (schistosomiasis, trachoma, intestinal helminth infections) was 2,213,000, about 1 million deaths due to malaria (United Nations World Water Development Report, 2003). In order to overcome water-related diseases (i.e. waterborne, water-washed, water-based, and water-related insect vectors) many approaches have been proposed over the past 200 years. In this book, the proposed approaches are known as "conventional sanitation".

Developing countries can be divided into two main categories, i.e. countries in transition with a higher growth rate and where industrialisation and urbanisation are taking place rapidly, and countries with a slower growth rate. Both categories have their own characteristics and should be discussed separately, especially for programmes or policy related to extensive capital investment, such as wastewater management.

1.1.1 Major trends

The chronological perspective on the trends of water pollution and control measures in developed countries are summarised in Table 1.1. However, due to limited information, the perspective presented is limited to modern civilisation. It is important to note that the earlier civilisations managed water and sanitation according to their needs and know-how. Copper (2001) illustrated that sewers have been widely utilised since the Mesopotamian empire (3500 to 2500 BC) and at King Minos' Royal Palace at Knossos, Crete, Greece (1700 BC). In Indus city of Mohenjo-daro, latrines and cesspools were widely used. The Ancient Greeks had public latrines which sewered to agricultural fields for irrigation and fertilising crops and orchards. Similar technologies, especially sewer networks, were also adopted in modern civilisations. For example, the Commission of Sewers in London was established in 1531. The utilisation of chemicals, which marks the demarcation line between earlier civilisations and modern civilisations was recorded in Paris in 1740, where lime was used for sewage treatment.

At the beginning of the 19th century, pathogenic microorganisms in sewage were the main public health concern that contributed to typhoid and cholera in Europe. Between 1848 and 1854, John Snow had proved the link between cholera outbreaks and water supply polluted by sewage. The counter measure was the sewer network. Sewage was properly collected from its generation points, sewered from the population and discharged to water bodies, mainly rivers and the sea. However due to high sewage loading, the aquatic systems became polluted with organic materials. As a result, sedimentation was introduced, followed by biological wastewater treatment, particularly using biological filters, to reduce the organic loadings. In 1916, the first full-scale activated sludge plant was built at Worcester, England. Similar plants were built in Soelleroed, Denmark and Rellinghausen, Germany in 1922 and 1926, respectively (Copper 2001). In the 1950s, public concern shifted to industrial wastewater, particularly heavy metals after the Minamata issue. The counter measure was to control the pollution on-site using various chemical treatments.

The trends focussed on eutrophication, particularly in Europe which required nitrogen and phosphorus control in sewage treatment plants. In the first decade of the 21st century, the issues of eco-hazards and micro-pollutants, such as endocrine disrupting chemicals (EDCs) became prominent in developed countries. EDCs, especially oestrogen mimics and androgen mimics, are harmful to the development of reproductive organs and function (Matsui *et a.l* 2001). The technical solutions are still in the infancy stage, but are heading towards membrane technology.

Table 1.1 Major trends of water pollution and its counter measures in developed countries.

Era	Issues/environmental impacts	Counter measures
Pre-1900s	Pathogenic organisms	Sewer network
1910s	Organic pollutants from sewage	Biological sewage treatment plant
1950s	Industrial waste; inorganics	Waste treatment on-site before discharge to sewer
1960s	Other organics: pesticides, fat & grease, colour, solvents	Advanced biological and chemical treatments
1970s	Eutrophication	Nitrogen and phosphorus removal in sewage plant
1980s	Odour, taste, colour	Membrane technology; activated carbon
1990s	Green house gasses	Biotechnology
Early 2000s	Micro-pollutants, eco-hazard	Membrane technology

1.1.2 Conventional sanitation approach

In general, the major trends of water pollution control have significantly contributed to the development of a "conventional sanitation" approach in terms of legal and financial frameworks, as well as technological enhancements. Literally, sanitation means the use of practical measures for preservation of public health associated with water supply, sewage and solid waste management, and other public health services. In the past few decades, sanitary engineers or public health engineers were mainly responsible for sanitation facilities, as well as public health medical practitioners. However, the rapid expansion of the subject matter has broadened the responsibility of the engineering disciplines, thus since the 1980s it has been better known as "environmental engineering". In general, the "conventional sanitation" approach, which has been fully developed in industrialised countries, consists of a few characteristics, as shown in Table 1.2.

Table 1.2 Characteristics of water and wastewater management systems in industrialised countries.

Items	Characteristics / Conventional practices
Water resources	Extraction of fresh water from protected areas
Water quality	Treatment of sufficient quality and quantity
Treatment objectives	Based on public health and risk assessment guidelines and standards
	Piping system with chlorine residual
Water distribution	Toilets, kitchen, bathrooms, washing
Generation of sewage	By-product of industrial processes, utilities, cleaning operations
Generation of industrial wastewater	Centralised system for sewer networking: combined or separate
Collection of wastewater	system
	Removal of pollutants: solids, organics, nutrients
Centralised wastewater treatment	Removal of micro-pollutants, eco-hazard
Advanced treatment	Sludge conditioning and disposal
Sludge management	Utilisation in landscape and agricultural irrigation (option)
Wastewater reuse (option)	Generally from grants from governments / taxpayers' money
Capital cost	Local governments' collection / local tax
Operating cost	

The scheme of the water and wastewater management system in industrialised countries, which is now considered the "conventional" approach as shown in Figure 1.1, is centralised in nature, using extensive sewer networks and big sewage treatment plants. Many cities and townships in developed countries have successfully implemented this "conventional" approach. This is also the approach which has been propagated in many manuals and textbooks as the "standard" to be followed, including by developing countries. It is based on the understanding that water should be supplied by a pipeline, and wastewater should be collected, transferred and treated before it can be discharged into the aquatic environment. The key factor in this centralised system is "control" using various water, wastewater and sludge treatments, as well as automation technologies.

However, such conventional approaches have at least four basic requirements:

- High capital cost to provide extensive sewer networks and wastewater treatment plants. The World Bank estimated that sewerage investment costs in eight developing countries ranged from USD 600 to USD 4000 per household, with total annual costs ranging from USD150 to USD 650 per household (1980 prices) (Mara 1996)
- High degree of legal enforcement to ensure proper connection from generation point to sewer networks

- High operating and maintenance cost. For example, the rehabilitation cost for a piping system in Germany was estimated to be in the range of 100 billion euros (Wilderer, 2001)
- In-house water and sewer connections in all premises in that particular catchment.

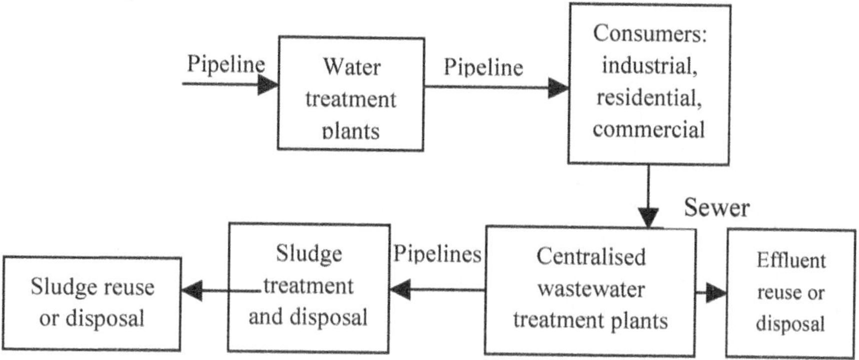

Figure 1.1 Water and wastewater management system in industrialised countries 'conventional approach'.

Despite the promotion and enhancement of science, engineering and legal frameworks on the conventional approach, 95 per cent of the wastewater in the world is released into the environment without treatment (Niemczynowics 1997). Only five per cent of the global wastewater was properly treated using "standard" sanitation facilities, mainly in developed countries. As a result the rest of the world's population are still at risk from water-borne diseases, and the quality of water resources has rapidly degraded, particularly in poor developing countries.

1.1.3 The challenge

The challenge now is to provide the world's population, especially the poor, with adequate water and sanitation facilities. Despite billions of dollars of investment spent every year, billions of poor people are still suffering and dying because of poor sanitation. At the beginning of this century, about 1.1 billion people lived without access to clean water (compared to about the same number in 1990), 2.4 billion without appropriate sanitation (compared to 2.3 billion in 1990) and four billion without sound wastewater disposal (WHO and UNICEF, 2000).

The challenge was translated into the form of the Millennium Development Goals for Water and Sanitation (MDGWS) proposed by the world community on many occasions. The goals are as follows:

- By 2015 to reduce by half the proportion of people who are unable to reach, or to afford, safe drinking water. This target was adopted at the United Nations Millennium Summit.
- By 2015 to reduce by half the proportion of people who do not have access to safe sanitation facilities. This target was adopted at the World Summit on Sustainable Development.

The future scenario will be that water resources are further depleted by a growing world population, which will be coupled with environmental degradation due to poor pollution control, particularly in most of the developing countries. At present, numerous efforts have been initiated in many developing countries to overcome environmental degradation. In Malaysia, for instance, besides new regulations and policies, a significant amount of investment has been directed into proper sewerage facilities, centralised hazardous waste treatment and disposal, incineration and sanitary landfill for solid and hazardous wastes, and river rehabilitation schemes (Ujang and Buckley, 2002). Other countries such as South Africa and Thailand have similar programs with special emphasis on the implementation of sustainable environmental resources. Waste Minimization Clubs, initiated by the University of Natal, South Africa are a good example of university-industry collaboration on pollution prevention and cleaner production (Barclay and Buckley, 2002).

1.1.4 Dilemma of developing countries

At the moment, many developing countries are adopting the "conventional sanitation" approach, i.e. to replicate the experience of the developed countries, without taking into consideration the funding availability, the development of technological know-how and the progress of environmental awareness among the general public. For example, in Japan the chronological development in environmental protection was mainly due to the progress in understanding among the general public on the nature of environmental and health problems, and the characterization of various pollutants. Thus the introduction of environmental standards for various kinds of pollutants and pollution control measures were chronologically introduced, roughly once every decade, as shown in Figure 1.2 (a). To control organic pollution from domestic wastewater in the 1960s, for instance, technologies such as activated sludge and biofilm systems were developed and properly implemented. The subsequent pollutants such as industrial wastes, nutrients, and hazardous chemicals were taken into consideration later in terms of environmental quality standards and pollution control (Ujang and Buckley 2002).

Figure 1.2 (opposite). Major trends of various water pollution components in developed and developing countries

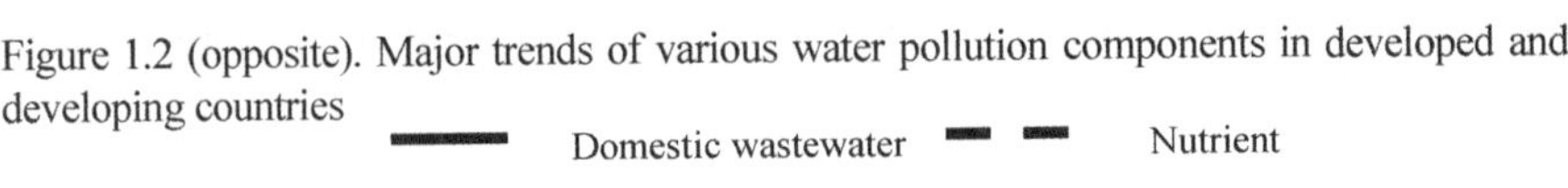

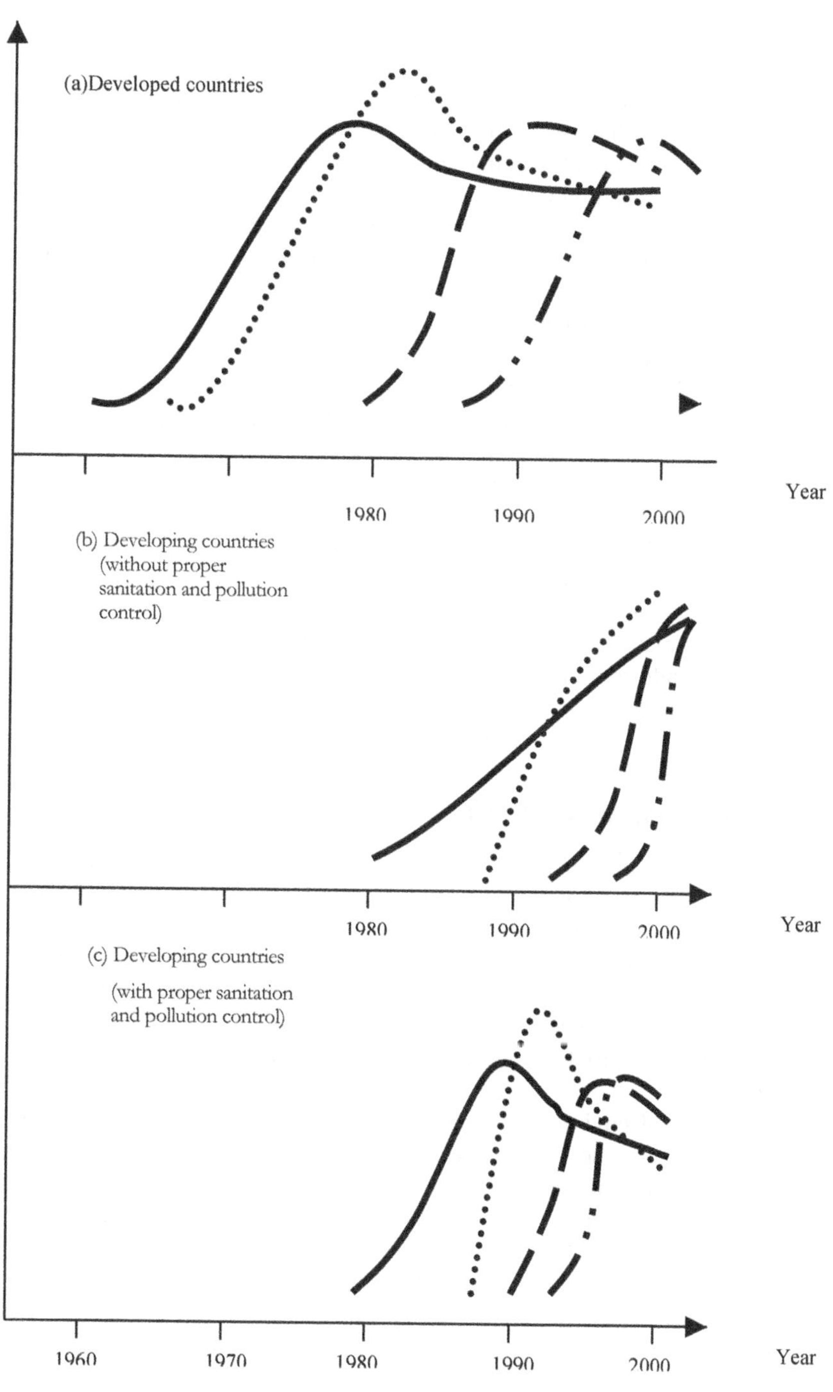
Strength of water pollution
(a)Developed countries
Year
(b) Developing countries
(without proper
sanitation and pollution
control)
1980
1990
2000
Year
(c) Developing countries
(with proper sanitation
and pollution control)
1960
1970
1980
1990
2000
Year

It is generally acceptable to say that the appropriate technologies are already available to treat the series of pollutants illustrated in Figure 1.2 (a), in a "package" form without waiting for the similar stages of implementation as experienced by Japan and other developed countries, i.e. "technology leap". At present many developing countries are experiencing water pollution phenomena as shown in Figure 1.2 (b), where the growth rate of urbanization and industrialization has rapidly increased from the 1980s to the present. The four classes of pollutants (domestic, industrial, nutrients and toxic) are occurring simultaneously and are not being detected sequentially. In this case, the environmental pollution control measures should not be limited to the chronological approach. The effluent treatment sequence should not necessarily mirror the historical trend in the developed countries. Many "advanced" technologies are not expensive if measured using a cost-benefit analysis taking into account the cost associated with public health, environmental quality, the eco-tourism potential and the sustainability of water resources as shown in Figure 1.2 (c). As the advantages of "cleaner production" are becoming more widely known, the link between economic development and environmental degradation is being broken. Thus many developing countries are heading towards "hi-tech" solutions especially in industries such as developing and applying membrane technology for the removal of micro-pollutants (heavy metals and EDCs) and ultra-pure water production for the electronic industry.

1.2 POOR DEVELOPING COUNTRIES

Many countries are facing a situation where economic activities are declining, leading to political instability and environmental degradation. In many situations, water resources are limited and the water quality is deteriorating, particularly in the case of Africa and South Asia. Water pollution issues, however, are not the main concern because other issues are more pressing such as national or racial security, food availability and epidemic control.

In general, the problems in these developing countries can be summarised as follows:

- Lack of environmental awareness among the majority of policy makers and the general public create a situation where water and wastewater management sectors are perceived to be less important than other sectors such as military empowerment, road improvement, electricity, mass education and health care facilities.
- Insufficient expertise, leading to gaps between ideal policies and implementation.

- Inappropriate policies on the conservation of water resources such as no legal requirements for the prohibition of deforestation activities in water catchments areas.
- Insufficient funding for water supply and sanitation programmes because of competing public expenditures due to rapid urbanization and population growth rate.
- Insufficient exploitable water resources especially in arid and urban areas.
- Inappropriate management systems and institutional support for providing water supply and sanitation facilities.

1.3 COUNTRIES WITH HIGH GROWTH RATES

Countries with high growth rates among developing nations have experienced rapid industrialisation and urbanisation, particularly since the 1980's. Both processes have been closely associated with environmental degradation, much of it caused directly by poor wastewater management. At present 48% of the world's population lives in urban areas; and it will be 60% in 2030 (United Nations World Water Development Report, 2003). Urbanisation is closely associated with high economic growth. In general urban areas with high rates of industrialisation provide the economic resources for water supply and sanitation services, as well as generating wastes.

In many cities in developing countries, the provision of a water supply and sanitation is dependent on the types of residential areas. Properly planned residential areas normally have sufficient levels (quality and quantity) of water supply, sanitation, road, electricity and other utilities. However, a densely populated squatter community is a common phenomenon in major cities, where both water supply and sanitation services are inadequate. People normally share toilets and pit latrines that are poorly maintained and dirty. The situation is worse in water-poor countries.

1.4 SOCIO-ECONOMIC CATEGORIES

In the past, many studies and publications have lumped developing countries together as low-income societies and most of the proposed solutions are only suitable for "poor" people. Thus many suggestions and proposals on the improvement of sanitation programme were not materialised, or partly failed because of several reasons as follows:

- The need for proper sanitation facilities is not only limited to poor people, but "rich" and government servants (who can be categorised as middle-income group) also demand such facilities. In many cases they are the policy makers and implementers of infrastructure projects.

Without taking their concern into account, the sustainability of any projects will definitely be reduced.

- Most of the publications, particularly those published by academics and international agencies are not representative in describing the structure of the sanitation programmes in developing countries. They are mostly targeted into (a) low-income rural or (b) slump areas in mega cities. No doubt that these are critical targets, however, there are also segments where sanitation programs are already available through private or public funding, but the framework is not sustainable. As a result, there is no model to follow.
- Many solutions designed for low-income societies propose free services – for both water supply and sewerage facilities. To support the services, the public is expected to maintain the facilities themselves. This framework is an ideal situation for covering capital cost, but such projects will be difficult to sustain because of covering operating and upgrading costs. Such frameworks also create internal conflict, especially with respect to "who should be getting what?" (Distributions, consumption, availability etc.).

In order to address the issue of water and wastewater management in developing countries it is necessary to take into consideration the segments of the society itself, particularly the types of housing areas. These segments will indicate the level of socio-economic, mentality and knowledge, which is important for any planned changes in life style and social engineering. It is also important to segregate the funding framework of any proposed projects. High-income urban communities, for instance are generally willing to pay for sewerage services and a higher water supply tariff, therefore a designated system can be accordingly provided. In Malaysia for instance, all new housing estates developed by the private sector (more than 98% of the total housing schemes) are required to provide sewers and centralised sewage treatment facilities, according to categories proposed by the Ministry of Housing and Local Authorities, the agency responsible for managing sewerage services. These categories reflect the sewerage tariff to be paid monthly (water supply is to be paid according to the level of consumption).

In view of the shortcomings in addressing the policy and operational framework of many proposals and solutions on sewerage services and management in developing countries, it is appropriate to categorise the communities in developing countries based on the types of residential areas (more discussion will be given in Chapter 3), as follows:

- Rural community
- Low-income peri-urban community

- High-income peri-urban community
- Low-income urban community
- High-income urban community

1.5 SUSTAINABILITY AND SANITATION

Sustainability means providing, supporting, encouraging and managing something to meet the needs of the present, and for future generations. In water and wastewater management, it is a need for "sustainable sanitation" because many professionals and policy makers believe that the existing framework with emphasise on the "conventional sanitation" approach will not be sustainable in nature, not only for developing countries, but also for developed countries. It is very costly, too extensive as a system, uses a lot of water, therefore it is only appropriate for water and money-rich developed countries.

Over the past 10 years, serious criticism has been given to the "conventional sanitation" approach, consequently many definitions, concepts and characteristics have been proposed for "a sustainable sanitation". Sustainable sanitation could be defined as a sanitation system which is technically manageable, socio-politically appropriate, systematically reliable, economically affordable that utilises minimal amounts of energy and resources with the least negative impact, and recovery of useable matters. The concept of sustainable sanitation emphasises three major components, i.e. (a) separation of pollutants at source, (b) decentralisation of facilities and (d) reuse of by-products, especially treated wastewater and sludge. The characteristics of sustainable sanitation is summarised in Table 1.3.

In general, the scheme for sustainable sanitation is illustrated in Figure 1.4. The emphasis is on separation at source, decentralisation and reuse of both effluent and sludge. It is also important to note that this scheme is also appropriate to treat and dispose of domestic solid waste together with sewage. In order to understand and benefit from the system, it is important to re-characterise domestic wastewater. The wastewater components, variation in person load and naming are illustrated in Tables 1.4, 1.5 and 1.6 respectively. The variations are due to the lifestyle and level of facilities available for water and sanitation systems. In many design manuals, person load is mixed up with the concept of population equivalent (PE). Normally the PE values widely used are: 1PE = 0.2 m^3/d and 60 g BOD/d, irrespective of the local person load (Henze *el at.*, 2002). Naming of wastewater is important to indicate the fraction of the household components that can be separated, treated and reused in a sustainable sanitation scheme.

Table 1.3. Characteristics of water and wastewater management systems within a sustainable sanitation framework.

Items	Characteristics
Water resources	Extraction of fresh water from protected areas; reuse from treated wastewater
Water quality	Treatment of sufficient quality and quantity; novel and decentralised facilities
Treatment objectives	Meeting health risk requirements
Water distribution	Piping system with chlorine residual; bottled drinking water
Generation of sewage	Toilets, kitchen, bathrooms; separation at source
Collection of wastewater	Centralised system for sewer networking: combined or separate system; on-site decentralised wastewater treatment requires no sewer network
Centralised wastewater treatment	Removal of pollutants: solids, organics, nutrients; reuse of effluent and sludge
Advanced treatment	Removal of micro-pollutants, eco-hazard; reuse of effluent and sludge
Sludge management	Sludge conditioning and disposal; reuse
Financial mechanism	Providing facilities based on the capability of the general public and industries
Capital and operating cost	Emphasis on cost-benefit analysis with reuse and recycling as requirements
Socio-cultural	Institutional requirements, acceptance and local expertise
Management	Clear policy on pollution control and water sustainability, human resource

However, in general the concept of sustainable sanitation is still in the development stage. Many ideas and arguments have been published and debated, but a concrete and standard definitive concept is yet to be agreed upon. The critical and ever-lasting questions are as follows: What are the detailed components of sustainability in terms of water supply and sanitation? What are the appropriate legal and institutional frameworks for sustainability, especially in countries where the political system is always unstable? How to embark on sustainable sanitation programmes in poor developing countries where the concept of self-financing is not well defined? Which types of technology could be considered as user-friendly and cheap but reliable and effective to control pollution? Can high-end and latest technology such as membrane bioreactors be considered as part of sustainable sanitation, as argued by DiGiano *et al.* (2004)?

The components of sustainability in water supply and sanitation can be divided into five categories as shown in Table 1.7, i.e. (a) policy and institutional, (b) economic, (c) environmental, (d) technical and (e) socio-political aspects. In order to achieve high levels of sustainability, it is a requirment to maintain high scores in all categories by providing a proper framework at least at 3 levels: community, Local Authorities and Federal Government. The ideal is that the three contribute according to their responsibilities. For example, community should maintain the sanitation infrastructures according to specifications, and pay the bills. Local Authorities should be responsible for authorising the systems and providing the necessary facilities such as land, electricity and water supply. Federal Governments should plan for the project, normally by providing capital investment. However, many facilities in developing countries, especially sanitation are provided by the communities themselves, without capital investment or support from Federal Government.

Table 1.4 Components of domestic wastewater (combined domestic and municipal sources; no industrial).

Major components	Specific items / interest	Environmental and public health effects
Micro organisms	Pathogenic bacteria, viruses and worm eggs	Health risk when bathing and eating shell fish
Biodegradable organics	Oxygen depletion in aquatic systems	Fish death, odours
Other organic components	Detergents, pesticides, fat, oil and grease, colouring materials, solvents, phenols, cyanides	Toxic, aesthetically unacceptable, bioaccumulation in food chain
Nutrients	Nitrogen, phosphorus, ammonium	Eutrophication, oxygen depletion in aquatic system
Metals	Hg, Pb, Cd, Cr, Cu, Ni	Toxic, bioaccumulation in food chain
Other inorganics	Acids, bases	Corrosion, toxic
Thermal effects	Heating water	Change aquatic conditions for flora and fauna
Odour and taste	Hydrogen sulphide	Aesthetically unacceptable, toxic
Radioactivity		Toxic, accumulation

Table 1.5 Variations in person load of domestic wastewater.

Components	Units	Level
BOD	g/capita.d	15-80
COD	g/capita.d	25-200
Nitrogen	g/capita.d	2-15
Phosphorus	g/capita.d	1-3
Wastewater	m^3/capita.d	0.05-0.40

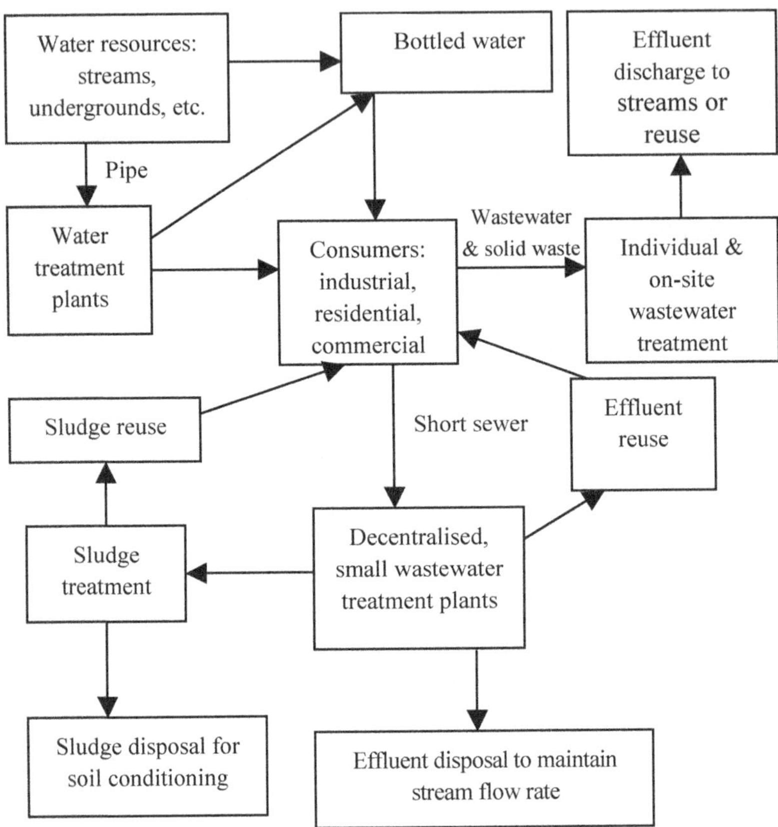

Figure 1.4 Sustainable sanitation approaches.

Table 1.6 Naming of components of domestic wastewater (from households) (Henze and Ledin, 2001).

Types	Contents / activities
Conventional sewage	Toilet, bathing, kitchen, washing
Black wastewater	Toilet (faeces and urine)
Grey wastewater	Bathing, kitchen, washing
Light grey wastewater	Bathing, washing
Yellow wastewater	Urine
Brown wastewater	Faeces

Table 1.7 Sustainability components for water supply and sanitation.

Sustainability components	Details	Example
Policy and institutional *To provide policy framework, strategy and motivation*	Federal implementing agency Local authorities Developers / Implementers Community	Ministry of Housing and Local Governments City council Water and sanitation companies Residential associations
Economic *To plan for capital investment, maintenance and cost recovery*	Investment of capital cost Affordability for operating cost Land usage Energy requirement	Direct Federal investment; private investment Water bills Public or private lands? Reliability of electricity supply
Environmental *To protect water resources and enhance pollution control*	Sufficient water resources River classification Effluent standards Advanced standards for recovery	Upstream of rivers; ground water Classes: Clean, moderate, poor Effluent for disposal Effluent quality for various options of re-use
Technical *To provide system and ensuring integrity*	Technical specifications, guidelines System approval procedures System reliability System complexity	Code of practice Based on standard design, or novel system Expected life-span of the system Adequate training for engineers, technicians
Socio-political *To involve local design, local supplier and contractor, and suitable with local culture*	Participation of local professionals Cultural perspective on sanitation Local taboos	Members, Association of Environmental Professionals Muslims require water for cleaning faeces and urine Faeces are taboo for any public debate amongst Asians

1.6 SUSTAINABLE SANITATION FOR DEVELOPING COUNTRIES

Sustainable sanitation is a relevant concept in order to achieve the goal of MDGWS by 2015 for providing adequate water supplies and sanitation for developing countries.

Sustainable sanitation is flexible in any community – poor or rich, urban or rural, water-rich or water-poor, and requires a lower investment cost compared to the conventional sanitation approach. It is also important to note that the framework of sustainable sanitation is much easier to adopt in developing countries where water supplies and sanitation infrastructures are still in the development stage. In some developing countries, no public facilities are available, and therefore it is an ideal situation to start a new infrastructure with a new framework.

REFERENCES

Barclay, S.J. and Buckley C.A. (2002) Promoting sustainable industry through waste minimisation clubs. *Wat.Sci.Tech.***46**, 79-86.

Copper, P.F. (2001) Historical aspects of wastewater treatment. In: *Decentralised Sanitation and Reuse*, P. Lens, Zeeman G. and G. Lettinga (eds). IWA Publishing, London.

DiGiano, F.A., Andreottola G., Adham S., Buckley C., Cornerl P., Daigger G.T., Fane A.G., Galil N., Jacangelo J.G., Pollice A., Rittman B.E., Rozzi A., Stephenson and Ujang Z. (2004) Safe water for everyone. *Water Environment & Technology*, June 2004, 30-35.

Feachem, R.G., Bradley, D.J., Garelick, H. and Mara D.D. (1982) *Sanitation and Disease: Health Aspects of Excreta and Wastewater Management*. John Wiley & Sons, Chichester.

Henze, M, Harremoës,P.,la Cour Jansen, J. and Arvin, E. (2002) *WastewaterTreatment-Biological and Chemical*. Springer Verlag, Berlin.

Henze, M. & Ledin, A. (2001): Types, characteristics and quantities of classic, combined domestic wastewaters. In: *Decentralised Sanitation and Reuse: Concepts, Systems and Implementation.* Lens, P., Zeeman, G. & Lettinga, G. (eds.), IWA Publishing, London, pp.59-72.

Magara (2001) Keynote address (slide presentation). *Asia WaterQual 2001*, September, Fukuoka, Japan.

Mara, D. (1996) *Low-Cost Urban Sanitation*. John Wiley & Sons, Chichester.

Matsui, S., Henze, M., Ho, G. and Otterpohl R. (2001) Emerging paradigms in water supply and sanitation. In: *Frontiers in Urban Water Management: Deadlock or Hope*, Maksimovics C and Tehada-Guibert, J.A.(eds). IWA Publishing, London.

Niemczynowics, J. (1997) The water profession and agenda 21. *Wat.Qual.Int.***2**, 9-11.

Ujang, Z. and Buckley, C.A. (2002) Water and wastewater in developing countries: present reality and strategy for the future. *Wat.Sci.Tech.* **46**, 1-9.

United Nations World Water Development Report (2003) *Water for People, Water for Life*. UN World Water Development Report and Berghahn Books, London.

Wilderer, P. (2001) Decentralised versus centralised wastewater management. In: *Decentralised Sanitation and Reuse*, P. Lens, Zeeman G. and G. Lettinga (eds). IWA Publishing, London.

World Health Organisation and UNICEF (2000) *Global Water Supply and Sanitation Assessment 2000 Report*.

2

Setting effluent quality standards

R. Hughes, G. Ho and K. Mathew

2.1 INTRODUCTION

This chapter is designed to provide information on the types of standards enforced in developing and developed nations for effluent discharge, water quality control, technological choice and subsequent protection of public and environmental health. It will first describe the types of standards used for pollution control and the problems encountered when implementing these in developing nations, as well as discussing approaches that are applicable to developing nations.

The classes of water and beneficial uses of water will be further discussed and the importance of maintaining standards for adequate sanitation and protection of water quality described. The final section will focus on the reuse of wastewater for water quality control and water resource management. This section will discuss standards important for reuse when compared to discharge.

Effluent quality standards are the basis by which authorities protect public and environmental health, by ensuring pollution control at the point of discharge and by allowing licensing of discharge (Enderlein *et al.*, 1997; Bishop, 2000). These standards are developed based on numerable years of research, compiled to produce values for effluent quality, which reduces environmental and public health impacts. The potential environmental and public health impacts are often evaluated based on the effluent inputs, such as the wastewater constituents, and the treatment technologies employed. In many developed nations the effluent standards are comprehensive in detail for particular environmental conditions and receiving environments, such as different soil and water body types (Bishop, 2000). Presently standards are available to regulate a variety of effluent reuse purposes (as opposed to traditional disposal) such as for food production and aquifer recharge (Bishop, 2000). Standards for reuse will be expanded further in Section 2.4, below.

The use of different standards, such as technology and discharge standards, can in some cases allow the enforcement of wastewater treatment systems at numerous stages, which ensures long-term pollution control. For example, wastewater systems may be firstly assessed during the feasibility development stage where they are scrutinised by the regional, state or national protection agency to ensure they can provide treatment to the standard discharge level. This entails assessing the capacity of a treatment system to provide effluent to the standard level when considering the influent composition and load. If a system passes the accreditation stage, it may be secondly enforced when installed by regular monitoring of the system to show if it is performing at an appropriate level. The monitoring of the system may include assessments of surrounding environments, such as groundwater, surface water, biota and vegetation. Typically, monitoring will show any site specific variations that may not have been accounted for in the design, or may show if the system is performing inappropriately due to poor maintenance or operator incompetence (USEPA, 1991).

The use of standards in many developed nations assumes many constraints, which are not uniform globally. These assumptions include:

- The availability of technologies to reach the standard values;
- The education of the designer to choose an appropriate system;
- The education of the personnel designing the monitoring program to ensure it is adequate to intercept any potential design faults or maintenance problems; and
- The education level of the enforcement personnel to assess the system and ensure it is monitored appropriately (Matsui *et al.*, 2001).

In most countries, standards have been developed by experts within the fields of chemical, environmental and engineering sciences to ensure comprehensive pollution prevention. This also ensures standards can only be implemented and enforced by highly educated personnel (UNEP, 2004). These factors may be missing in some developing countries (Matsui *et al.*, 2001), although there is often enough capital, either from grants or loan funds, to design wastewater systems, which can initially discharge at standard levels, using contracted personnel (Ali, 2002).

2.1.1 Current conditions

There are numerous reasons pollution control through appropriate sanitation systems is inadequate in developing nations. These may include a lack of political will and/or stability, poor technology choice leading to high operational and maintenance costs, and inadequate enforcement and adherence of standards (UNEP, 2002). One of the major factors leading to inappropriate standard adherence is the inability of authorities to develop standards appropriate for a country or region, this including local availability of resources (UNEP, 2004). In some regions where localised conflicts and civil war destroy any pollution control attempts, pollution control is a secondary priority for authorities and may be important during the rebuilding process.

In many developing nations the extrapolation of standards from developed nations, such as those used by the United States Environmental Protection Authority (USEPA) and the World Health Organization (WHO), is currently employed as a means to ensure that their public and environmental health is protected (Stephenson, 2001). Due to the high cost of standard production, lower socio-economic countries may not develop their own standards and instead use the available standards from other nations. However, the direct adoption of standards has been found to cause some significant problems, besides the obvious notion that these standards may not represent the local environment of these regions e.g. tropical compared to temperate (UNEP, 2004).

The standard values from many developed nations are a reflection of the socio-economic values of those communities and the desire to have a high standard of living, including low incidence of disease transmission and/or toxic effects of pollution. The standards in many developed nations may have an ecological focus by attempting to protect local ecosystems in a highly conservative manner e.g. 95% of all species. This has allowed many of the standards to have stringent values for wastewater constituents e.g. nutrients (UNEP, 2004). In most developing nations public health (e.g. potable water supplies) is of primary concern, and the use of standards aimed at conserving a high proportion of species within an ecosystem may not be appropriate, where the socio-economic status cannot afford such luxuries e.g. see Figure 2.1 below. Nevertheless, the allowable level of ecological impact should be assessed within local regulatory budgets because environmental remediation may have associated long-term costs and health risks (Enderlein *et al.*, 1997).

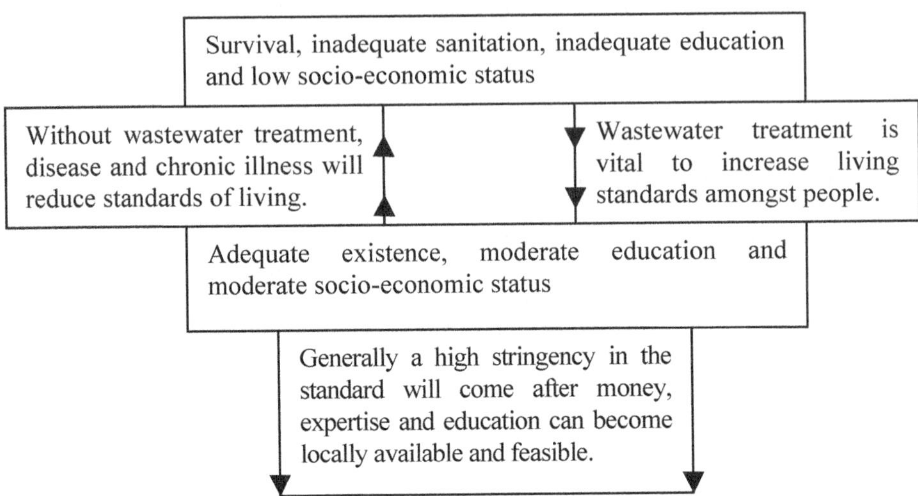

Figure 2.1 The depiction of problems when implementing standards for many developing nations.

An example of the high cost of wastewater treatment to meet the stringent standards from developed nations is shown with nutrient removal through conventional treatment. In this example the first 50% of nutrient removal is often cost-effective, the next 30-40% may or may not be cost-effective depending on the technology and processes involved, but the last 20-10% is extremely expensive and generally not cost-effective. In many developed nations (e.g. Germany, England and America) governments spend more money on wastewater treatment than many other amenities. Commonly, these costs are not fully recovered through water rates, and wastewater treatment is covered by subsidisation (Hoehn and Krieger, 2000). In this sense, understanding the importance of sanitation to a country and a government is highly important.

Another problem in developing nations which should be pointed out, is pollutant discharge due to an inability to enforce standards through monitoring numerous wastewater constituents (Dickens and Graham, 1998). As the standards have been developed by highly specialised personnel and assume that the personnel adhering to the standard have appropriate training, they can be impractical for many developing nations, which may not be able to support or supply such personnel. This problem becomes worse where a wastewater treatment plant is developed using loaned capital and an upgrade is not financially possible for many years. As the wastewater system degrades, either through time or by increasing populations and discharges from urban centres, the wastewater discharged will become more toxic. Therefore if little or no monitoring takes place the wastewater will be initially discharged at the standard level when the plant is built, but with time it will become worse. Hence, with inadequate monitoring the poorly treated discharge will not be detected until the environmental impact becomes significant e.g. disease transmission (Matsui *et al.*, 2001; Stephenson, 2001).

2.1.2 Setting realistic quality standards based on available resources

The approach of using unchanged standards from the developed world is detrimental to developing nations, as enforcement and adherence of the high stringency in the standards can be beyond the socio-economic status of the people. Due to this, standards in low socio-economic regions are commonly neglected (UNEP, 2004). In many developing countries, it may not be possible to develop independent standards for all site-specific situations. This is because the information is likely to be unavailable and is costly and time consuming to obtain through research. Hence, the development of standards in many developing nations should take an approach which gathers the relevant information from other regions of the world, and makes this information available to suit specific national criteria. This has been done in some developing nations (e.g. Nigeria) across the world where appropriate standards from countries with similar climates and environments have been used (Enderlein *et al.*, 1997). This approach is also currently undertaken by many developed nations with smaller populations, such as Australia, although these standards are very conservative in nature (Markich and Camilleri, 1997).

In developing nations, it is imperative that the standards must reflect the standard of living experienced by the people and appropriately incorporate realistic values that can be attained by the majority. These values need to be within the economic capacity of environmental protection and public health authorities that enforce them. The standards can therefore make available the greatest public and environmental health affordable to a community (UNEP, 2004). A further point that also needs to be addressed is the ability of technologies to meet these standard values, as without appropriate technologies, that can be maintained within locally available budgets, standards are likely to be unattainable in the long-term (Ali, 2002).

The development of standards may be done by following specific criteria such as:

- **The assessment of standard values to provide adequate protection.**
  The development of standard values to provide adequate protection involves assessing the criteria, which will be easily implemented by the population whilst providing adequate protection. The development of a national standard that has flexibility for changes in values due to different environmental classifications may offer an approach that is more affordable

for many developing countries. The national standard may include criteria for specific environmental classifications, either through the designated use of the environmental as is the case for water bodies (e.g. potable water supplies) or through the ecological significance of the area. Generally, the criteria based on the use of the area, is more easily implemented, as ecological classifications are highly conservative but offer the best level of protection. Where site-specific standard criteria are required the assessment may be carried out by industrial groups that place increasing pressure on wastewater treatment and which are profiting from the fast industrialisation of many developing nations (Gruttnera, 1997).

The standards should possibly highlight hazardous chemicals for specific areas if the information is available, which will allow more accurate monitoring of high-risk substances. Currently there are numerous inventories on hazardous and priority listed chemicals such as the International Register of Potentially Toxic Chemicals (IRPTC) developed by the United Nations Environmental Program (UNEP). These inventories help to identify potentially hazardous chemicals and can show the best routes of disposal as well as their environmental and human health effects. The use of inventories for industries within specific locations is an approach becoming more popular across the world especially where they discharge to municipal treatment systems. This will allow greater regulation of hazardous chemical use before they reach the wastewater system and subsequent receiving environment (Chave, 1997). Delisting of certain chemicals before they reach the sewerage system may be warranted in cases where technologies are not available to treat them adequately (Enderlein *et al.*, 1997). For example, it has been found that hazardous substances, such as branched alkylbenzene sulfonates which has been banned in Europe and withdrawn from the US market, have been found to persist in the Phillippines due to inadequate treatment technologies (Eichhorn *et al.*, 2001). Eichhorn *et al.*, (2001) noted that replacing such chemicals with readily biodegradable and environmentally friendly compounds would lessen any environmental impacts.

- **Developing standards within the enforcement means of an area.**
  The level of standard enforcement will be important in ensuring industry and private sectors, as well as the public sector are monitored for standard adherence by an enforcement body. The enforcement body may be a regional, state or national government department or a number of departments, which enforce different aspects of the standard e.g. department of health for microbiological aspects and environmental protection agency for chemical aspects. The level of training will obviously be a critical part of this assessment, as the technical expertise as well as the

equipment involved in enforcement can be hard to obtain (Ingallinella *et al.*, 2002). The improvement of enforcement should also be part of this assessment and may include a long term plan which aims to increase institutional development to a level that can provide the best standard implementation given the available monetary funds (von Sperling and Chernicharo, 2002). The use of different levels of government (e.g. regional, state and national) and different departments may also allow for an integrated approach. The integrated approach could involve bodies that are equipped with locally available personnel and equipment in the field of research such as tertiary institutions.

- **The feasibility of adherence by people living within an area.**
  This assessment is critical for standard adherence, as standard failure may be due to an inability to use treatment systems that reach the desired treatment level (Kendie and Johnson, 1999; von Sperling and Chernicharo, 2002). This factor involves the socio-economic status of an area and is linked with the cost of designing and maintaining appropriate treatment systems (Hoehn and Krieger, 2000). For example, the cost of private wastewater treatment systems (e.g. onsite systems) might be beyond local income for many lower socio-economic people and other factors might be priorities e.g. food, shelter and clothing. An example of this for local small-scale commercial enterprises may occur where the cost of commercial wastewater treatment is beyond the earning capacity of small business and therefore other means of disposal may be preferred e.g. illegal discharge into the sewerage system. Future industrial increase within an area and the potential for industrial wastewater discharge into a municipal treatment plant may also be included in this assessment. When it is known the standard values are within the economic capacity of businesses within an area, closer monitoring and licensing of industrial discharges may be warranted (Gruttnera, 1997; Kendie and Johnson, 1999).

- **Other factors that need to be addressed for future adherence and quality assurance**
  Educational activities are an important factor rather than pure enforcement, to ensure pollution control measures (e.g. sanitation) are valued within the community. The use of a demand driven approach versus supply driven approach is an ideal concept in view of this, as its basis is on education of the governing bodies and local community. The participation in development and maintenance of a sanitation project can be a major factor in the future maintenance and operational success of the scheme. Education on public health and the importance of appropriate sanitation levels may in some cases place a greater importance on environmental assurance and increase the money involved in developing appropriate treatment systems.

The future use of opportunity-driven approaches may be successful after demand approaches have been used to exemplify that sanitation is one of the most influential factors on the standard of living within a region e.g. catchment, state or nation (UNEP, 2004).

The use of educational institutions to specifically train engineers with critical local expertise is one approach that can increase the specialisation in developing nations. This will allow professionals to be available with the ability to develop and implement socially acceptable standards. These specialists may be used to direct the wastewater industry to a more self-sustaining format that uses local knowledge and experience but also draws on world wide research to implement locally feasible directives (Ujang *et al.*, 2004).

Other approaches may be warranted such as regulation through the use of specific treatment technologies that are known to treat wastewater to an acceptable level within an area. A similar scheme that follows this methodology is the best available technology (BAT) approach, which has been applied across the UK. When applied to lower socio-economic regions the approach may be termed the best available technology not entailing excessive cost (BATNEEC). Unlike BAT, the BATNEEC approach ensure the technologies chosen are affordable to the user, even where better technologies are available but too exorbitant for the user (Chave, 1997). These approaches aim to provide the best treatment available and may include the use of locally available cost-effective technologies (Ingallinella *et al.*, 2002). It should be noted that with such approaches, the technology has to be proven and its long term performance understood (von Sperling and Chernicharo, 2002). It is likely that technological approaches, especially the use of on-site systems, needs more research and better management criteria to ensure they can perform over long-term periods, although they offer a more cost-effective approach to standard enforcement than end of pipe discharge limits and *in situ* monitoring.

The BAT approach also involves the use of incremental technological steps to improve pollution control and subsequent environmental conditions e.g. water quality (Wannera *et al.*, 1996; von Sperling and Chernicharo, 2002). An example of this includes the use of primary technologies such as wetland (Shrestha *et al.*, 2001) and pond based treatment (Senzia *et al.*, 2003). When industrialisation and urbanisation are rapid, the technologies involved (e.g. wetlands) may suffer under toxic stress from accumulation of toxic compounds (e.g. heavy metals) and may not provide adequate treatment with the wastewater produced (Gatts *et al.*, 2003). Therefore, where increasingly rapid industrialisation occurs, it is important that industry provide some resources for the municipal wastewater systems and the systems are upgraded to suite the new wastewater composition and load.

However, the original primary systems may still provide stormwater or wastewater treatment from domestic sources.

In examples where industrialisation and urbanisation are slower primary technologies may still be useful and provide an economical means to ensure sanitation (Meutia, 2001). The presence of disease vectors such as mosquitoes should be an important consideration in such approaches e.g. wetlands. Mosquito breeding in wetlands has become a critical factor when designing wastewater systems and is heavily regulated in many areas of the world, because the mosquitoes may offer direct exposure to a disease (Stephenson, 2001). The improvement of these open systems to subsurface wetlands may provide an extra technological step by reducing disease vectors such as mosquitoes (Stott *et al.*, 2003).

2.2 WATER QUALITY

Water quality is the crucial aspect of water management which impacts the environment through ecosystem degradation and public health problems e.g. pathogen transfer. Drinking water quality is the most important aspect of water management and when potable water supplies are maintained it often leads to increased human health within an area. Presently over 1.1 billion people in the developing world have no access to safe drinking water and up to 80% of disease in developing nations is due to the contamination of drinking water. Providing these people with clean water is obviously the key to successful water management in developing nations.

2.2.1 Water quality criteria and standards

Water quality standards have been developed primarily to ensure that contaminants such as those present in wastewater are present in the environment at concentrations that do not adversely impact water quality and subsequent environmental features e.g. fish impairment. Numerous criteria are contained in standards for protection of different water classifications. These classifications can include surface waters such as aquatic, marine and estuarine ecosystems, aquaculture systems, and potable and industrial water supplies. Values for ground waters are also included within standards. Typically, water quality is measured with the use of an ambient (also termed 'environmental') standard.

Ambient standards relate to the maximum concentration of a particular constituent within a particular environment, such as the biochemical oxygen demand (BOD) within a river. The ambient standard is often used to ensure water quality and aquatic health is protected in an environment that receives multiple point source discharges, or has a high significance value for the population e.g. potable water supplies. Moreover, ambient standards may be used where there are unknowns in respect to the assimilation capacity of an environment or where an ecosystem has been degraded through pollution. They can further be used where diffuse source pollution is reducing the assimilation capacity of an environment in conjunction with point sources (USEPA, 1991).

The original standards such as the US Water Protection Act specify protection using appropriate criteria such as nutrient levels, dissolved oxygen (DO), suspended solids (SS) and faecal coliform counts. The initial criteria have changed overtime to include assessment of the presence of persistence organic compounds including pesticides, fungicides and herbicides, many of these such as DDT have been banned due to there highly persistent nature. Currently newer criteria are being developed such as those for pharmaceutically active compounds, bisphenol A and hormones such as estradiol. These micro-contaminants have known endocrine disrupting properties and can cause organism impairment in the part per trillion range (Ternes, 1998; Xia *et al.*, 2001; Ternes *et al.*, 2003).

2.2.2   Scientific basis for development of criteria and standards

The science used to derive the figures in water quality standards is from experimental research that is aimed at finding the concentration (or unit) of a water contaminant that does not affect an aspect of the environment e.g. human and biological health, and ecosystem functioning. The concentration of the contaminant within the standard is often related to the exposure pathway, such as through secondary contact (e.g. recreational use of water) or direct uptake (e.g. drinking water) (Anon, 2000; Stephenson, 2001). For many of the pathogenic organisms other factors besides the type of exposure are important in determining the human health risk. These may include the ability of the pathogen to survive within a particular environment and the method in which they are best transferred. The detection of pathogens due to the size of the bacteria involved in disease transfer is commonly identified with the use of an indicator such as *Escherichia coli* (Stephenson, 2001).

The science used to derive non-pathogenic contaminants has placed an importance on the integration of humans within the environment, as often changes in environmental conditions can lead to human health disruption, even where the concentration of a parameter has no direct impact on human health. This can occur at a miniscule concentration, as for instance with low loadings of inorganic phosphates (primarily orthophosphate) into oligotrophic lentic water bodies, resulting in subsequent eutrophication and blooming of toxic cyanobacteria. Such a problem may be spurred on in some regions by the spring bloom phenomenon.

The types of experimental research used to derive non-pathogenic standard values for water quality criteria may include; acute and chronic toxicological studies either within laboratories or within the environment; site specific assessments using biological indicators or physico-chemical parameters; and assessments of the assimilative capacity of surrounding environments. Where little or no information exists for a particular contaminant standard values are derived with reference to experts within the medical and environmental fields. Typically, standard values that are derived without considerable research are conservative in nature (UNEP, 2004). Similarly, standard values that are not site specific, which account for natural changes in different environments are also conservative. Presently standard values may include some of the information that can account for the transformation of toxic constituents within the environment, such as changes in a constituents form (e.g. metal species) with pH and organic matter concentrations. Standards are starting to place an importance on the toxic forms (bio-available form) that a constituent occurs in, rather than total values that may be over conservative.

2.2.3 Classes of water

There are numerous classes of water. Commonly each class of water is specified based on indicators of water quality. The indicators include faecal coliform counts (some include more specific indicators e.g. *E. coli*) and BOD and/or COD, dissolved oxygen, nutrient, SS and chlorophyll 'α' concentrations. The number of classifications may represent the types of uses of water within a region e.g. recreation, agriculture, aquaculture, drinking water etc (Stephenson, 2001). Table 2.1 below has a brief description of the values for each class of water.

Table 2.1 An example of the parameter guidelines used to classify different types of water.

Water Class	TN (µg/l)	TP (µg/l)	pH	DO (mg/l)	BOD (mg/l)	SS (mg/l)
Class 1	<300	<10	6-8	8	2-3	10
Class 2	300-750	10-25	6-8	6	4	30
Class 3	750-1,500	25-50	6-9	4	7	80
Class 4	1,500-2,500	50-125	6-9	3	20	100

Source: (SoER, 2001).

Class 1 is the highest level of water quality and has been set aside for drinking purposes and high conservation natural reserves. Discharge of wastewater into class 1 water is generally not acceptable, although in some locations discharge may be allowed where the wastewater is treated to the same level as the drinking water e.g. tertiary treatment. This type of water represents a natural water body that has not been impacted by anthropogenic activities e.g. diffuse and point source pollution.

Class 2 refers to water that can be used for secondary contact purposes such as recreation. This level of water quality may also be useful for maintaining aquatic health in areas of conservation although a higher classification may be warranted where lotic oligotrophic water bodies are present. Class 2 water is suitable for most food production purposes such as agriculture and aquaculture.

Class 3 refers to water that can be used for agriculture and industrial purposes. For agricultural purposes, the water may have to be assessed for ammonia concentrations and pathogen transmission possibilities. Disinfection should be provided at this level. Class 3 water is typically not suitable for sustaining aquatic health.

Class 4 refers to water that is substantially polluted and cannot support much aquatic health. Preferably, water should not be allowed to reach this level of pollution. Water at this level may have some beneficial uses (e.g. non contact plantation irrigation) but should be assessed for risks, and disinfection and/or treatment before use is advised.

2.3 CRITERIA FOR DISCHARGE

Discharge criteria have been devised to stop impacts on receiving environments deemed unsuitable by the enforcement body. These criteria are contained in standards, which allow assessment of the treatment process to reach specific discharge criteria and assessment of actual discharge based on load. Common types of discharge standards with specific criteria are detailed below:

Technological Standards (treatment based)
The use of technological standards is often employed to ensure wastewater treatment systems can provide a minimum level of treatment within an area. They can also be used to assess which technologies are suitable for an area by having a set treatment level (UNEP, 2004). An example of this is the use of secondary treatment in the US as the minimum treatment level for all municipal wastewater systems. This ensures that the treatment process employs the minimum treatment process (e.g. stabilisation pond) to remove SS and BOD. Similar to load based discharge standards, the level of treatment (e.g. primary, secondary and tertiary) is linked to the concentration of wastewater constituents

after treatment. For example, at the secondary treatment level BOD and SS should not exceed 30mg/l over a 30-day period and the pH of the wastewater should be between 6 and 9.

Discharge Standards (load based)

The standards are derived by assessing the assimilation capacity of receiving environments (e.g. class of surface water), disease transmission probabilities and the potential impacts to abiotic and biotic factors (Stephenson, 2001). The discharge standard will measure the quantity and quality of effluent discharged into the environment, thus allowing monitoring of pollutant loads over a particular period of time. The load will then be assessed against tolerable limits for a receiving environment and any pollution outside the tolerable limit for that ecosystem calculated. If the discharge is outside the tolerable limit, fines may be implemented to the polluter (see 2.3.4 Discharge licences for water quality control for more details).

These standards originally measured the organic loading to water bodies by assessing the BOD and SS. Currently standards may incorporate values for nutrients such as ammonia, total nitrogen (TN) and total phosphorus (TP), and total dissolved solids. Nutrients are measured because they are known to cause significant water quality problems, such as eutrophication, which can deoxygenate water bodies (Enderlein *et al.*, 1997; Bishop, 2000). Where considerable industrial effluent is discharged to sewer without adequate treatment, an emphasis may be placed on constituents that are highly concentrated due to production or refinement of a product e.g. chromium in tanneries. Often industrial effluents will be assessed for heavy metals and hazardous organic compounds. Effluent toxicity may also be assessed using whole effluent toxicity or inhibition testing with a suitable bioassay species.

2.3.1 Assimilative capacity of receiving environments

The assimilative capacity of an environment refers to its ability to receive foreign substances and remain in a relatively healthy state. In terms of municipal wastewater management, the receiving environment is commonly a surface water body and its ability to assimilate treated wastewater will be detected through changes of water quality and aquatic health. Changes in the environment are assessed against the natural variations that an area experiences (Dickens and Graham, 1998). The assimilation of wastewater in a surface water body will change in reference to the discharge point, where the discharge point will show greater change (and less assimilation) due to higher concentrations of wastewater. Beyond the discharge point, the wastewater will form a mixing zone. The mixing zone then extends in a gradient (i.e. dilution gradient) from the discharge point and finishes where the original surface water and effluent no longer show any differentiation with respect to natural water quality and aquatic health. The wastewater will have been fully assimilated where no further natural change from the discharged wastewater is present.

The physical assimilative capacity of an environment can be determined by assessing the types of receiving environment e.g. ocean, river, lake. Different receiving environments have different assimilation capacities, which are often related to the natural level of a particular substance the environment has adapted to over time. For example, coral reef ecosystems have adapted to very low nutrient and suspended solid levels and rely on clear water for the corals and associated zooxanthelae to survive. Therefore, discharging low levels of either nutrients or suspended solids into a coral reef ecosystem is likely to cause singificant impacts through algae blooms and subsequent smothering through suspended solid and algal deposition.

An important factor in assessing the assimilative capacity of an environment is the types of constituents in the wastewater, as the higher the toxicity of the wastewater, the greater its affect on receiving ecologies. The toxicity of the wastewater and its assimilation in the environment can be assessed using biological organisms. For example, it is common for macroinvertebrate assemblages to be dominated by polychaete species, such as *Capitella capitata* and *Mediomastis californiensis* in the immediate areas of heavy metal contaminated effluent discharge. However, further away from the discharge point it is common that other species start to show a presence until the point is reached where other species begin to dominate and the assemblage becomes more diverse. The point where other species begin to dominate and there is a diversity of species is typically the area where appropriate assimilation of heavy metals has occurred (Warwick and Clarke, 1998; Warwick *et al.*, 2002).

The assessment of the receiving ecologies requires considerable effort, such as the assessment of benthic communities, aquatic ecology and chemical parameters of the water. This requires high costs and expertise to be carried out effectively and other means such as rapid bioassessment using whole effluent toxicity (WET) tests or bioindicators of aquatic health may prove worthwhile (Dickens and Graham, 1998). These indicators show changes in the mixing zone and can show where aquatic health is impaired. Dickens and Graham (1998) noted that in southern Africa the use of bioindicators such as macroinvertebrates was the only appropriate means to assess riverine health and showed that the discharge of inappropriately treated wastewater lead to poor stream quality. The use of some species, such as the mussel *Mytilus edulis*, can also show variations in wastewater constituent loading over a period of time (Chafika *et al.*, 2001).

2.3.2 Relating discharges to assimilative capacity of receiving environments

Relating the effluent discharge to the assimilative capacity of receiving environments is the most appropriate method to ensure that discharge limits are not over or under conservative (USEPA, 1991). Relating the discharge and assimilative capacity of an area requires the use of knowledge on the assimilative capacity of the receiving environment. This knowledge is derived from research, which shows what level of a contaminant significantly affects water quality and aquatic health within a particular ecosystem. For example, research has shown that some ecosystems may need wastewater to be treated to a tertiary level, due to their highly sensitive nature (e.g. coral reef ecosystems or clear oligotrophic lentic water bodies), whereas other ecosystems (e.g. muddy mesotrophic lotic water bodies) may not require such high standards of treatment and may require only secondary treated wastewater. Standards for discharge often contain information on discharge limits for particular environments, which have been produced to allow licensing of discharge.

The loading of wastewater into a particular environment will change over a period of time however and it is important that the overall discharge is factored into the assimilative capacity of a receiving environment. In this case, the assimilative environment may be given a particular load (within a standard as described above) that it can assimilate over a period of time. The measurement of discharge over a period of time is then modelled using software such as the common 'wasteload programme'. The modelling of the wastewater discharge in this way shows whether the discharge is above or below the assimilative capacity of the receiving environment (Mahajana *et al.*, 1999).

The long-term assimilation of persistent wastewater constituents, such as heavy metals, within the environment is another factor to be considered in some wasteload applications. These substances may persist in the environment (e.g. sediments) until major disturbances (e.g. dredging) lead to significant water pollution.

2.3.3 Setting effluent standards with multiple discharges

Setting standards for multiple discharges ensures the assimilative capacity of an environment is within the limits of all discharges. It is important for multiple dischargers to assess whether the mixing zone of the effluent representing the area where the wastewater is being diluted overlaps. If there is no overlap of mixing zones and the wastewater from each discharge is assimilated into the environment, regulation as would occur in a single discharge situation maybe warranted. However, when the mixing zones overlap, other methodologies are needed, to ensure pollution does not exceed the receiving environments limit. Such a problem may be compounded by diffuse pollution in an area.

The use of technological and discharge standards to reduce problems of water pollution can be commonly applied to ensure all dischargers meet the required level of treatment for an area. The use of a discharge standard may also allow the licensing of dischargers to ensure they meet the required environmental loading criteria. Monitoring the load of discharge for all dischargers will show where any pollution sources come from and if the load is above the allowable limits.

The use of *in situ* instrumentation may be used to assess the site-specific environmental effects within the mixing zones. This can show whether the water body is receiving loads of wastewater toxic to the aquatic health in the area. The use of WET tests can also be used for measurement of water quality in the mixing zone areas and can rapidly show whether the wastewater discharged into an area is likely to cause any significant aquatic health problems.

2.3.4 Discharge licences for water quality control

The use of discharge licences is common practice in municipal wastewater management and has principally been established due to pollution prevention acts, which specify that toxic or noxious wastes should not be discharged into the environment or controlled waters in a way that impairs the receiving water body. Examples of these acts include the Canadian Ontario Environmental Protection Act and the UK Rivers Act. As the discharge of wastewaters contravenes such laws, a licence for discharge is often required for discharge into receiving environments (Chave, 1997).

The discharge licences for water quality control are used to ensure the discharger does not exceed environmental tolerance limits and allows enforcement of tolerance limits. The limits of discharge are usually calculated on a load approach rather than concentration units, as it can show the overall load discharged into receiving environments (Mahajana *et al.*, 1999). Discharge loads can be measured using a number of modelling approaches that show the load over a period of time, including fluctuations e.g. peaks. The models are based on software such as wasteload described above. The models show the daily allowable limit and monthly allowable limit, which can show the acute and chronic discharge concentrations.

Regulation based on monitoring discharges can be used to implement enforcement, by allowing the authorities to assess if the discharge is within an acceptable load for the assimilative capacity of the receiving environment. If the discharge has been shown to be above the acceptable load (including tolerance for discharge load variations), then fines may be charged, allowing a 'polluter pays system'. The fine amount and extent of enforcement may depend on the level of the offence and the classification of the receiving environment as well as the likely remediation costs brought about by the offence.

The modelling of wastewater loads requires specific criteria to be in place to allow accurate measurements such as the design of the monitoring program and the equipment used for monitoring and modelling the wastewater flow and composition. The types of constituents monitored will include traditional parameters such as SS, BOD and pH, but may also include nutrient, heavy metal and organic compound concentrations depending on the types of sources that discharge into the wastewater treatment system. The use of WET testing may also be incorporated in the discharge licence to ensure the wastewater does not exhibit greater than expected toxicity, as measured by a standard endpoint e.g. no observed effect concentration (NOEC) or lethal concentration (LC). The WET tests are often employed where there are unknowns with respect to wastewater constitution and concentration, as many wastewaters are complex in nature and may have numerous sources of influent e.g. small industries.

2.4 CRITERIA FOR REUSE

Wastewater reuse is becoming a popular approach to turn the useful components of wastewater into a resource. Wastewater can be reused for domestic, industrial, agricultural, aquacultural and natural flow allocation purposes (Nhapi *et al.*, 2002). The high value of clean freshwater ensures that reuse is a viable option for pollution control (Emongor and Ramolemana, 2004). Using clean water for purposes that do not need potable water is wasteful. Wastewater may have greater benefits than clean water due to the higher nutrient levels, which can bolster crop yields or increase the efficiency of the aquacultural process. It is estimated that up to 10% of the world population consumes food grown using wastewater (Smit and Nasr, 1992). Applying the useful wastewater components for such applications can protect surface waters, where technology is unavailable (e.g. too expensive) to treat the wastewater to acceptable nutrient loads before discharge. The use of wastewater can also reduce fertiliser demand in agriculture and effectively close the nutrient cycle, by allowing nutrients from food to be recycled back on to agricultural lands (Nhapi *et al.*, 2002).

The reuse of wastewater is obviously more appropriate than discharge at resource utilisation, but has a number of constraints. These include:
- The availability of technologies to treat the wastewater to the desired reuse standard (e.g. pathogen inactivation levels)
- The available space or industry to reuse the wastewater
- Knowledge on the appropriate reuse of wastewater, especially when using it for multiple reuse purposes (Lawrencea *et al.*, 2002)

- Availability of standards and criteria to appropriately reuse the wastewater in different environments (Gallegosa *et al.*, 1999).

Wastewater treated to the appropriate level is termed 'reclaimed water'. In most cases it is important to match the desired treatment level with the desired usage (Bahri, 1999). For example, it may be appropriate to inactivate more pathogen before urban reuse on public open space, than it would be to reuse wastewater for agriculture in a rural area, where low contact is evident. Hence, the desired treatment level should leave the important components at the right concentration for that use and remove the concentrations of components that may lead to pollution or potential disease transmission. Such an approach will reduce the risk in any reuse scenario (Mara, 2001). Currently, however, in many parts of the world wastewater may not be treated properly, if at all, before reuse, especially in peri-urban areas (Parkinson and Tayler, 2003).

2.4.1 Reuse as a means of water quality control

The reuse of wastewater for water quality control is aimed at providing best practice water resource management. In this context, the wastewater that matches the criteria of the reuse purpose can be applied to that sector of water management (Bahri, 1999). This involves assessing the constituents of the wastewater with the reuse purpose requirements and available treatment technologies (Mara, 2001). In some cases, a low level of treatment may be required for the reuse purpose, although this is generally not recommended for long-term situations. However, where there are available technologies the wastewater may also be applied to any area of wastewater reuse, especially where there are restrictions on potable water supplies and the economy can afford to reuse the wastewater.

The best approaches for water quality control involve onsite or source treatment, before mixing of different wastewater streams (Iwugo *et al.*, 2002; Jonsson, 2002; Maurer *et al.*, 2003). After mixing, the treatment required is often more costly and less effective, especially within some processes such as tanneries which may mix chromium contaminated wastewater into the main stream (Konrad *et al.*, 2002). In these cases, a cleaner production approach that assesses each process and finds methods to recycle the wastewater before the next process may be required. Such an approach allows wastewater recycling within an industry itself and thus reduces potable water consumption (Gumbo *et al.*, 2003).

An example of this was shown in Bulawayo, Zimbabwe by Gumbo *et al.*, (2003) who studied 3 different industries, a wire galvanising company, a soft drink manufacturer and a sugar refinery. The authors found that all industries could save costs with water reuse within production cycles. For the wire galvanising company it was found that 17% of the water could be recycled through the hot quench water-cooling system. For the soft drink manufacturer, water intake could be reduced by using filter backwash through the water treatment plant, and for the sugar refinery the use of a heat exchanger system could save 57 m^3 of water per 100tonne of sugar. Gumbo *et al* (2003) noted that although developing countries can often use outdated technologies updating them with cleaner production processes, can lead to greater environmental and financial benefits.

Separating the most toxic fractions of wastewater may allow recovery of important constituents, and ensures that technologies used for treatment of the toxic fractions can be more technologically advanced due to a reduction in quantity of influent. The technology may also be less expensive to maintain. Additionally, the separation of the toxic fractions ensures larger municipal plants, that rely on biological treatment, experience less biological inhibition (Tare *et al.*, 2003).

The reuse of wastewater is effective at reducing environmental pollution, but may in some instances create problems when the reuse system is designed inappropriately for the types of environment. This can be especially the case for natural environments that involve soil and water processes. For example, if a designed agricultural reuse system has been overloaded with nutrients and water it can lead to groundwater contamination and pollution of surrounding environments. To ensure best management practice, standards using appropriate buffer distancing, environmental condition constraints, including loading rates and risk assessment, need to be in place.

2.4.2 Water quality criteria for reuse

The main objectives of reuse criteria are to maintain water quality for the desired application that reduces the risks to the users of the effluent (e.g. farmers, transporters and sales) and consumers of the product. The risks involved may include pathogen transfer via the food source or direct uptake from the reuse application. The standards for reuse are newer developments than typical discharge standards with the first comprehensive reuse standard being completed by WHO in (1989). Table 2.2 below exemplifies the criteria associated with reuse of water in a few different types of applications. This example is a simplified version that is used to illustrate the principles behind reuse and more quantified values may be located in actual guidelines. The US EPA has recently developed a new manual reuse for wastewater in 2004 and WHO are currently about to realise the next reuse guidelines following the revision of the original 1989 document.

Table 2.2 The level of risk when different levels of treatment are applied to wastewater (from WHO 1989).

Control measures	Field or pond	Crop	Inhibitory measure between production and consumption	Consumer
	Level of contamination			Level of risk
No protective measures	High	High		High
Crop restriction	High	High		Safe
Application measures	High	Safe		Safe
Human exposure control	High	High		Low
Partial treatment in ponds	Low	Low		Low
Partial treatment by conventional methods	Low	Low		Low
Partial treatments in ponds, plus crop restrictions	Low	Low		Safe
Partial treatment by conventional methods, plus crop restrictions	Low	Low		Safe
Partial treatment, plus human exposure control	Low	Low		Low
Crop restriction, plus human exposure control	High	High		Safe
Full treatment	Safe	Safe		Safe

The types of criteria for reuse are often similar to water use classifications. This has involved allocating different treatment standards for different reuse approaches, such as the different treatment requirements between surface and subsurface irrigation. Buffer distances between the reuse area and public access, drinking wells, bores and natural areas are also maintained to ensure pollution does not diffuse into these areas. Reuse criteria typically specify monitoring of surrounding surface and groundwater areas to ensure that wastewater constituents are not transmitted beyond the application zone to areas where pollution may occur.

The criteria for reuse for some applications can be different from disposal standards for particular environments, especially for soil. The soil environment tends to act less as a dilution substance and more as an aerobic filter for many wastewater contaminants and has the capacity to lock up phosphorus and remove nitrogen with associated micro-organisms and invertebrates e.g. earthworms. The reuse of wastewater on soils has been derived using classification of soil types corresponding to different grain shapes and sizes, infiltration rates, nutrient immobilisation capacity such as the phosphorous retention index (PRI) and cation exchange capacity (CEC).

The infiltration capacity of the soil is assessed in most site assessments. This allows application rates to be set below infiltration rates when wastewater is applied to stop ponding and surface runoff. An understanding of the soil chemistry with the wastewater composition is also required. For example, soils with a high CEC may have impaired infiltration rates when wastewater with high sodium ion concentrations is applied. In this example the soil absorbs sodium ions from the wastewater to saturation, causing the soil to swell and clog pore spaces (Patterson, 2001).

2.4.3 Reuse for forestry, agriculture (including hydroponic systems), horticulture, aquaculture, polyculture

Designing a scenario specific wastewater reuse system requires some careful consideration. A site and soil assessment using specific criteria such as those mentioned above is essential to ensure no public health risks occur or surrounding environments are affected. In addition to this, the design should ensure the wastewater can supply enough nutrients or desired wastewater contaminants at the right area for efficient uptake and use by the system (Gallegosa *et al.*, 1999).

The design of plant based reuse systems involves the same aspects of design as would occur for irrigation with other waters (e.g. bore water), although the use of additional fertiliser either through fertigation or ground application is reduced in a wastewater reuse scenario. The factors that need to be considered are the water requirements of the plants, the evapotranspiration and precipitation rates of the area e.g. water balance approach. The reuse of wastewater for agriculture and horticulture will generally involve either the use of subsurface (e.g. drip) or surface irrigation. In both instances, the wastewater should be treated to at least a secondary standard with surface irrigation requiring disinfection, unless additional tertiary treatment is available.

The reuse of wastewater for forestry can involve wastewater treated at the primary level, but largely depends on the distance of the forest to human contact zones, surrounding water bodies and ground waters, and the soil classification. The application of primary treated wastewater to a forest can involve the use of trenches, designed to carry significant loadings of BOD and SS. Subsurface irrigation through drip lines and trenches (e.g. 100mm piping surrounded by coarse aggregate) may be applied to forests where secondary treatment is available and human contact is possible. Surface irrigation may also be warranted in some situations.

Other factors that need to be considered with wastewater reuse on plant systems are the nutrient requirements of the plants, the nutrient adsorption capacities of the soil and nutrient imbalances that may occur from different plant requirements and nutrients in the wastewater. This can be found by assessing the plants essential requirements, including changes with season due to flowering and fruiting (e.g. high nutrient application may reduce fruit yields if applied at the wrong period) and the ability of the soil to store nutrients. Amendments with nutrient adsorbing materials, such as red mud, bentonite and zeolite, should be considered if possible to ensure the nutrients remain within the application zone.

The aquacultural reuse of wastewater has been developed around an aquatic based ecosystem with few trophic levels that thrives on high nutrient inputs. The simplified version of this approach to produce fish is detailed below (Figure 2.2). An aquacultural system is not only limited to fish production however and may involve the production of aquatic macrophytes such as water spinach for food, and the production of freshwater and marine crustaceans and molluscs.

In some countries, such as in Bangladesh, the product of one aquaculture system, such as duckweed, may be the food for the next aquaculture system e.g. carp production. This obviously reduces risks in pathogen uptake by the second system and may allow the first system to have greater influent loading, as macrophytes such as duckweed are more resilient and thrive under high nutrient loadings when compared to fish.

When designing a pond based aquaculture system, it is important to understand the capacity of the fish to survive under different wastewater loadings. Although many of the aquacultural species such as grass carp can withstand DO concentrations as low as 3-4mg/l, the system may not perform as well under stressful conditions, leading to poor output and also contaminated water e.g. high ammonia concentrations. Pond based systems are well maintained with inputs of 10-30 kg of BOD/ha/day and 4kg TN and 1kg TP/ha/day.

The performance of pond aquaculture in many developing countries is assessed through farmer experience. The aspects showing poor water quality may be assessed through changes in the colour of the water. The best colour of pond water is a medium green, showing good phytoplankton production. Other factors that are assessed include fish behaviour such as gulping for air at the surface (signifies low DO) and light penetration from the surface, which should be at 10-20cm from the surface (UNEP, 2002).

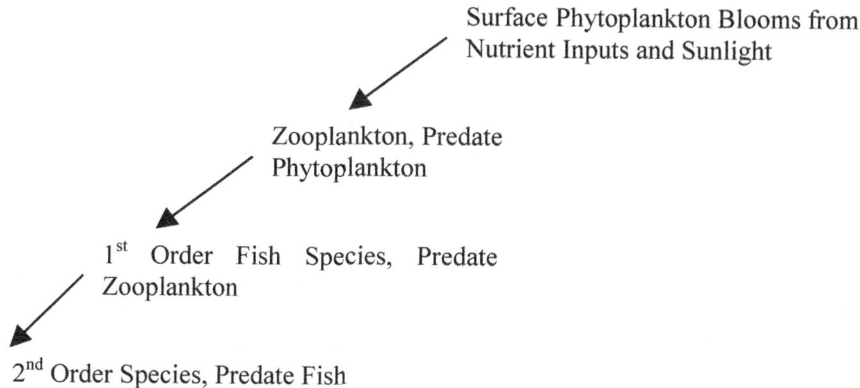

Figure 2.2. The food web dynamics of a fish based wastewater aquaculture system.

2.4.3.1 An example of wastewater reuse for Aquaculture in a developing country

The Calcutta wastewater based aquaculture system is an example of applying appropriate reuse techniques within the social, economic and environmental constraints.

The Calcutta area has inadequate centralised wastewater treatment systems for the quantity of wastewater produced on a daily basis. Much of the areas wastewater and stormwater flows through main sewers into channels that convey it into the nearby estuary. Along the dry weather flow channel (sewage), and stormwater channel there has been the establishment of fishponds that rely on wastewater to increase fish production.

The ponds are drained and prepared in the coolest months and are initially filled with the untreated wastewater. The ponds are then allowed to stabilise for 20-30 days before fish stocking. After this period, small doses of wastewater are added, when required, to boost phytoplankton production and increase fish growth.

The area provides economic incentives for many poor farmers and a source of cheap protein for many poor villagers. The area also offers the potential to treat wastewater to a higher standard before removal from the system. The area is, however, reducing its size of production due to urbanisation and imminent threats from industrial pollution of wastewater supplies (UNEP, 2002).

2.5 CASE STUDIES OF A DEVELOPING COUNTRY

2.5.1 Vietnam

Vietnam is a country in the south east of Asia and has a population of approximately 70 million people. Like many other countries within the region, it is experiencing rapid urban growth. Vietnam is also experiencing greater demand on water supplies as city sizes grow and with greater water demand, wastewater production is increasing. In many of the larger cities, such as Ho Chi Minh City, public water supplies are supplemented with private household bores.

The use of flush toilets in urban centres in Vietnam, either pour flush or cistern, is considerable. The table below (Table 2.3) shows the percentage of flush toilets. The percentage of onsite treatment through septic tanks (flush toilets) in Vietnam is high, especially in the larger cities e.g. city class 1. Due to the demand for space in many urban centres, many of these onsite systems do not have enough space for adequate wastewater infiltration and they discharge into streets, drainage ditches and natural water courses (UNEP, 2002).

The smaller urban centres of Vietnam have lower sanitation standards. The people may rely on bucket latrines or other simple facilities, while a small percentage (e.g. 14%) may be without any adequate sanitation. Many of those without sanitation may be living in temporary houses and could be served by communal facilities.

Table 2.3 Disposal of domestic wastewater, % of households in Vietnam.

	Flush toilets	Disposal to public sewers/drain	On-site disposal from septic tanks
City Class I	84	48	37
City Class II	75	44	31
City Class III	14	25	24
Average North	41	27	14
Average South	91	52	39
Average All	76	44	32

Source: Vietnam National Urban Wastewater and Sanitation Strategy Nov. 1995

2.5.1.1 Standards

Sewerage systems in Vietnam are classified as basic urban facilities and function to conserve the water quality of public waters. The Sewer Law governs most of the aspects of the sewerage system in Vietnam including control of water quality from sewerage systems. The Basic Law of Pollution Control and Water Pollution Control are important laws relative to the Sewer Law. The types of codes available for different aspects of sewerage and water pollution control are detailed below (Table 2.4).

The government has a water management policy that highlights the need for adequate quantity and quality of water for all beneficial uses, as well as control of point and diffuse sources. The government has a long-term plan for the development and management of water resources and an expected reduction of impacts from pollution within the Mekong River delta. Criteria for water quality of the delta in countries with riparian rights along its length are part of this plan to ensure pollution control throughout the river basin (Enderlein *et al.*, 1997).

Vietnam currently has national water quality criteria for potable water, aquatic ecosystems and irrigation (ESCAP, 1990). The criteria cover ambient factors, such as pH, DO, nutrient, heavy metal and dissolved solid concentrations. Local authorities are now also starting to prepare their own criteria such as those developed by Ho Chi Minh City for pesticide concentrations in water supplies.

Table 2.4 Water supply, sewerage and drainage systems standards in Vietnam

Code	Title
TCXD 70-77	Maintenance and operation rules for sanitation facilities of industry and civil construction works
TCVN 5576-91	Water supply and sewerage systems, technical management rules
TCVN 3988-85	Design documents and work drawing standards for construction of water supply and sewer networks
TCVN 4038-85	Terminology and definitions for water sewerage and drainage systems
TCVN 2622-78	Fire fighting water
2OTCN 51-84	Standard branch sewerage and drainage system and works. Standard design
2OTCN 66-91	Operator of water supply and sewer system safety
Sector Standard	Design criteria for water distribution systems and structures
Sector Standard	Drinking water quality standards

2.5.1.2 Governance

The government mostly administers the governance of water and sanitation in Vietnam. For example, in the city of Halong the Department of Science,

Technology and Environment promote new environmentally friendly technologies and monitor the environment for compliance with standards. The Quang Ninh Hygiene and Epidemiology Institute monitor the water quality of rivers and lakes to assess public health risks.

Problems with governance within Vietnam may include inappropriate technology choice leading to pollution. For example, the high use of septic tanks in urban centres, as earlier mentioned, can lead to pollution of local receiving environments. This problem is also compounded in some areas (e.g. Haiphong), where septic tanks and adsorption trench systems are used on high content clay soils. The low infiltration rate of these soils can lead to ponding and wastewater runoff.

The other problem with governance experienced in Vietnam is the ignorance of wastewater system designers and builders towards standards. In some cases, resources and time can be lost when approval is open to influence by manipulation and abuse. This can lead to inappropriate wastewater systems being developed (UNEP, 2002).

REFERENCES

Ali A. (2002). Operational Problems of Waste Water Treatment Plants in Developing World. In: *WAPDEC: Water and Wastewater Perspectives of Developing Countries*, New Delhi, India, pp. 933-937.

Anon. (2000). Global water supply and sanitation assessment. World Health Organisation, UNICEF, Water supply and sanitation collaborative council.

Bahri A. (1999). Agricultural reuse of wastewater and global water management. *Water Science and Technology*, **40**(4-5), 339-346.

Bishop P. (2000). *Pollution Prevention: Fundamentals and Practice*. McGraw Hill International Editions, Boston, USA.

Chafika A., Cheggourb M., Cossa D. and Sifeddinea S. B. M. (2001) Quality of Morrocan Atlantic coastal waters: water monitoring and mussel watching. *Aquatic Living Resources*, **14**(4), 239-249.

Chave P. A. (1997) Legal and Regulatory Instruments. In: *Water Pollution Control - A guide to the use of Water Quality Management Principles*, World Health Organization, London. Available online: http://www.who.int/docstore/water_sanitation_health/wpcontrol/begin.htm.

Dickens C W S. and Graham P M. (1998) Biomonitoring for effective management of wastewater discharges and the health of the river environment. *Aquatic Ecosystem Health and Management*, **1**(2), 199-217.

Eichhorn P., Flavier M. E., Paje M. L. and Knepper T. P. (2001) Occurrence and fat of linear and Branched alkylbenzene sulfonate and their metabolites in surface waters in the Philippines. *Science of the Total Environment*, **269**(1-3), 75-85.

Emongor V. E. and Ramolemana G. M. (2004) Treated sewage effluent (water) potential to be used for horticultural production in Botswana. *Physics and Chemistry of the Earth, Parts A/B/C*, **29**(15-18), 1101-1108.

Enderlein U.S., Enderlein R E. and Williams W.P. (1997) Water Quality Requirements. In: *Water Pollution Control - A guide to the use of Water Quality Management Principles*, World Health Organization, London. Available online: http://www.who.int/docstore/water_sanitation_health/wpcontrol/begin.htm.

ESCAP. (1990) *Water Quality Monitoring in the Asia and Pacific Region*. United Nations, New York.

Gallegosa E., Warrenb A., Roblesa E., Campoya E., Calderona A., Sainza M.G., Bonillaa P. and Escoleroc O. (1999) The effects of wastewater quality on groundwater in Mexico. *Water Science and Technology*, **40**(2), 45-52.

Gatts C.E.N., Faria R.T., Vargas H., Lannes H., Aragon G.T. and Ovlle A.R.C. (2003) On the use of photo thermal techniques for monitoring constructed wetlands. *Review of Scientific Instruments*, **74**(1), 510-512.

Gruttnera H. (1997) Regulating industrial wastewater discharge to public wastewater treatment plants - a conceptual approach. *Water Science and Technology*, **36**(2-3), 25-33.

Gumbo B., Mlilo S., Broome J. and Lumbroso D. (2003) Industrial water demand and cleaner production potential: a case of three industries in Bulawayo, Zimbabwe. *Physics and Chemistry of the Earth*, **28**(20-27), 797-804.

Hoehn J.P. and Krieger D.J. (2000) An economic analysis of water and wastewater investments in Cairo, Egypt. *Evaluation Review*, **24**(6), 579-608.

Ingallinella A.M., Sanguinetti G., Koottatep T., Montangero A. and Strauss M. (2002) The challenge of faecal sludge management in urban areas - strategies, regulations and treatment options. *Water Science and Technology*, **46**(10), 285-294.

Iwugo K.O., Andoh R.Y.G. and Feest A.F. (2002) Cost-effective integrated drainage and wastewater management systems. *Journal of the Chartered Institution of Water and Environmental Management*, **16**(1), 53-57.

Jonsson H. (2002) Urine separating sewage system - environmental effects and resource usage. *Water Science and Technology*, **46**(6-7), 333-340.

Kendie S.B. and Johnson J.W.K.D. (1999) Some polluting effects of small-scale industrial production in the central region of Ghana. *Singapore Journal of Tropical Geography*, **20**(2), 131-147.

Konrad C., Lorber K.E., Mendez R., Lopez J., Munoz M., Hidalgo D., Bornhardt C., Torres M. and Rivelca B. (2002) Systematic analysis of material fluxes at tanneries. *Journal of the Society of Leather Technologists and Chemists*, **86**(1), 18-25.

Lawrencea P., Adhamb S. and Barrota L. (2002) Ensuring water re-use projects succeed - institutional and technical issues for treated wastewater re-use. *Desalination*, **152**(1-3), 291-298.

Mahajana A.U., Chalapatiaoa C.V. and Gadkaria S.K. (1999) Mathematical modelling - a tool for coastal water quality management. *Water Science and Technology*, **40**(2), 151-157.

Mara D.D. (2001) Appropriate wastewater collection, treatment and reuse in developing countries. *Proceedings of the Institution of Civil Engineering-Municipal Engineer*, **145**(4), 299-303.

Markich S. and Camilleri C. (1997) Investigation of metal toxicity to tropical biota: Recommendations for revision of Australian water quality guidelines, Supervising Scientist, Canberra, Australia.

Matsui S., Henze M., Ho G. and Otterpohl R. (2001) Emerging Paradigms in Water Supply and Sanitation. In: *Frontiers in Urban Water Management: Deadlock or Hope*, Maksimovic, C.and Tjada-Guibert,J.A. (eds). IWA Publishing, London, England, pp. 229-263.

Maurer M., Schwegler P. and Larsen T.A. (2003) Nutrients in urine: energetic aspects of removal and recovery. *Water Science and Technology*, **48**(1), 37-46.

Meutia A.A. (2001) Treatment of laboratory wastewater in a tropical constructed wetland comparing surface and subsurface flow. *Water Science and Technology*, **44**(11-12), 499-506.

Nhapi I., Hokob Z., Siebelc M. and Gijzenc H J. (2002) Assessment of the major water and nutrient flows in the Chivero catchment area, Zimbabwe. *Physics and Chemistry of the Earth, Parts A/B/C*, **27**(11-22), 783-792.

Parkinson J. and Tayler K. (2003) Decentralised wastewater management in peri-urban areas in low-income countries. *Environment and Urbanisation*, **15**(1), 75-90.

Patterson R.A. (2001) Water quality relationships with reuse options. *Water Science and Technology*, **43**(10), 147-154.

Senzia M.A., Mashuari D.A. and Mayo A.W. (2003) Suitability of constructed wetlands and waste stabilisation ponds in wastewater treatment: nitrogen transformation and removal. *Physics and Chemistry of the Earth*, **28**(20-27), 1117-1124.

Shrestha R.R., Haberl R., Laber J., Manandhar R. and Madar J. (2001) Application of constructed wetlands for wastewater treatment in Nepal. *Water Science and Technology*, **44**(11-12), 381-186.

Smit J. and Nasr J. (1992) Urban agriculture for sustainable cities: using wastes and idle land and water bodies as resources. *Environment and Urbanisation*, **4**(2), 141-152.

SoER. (2001) *State of the Environment Report: Water Quality Classification.* Ministry of Environment and Physical Planning. Skopje. Republic of Macedonia.

Stephenson D. (2001) Problems of Developing Countries. In: *Frontiers in Urban Water Management: Deadlock or Hope*, Maksimovic, C. and Tjada-Guibert, J.A. (eds). IWA Publishing, London, England, pp. 264-312.

Stott R., May E. and Mara D.D. (2003) Parasite removal by natural wastewater treatment systems: performance of waste stabilisation ponds and constructed wetlands. *Water Science and Technology*, **48**(2), 97-104.

Tare V., Gupta S. and Bose P. (2003) Case studies on biological treatment of tannery effluent in India. *Journal of the Air and Water Management Association*, **53**(8), 976-982.

Ternes T.A. (1998) Occurrence of drugs in German sewage treatment plants and rivers. *Water Research*, **32**(11), 3245-3260.

Ternes T.A., Stuber J., Herrmann N., McDowell D., Ried A., Kampmann M. and Teiser B. (2003) Ozonation: a tool for removal of pharmaceuticals, contrast media and musk fragrances from wastewater?" *Water Research*, **37**, 1976-9182.

Ujang Z., Henze M., Curtis T., Schertenleib R. and Bell L.L. (2004) Environmental engineering education for developing countries: framework for the future. *Water Science and Technology*, **49**(8), 1-10.

UNEP. (2002). *International Source Book on Environmentally Sound Technologies for Wastewater and Stormwater Management.* United Nations Environment Programme - International Environmental Technology Centre, Osaka and International Water Association Publishing, London.

UNEP. (2004). *Guidelines on municipal wastes management.* Hague, Netherlands.

USEPA. (1991) *Technical Support Document for Water Quality Toxic Controls.* Washington, USA.

von Sperling M. and Chernicharo C.A. L. (2002) Urban wastewater treatment technologies and the implementation of discharge standards in developing nations. *Urban Water*, **4**(1), 105-114.

Wannera J., Cechb J.S., Kosc M. and Barchanekd M. (1996) Municipal effluent standards in the Czech Republic. *Water Science and Technology*, **33**(12), 1-10.

Warwick R.M., Ashman C.M., Brown A.R., Clarke K.R., Dowell B., Hart B., Lewis R.E., Shilliabeer N., Somerfield P.J. and Tapp J.F. (2002) Inter-annual changes in the biodiversity and community structure of the macrobenthos in Tees Bay and the Tees estuary, UK, associated with local and regional environmental events. *Marine Ecology Progress Series*, **234**, 1-13.

Warwick R.M. and Clarke K.R. (1998) Taxonomic distinctness and environmental assessment. *Journal of Applied Ecology*, **35**(4), 532-543.

WHO (1989). Health Guidelines for the use of Wastewater Agriculture and Aquaculture. Report of a Scientific Group Meeting. Technical Report Series, No. 778, World Health Organization, Geneva.

Xia K., Keller H.L. and Wagner J. (2001) Occurrence, distribution, and fate of 4-nonylphenol in Kansas domestic wastewater treatment plants. *Water Resources Update: Pharmaceuticals and Endocrine Disrupting Chemicals: Emerging Contaminants in Water*, **120**, 41-48.

3

Strategy and planning of sewerage infrastructures for developing countries: experience of Malaysia

Zaini Ujang

3.1 INTRODUCTION

Sustainable sewerage services require an appropriate strategy and planning. It is a common phenomenon in developing countries where many master plans have been expensively prepared for sewerage infrastructures involving high cost of investment, without preparing a proper legal and financial framework to implement the recommendations and action plan. As a result the master plans remained in bookshelves, without being able to be effectively implemented. This chapter will illustrate an example of sewerage strategy and planning in Malaysia, focusing on non-technical matters and implementation strategies.

Malaysia is a good case study because the country has experienced various stages of sewerage infrastructure development, ranging from individual on-site systems such as septic tanks and latrines to conventional centralised sewerage systems. The development of institutional and financial systems also could be used as a good case study for developing countries in improving public utilities. Examples will be given on specific measures and options, especially on the financial matters.

3.1.1   Malaysia in brief

Ever since Independence in 1957, Malaysia has been a country of a progressive multi racial society. The political system is based on parliamentary constitutional monarchy with a Federal Government structure, comprising 14 states. The constitution and parliamentary system is almost similar to the British Westminster model, except the members of Local Governments are appointed by the ruling parties at Federal and State levels. The location of both Peninsular Malaysia, and Sabah and Sarawak (in Borneo) lies entirely in the equatorial zone.

The climate is governed by the regime of the north-east and south-west monsoons which blow alternately during the course of the year. The average temperature throughout the year is 26°C with diurnal temperature range is about 7°C. Annual rainfall is about 2500 mm with high humidity (80%) due to the high temperature and rate of evaporation. In the year 2000 the population was around 22 million.

The main economic activities based on share of GDP in 2001 were service sector (including tourism, 54.6%), manufacturing (32.8%), agricultural, forestry and fishing (8.4%), mining (mainly petroleum, 6.8%), construction (3.4%) (Malaysia 2002 Yearbook). Under the Malaysian constitution, sanitation and public health are concurrent matters in which both Federal and State Governments have responsibilities. However, the constitution was amended in the year 2005 to give the Federal Government more power over water management (water supply and sewerage services) but the State Governments have control over water resources.

3.1.2 Sewerage development in Malaysia

The establishment of sewerage policy and facilities started with the formation of Sanitary Boards in major towns since the 19th century during the British occupation. The Sanitary Board of Kuala Lumpur for example, has been established since 1890. Despite limited control over finance, and could only make recommendations to the British Resident, the Sanitary Boards were responsible for providing sanitary services such as water supply, latrines, solid waste collection, roads, drainage and irrigation (Khoo 1996). The Boards also approved permits and licences for the construction of buildings, and collection of council or municipal tax.

Following Independence, the Boards have been upgraded to Municipal or Town Authorities. However, with the diversification and growth of urban services that need to be provided and the gradual transformation to the present day municipal councils, the emphasis on health and sanitation has been marginalized, compared to other public utilities and services.

In 1970, a survey was conducted on the status of sanitation facilities showing that only 3.4% of the population in Malaysia was served by the centralised sewerage systems, mainly in Kuala Lumpur, Georgetown and major towns in Sabah. For the majority of people in the urban areas in 1970, the sewerage facilities or technologies used for wastewater treatment were mainly septic tanks and bucket latrines. The people in rural areas were using pit latrines or resorting to the natural environment (Sewerage Services Department 1998).

It is important to note that after Independence, the sewerage services was initially administered by the Environmental Health and Engineering Unit in the Ministry of Health, formed in the mid-1960s. The Unit has initiated a two-pronged programme to improve sanitation facilities in rural and peri-urban, as well as urban areas. The programme for rural and peri-urban areas has been designed using on-site individual systems mainly by installing pour-flush latrines. The coverage and implementation of this programme was recognised as the most successful model in developing countries by the World Health Organisation, with a minimum of 90% coverage in 1995 (compared to 2.6% in 1970) (Sewerage Services Department 1998).

For urban areas, the Unit has adopted the "conventional sewerage" approach, as discussed in Chapter 1. The progress, however, was slow due to insufficient funding. In a bigger perspective, it was also due to an insufficient financial and institutional framework. In practice, the implementation of sewerage facilities in urban areas is under the jurisdiction of the Local Authorities, especially in developing new residential areas, industrial zones or townships. The function of the Unit was mainly to assist the Local Authorities in planning the sewerage facilities, including approving schemes proposed by private property developers.

In 1979 a new regulation was enacted on the standards of discharge quality for both sewage and industrial effluents, as shown in Table 3.1. The regulation, known as Environmental Quality (Sewage and Industrial Effluents) Regulations 1979 which was part of the Environmental Quality Act 1974, was very instrumental in controlling water pollution measures in Malaysia. It is also the first legal framework to regulate discharge quality into any inland water bodies, which require extensive upgrading of the sewerage facilities in Malaysia.

However, there is an exemption to contravene the standards that requires a special licence for contravention under five conditions, i.e.:

- No known practicable means of control to enable compliance with the acceptable conditions
- The estimated cost to be incurred for compliance will be prohibitive with regards to the nature and size of the industry, trade or process being carried out in the premises discharging the effluent
- The design and construction of any treatment plant or other control equipment and their commissioning require a longer period than the period for compliance with these regulations
- The imposition of the acceptable conditions as prescribed may result in circumstances which are not reasonably practicable or are contrary to the intent and spirit of the Act
- A new sewerage system is to be provided and later the effluent is allowed to be discharged into the sewerage system.

In 1980 the Federal Government adopted a policy for developers of all new housing developments of more than 30 units which required them to provide sewerage infrastructure with their own local sewage treatment plants. At the early stage, waste stabilisation ponds (WSPs) and Imhoff tanks have been recommended and built due to its simplicity in design, construction and maintenance. It is important to be simple because of shortage of operational budget and manpower. In the guidelines, after a year of completion of the sewage plants by the housing developers, the Local Authorities will take over the operation and maintenance. It was a massive and costly exercise since it was the period where the growth of development in Malaysia achieving two digit numbers with high growth of the property market, including residential and commercial areas.

The functions and funding mechanism of both the Ministry of Health and the Ministry of Housing and Local Government were limited in policy making at national level with less emphasis on the implementation at local level. On the other hand, not all Local Authorities fully complied with the requirements of the national policy due to insufficient funding and lack of expertise. As a result, by 1990, the coverage of centralised sewerage systems had only grown to 5.0% from 3.4% in 1970, while septic tank coverage in urban areas had grown to 37.3% from 17.2% in 1970 (Sewerage Services Department 1998).

Table 3.1 Environmental Quality (Sewage and Industrial Effluents) Regulations, 1979.

Parameters	Standard A	Standard B
Temperature, °C	40	40
pH value	6.0 - 9.0	5.5 – 9.0
BOD_5 at 20°C, mg/l	20	50
COD, mg/l	50	100
Suspended solids, mg/l	50	100
Mercury, mg/l	0.005	0.05
Cadmium, mg/l	0.01	0.02
Chromium, hexavalent, mg/l	0.05	0.05
Arsenic, mg/l	0.05	0.10
Cyanide, mg/l	0.05	0.10
Lead, mg/l	0.10	0.5
Chromium, trivalent, mg/l	0.20	1.0
Copper, mg/l	0.20	1.0
Manganese, mg/l	0.20	1.0
Nickel, mg/l	0.20	1.0
Tin, mg/l	0.20	1.0
Zinc, mg/l	1.0	1.0
Boron, mg/l	1.0	4.0
Iron (Fe), mg/l	1.0	5.0
Phenol, mg/l	0.001	1.0
Free chlorine, mg/l	1.0	2.0
Sulphide, mg/l	0.50	0.50
Oil and grease, mg/l	Not detectable	10.0

In 1993, a landmark development in the sewerage sector in Malaysia took place with the establishment of the Sewerage Services Act 1993, and the formation of the Sewerage Services Department as a new Federal agency. The functions of Sewerage Services Department are mainly to plan, regulate and enforce all laws and regulations related to the sewerage services sector. The Department is also responsible for ensuring smooth implementation of appropriate and modern sewerage systems and to encourage and develop the sewerage industry in order that it be managed efficiently in terms of cost, technology and manpower resources. Since 1993 the Department has been responsible for ensuring national sewerage privatisation project be successfully implemented.

In general, the types of wastewater treatment technologies that have been applied in Malaysia, its purpose and options can be summarised in Table 3.2.

The systems can be broadly divided into five classes, reflecting the nature of the served community (particularly the type of residential facilities) and its design objectives. Individual septic tanks are the most common, particularly for rural and urban poor residential areas, and probably the easiest to be fully constructed using local materials and expertise. In Malaysia for example, about 40% of the population (who are occupying rural areas) are mainly served by individual septic tanks. Communal septic tanks are also important in serving clusters of houses in properly planned residential or commercial areas.

Table 3.2 Common types of wastewater treatment technologies commonly used in Malaysia.

Systems	PE	Target populations	Objectives	Options
Individual on-site systems	1-10	-Rural areas -Urban poor	-SS removal -Partly organic removal	-Septic tanks -Ventilated improved pits
Communal on-site system	10-100	-Urban poor -Government-sponsored housing areas	-SS removal -Partly organic removal	-Communal septic tanks -Imhoff tanks
Household-centred low-cost sewerage systems	100-5,000	-Urban poor -Middle class urban	-SS removal -Organic removal	-Waste stabilisation ponds -Wetlands, land treatment -Sea outfall
Decentralised and small mechanical sewerage systems	5,000-50,000	-Low cost flats in urban -Middle class residentials -Resort areas -Government houses -Business areas -Properly planned housing areas	-SS removal -Organic removal	-Package plants: activated sludge systems -Package plants: biofilm systems -Package plants: hybrid system
Large-size, mechanical centralised systems	50,000-500,000	-Clusters of properly planned housing areas -Business areas -Middle class urban	-SS removal -Organic removal -Nutrient removal	-Conventional activated sludge plants -Sequencing batch reactors

Table 3.3 Major developments of the sewerage sector in Malaysia.

Era	Issues/Developments	Agencies
Before 1900s	Individual and communal latrines facilities introduced	Sanitary Board
1960s	Establishment of Environmental Health and Engineering Unit	Ministry of Health
1970s	Introduction of pour-flush latrines for rural areas	Ministry of Health
1970s	Introduction of "conventional centralised sewerage" facilities	Ministry of Health
1970	Survey on sewerage facilities nationwide	Department of Statistics; Department of Health
1974	Establishment of Environmental Quality Act 1974	Department of Environment
1975	Establishment of the Department of Environment	Ministry of Science, Technology & Environment
1979	Establishment of Environmental Quality (Sewage and Industrial Effluents) Regulations 1979	Department of Environment
1980	Policy on new housing development to provide local sewage treatment system, funded by developers	Ministry of Health & Ministry of Housing and Local Government
1991	Establishment of the *Malaysian Standard 1228: Code of Practice for Design and Installation of Sewerage Systems*	Standards & Industrial Research Institute of Malaysia (SIRIM)
1993	Sewerage Services Act	
1994	Formation of the Sewerage Services Department	Ministry of Housing and Local Government
1994	Concession to Indah Water Konsortium (IWK) to provide sewerage services in Local Authority operational areas for 28 years	Ministry of Housing and Local Government
1995	Publication of *Guidelines for Developers: Vol. 1 - Sewerage Policy for New Developments; Vol. II -Sewerage Works Procedures; Vol. III - Sewer Networks and Pump Stations; Vol. IV - Sewage Treatment Plants; Vol. V- Septic Tanks*	Sewerage Services Department and Malaysian Water Association
1996	Policy on sewerage capital contribution was approved by the Cabinet, i.e. 1.65% of the property value	Sewerage Services Department

2002	New regulation that all septic tanks must be desludged once every two years	Sewerage Services Department
2004	Formation of a new Ministry where water supply and sewerage services are integrated towards better and more sustainable management of water resources	Ministry of Energy, Water and Communications
2005	Constitutional amendment on the management of water services from State Governments to Federal Government	

It is important to note that decentralised and small treatment systems using package plants, complete with sewer networks have been successfully implemented in many planned cities in Malaysia and other developing countries. Package plants are relatively easy and cheap to build and assemble, and require minimum expertise and could be designed with low operating and maintenance costs.

Table 3.3 shows the overall chronological development for the sewerage services sector in Malaysia before the twentieth century, by the formation of Sanitary Boards in major towns. It is obvious that the developments during the 1990's are the most rapid and structurally significant to the modernisation of sewerage facilities, after the establishment of the Department of Sewerage Services and the Sewerage Services Act 1993. It was planned that the decade was mainly for reengineering and restructuring of the technical, financial and institutional framework, while the decades ahead are to implement the policies towards better environment and effective pollution control.

3.2 SEWERAGE POLICY

Sewerage policy in Malaysia has gradually developed in response to various infrastructure requirements and environmental issues. As shown in Table 3.3, before the 1980's, sewerage facilities were provided without proper financial, as well as insufficient legal and institutional support. As a result, the progress was slow and the industry was not lucrative enough to invite private investment on research and development, and product innovations.

However, the major landmark in sewerage policy and implementation in Malaysia was started after the establishment of Environmental Quality (Sewage and Industrial Effluents) 1979 and Government policy made it compulsory for developers to provide local sewage treatment plants for any residential or township development with more than 30 units. The policies have been translated into various guidelines, code of practices and regulations mainly established by the Engineering Unit at the Ministry of Health and Ministry of Housing and Local Governments. Later in 1993, the task was transferred to a newly established department, the Sewerage Services Department. In general the policies can be summarised into three major components, as follows:

- National framework
- Regional and Local Authorities
- Privatisation.

3.2.1 National framework

The coverage of sewerage systems and services in Malaysia is limited to human waste and household wastewater. Industrial and trade wastewaters are separately treated by individual on-site treatment plants. The plants are approved and regulated by the Department of Environment. In addition, the management of scheduled or hazardous wastes is also under the jurisdiction of the Department of Environment in which all wastes should be properly collected, labelled, transferred, treated and disposed according to a prescribed procedure. At the moment Malaysia has a centralised scheduled waste treatment and disposal plant, located in Bukit Nanas (Ujang 2003).

The Sewerage Services Act 1993 vests the Federal Government with the executive authority on all matters related to sewerage services in Malaysia. The Acts also provide the authority to the Federal Government to transfer all assets of Local Authorities and State Governments used for the purpose of sewerage systems and services. In general, the Act provides a national policy framework for the sewerage sector which includes:

- The authority for sewerage systems and services was fully transferred to the Federal Government
- The property for sewerage infrastructures was transferred to the Federal Government from Local Authorities
- Privatisation of sewerage services and systems
- Appointment of Director-General Sewerage Services to head the Sewerage Services Department
- Public sewerage system; its construction, management, operation, maintenance
- Private sewerage systems and septic tanks
- Charges on sewerage services
- Approval of plans and specifications of sewerage systems or septic tanks
- Licensing

After the establishment of the Sewerage Services Act 1993, the Sewerage Services Department has been responsible for planning, providing, operating and managing sewerage services in Malaysia. The responsibility was gradually transferred from Local Authorities. In general, the Sewerage Services Department is responsible for upgrading the sewerage sector by providing appropriate sewerage systems, which meet the treatment objectives for effluent quality standards and ensuring a more responsive and efficient sewerage service to the customers. In specific terms, the roles of the Department are as follows:

- To plan, regulate and enforce all laws and regulations to the sewerage sector in Malaysia according to the Sewerage Services Act 1993
- To ensure the smooth implementation of appropriate and modern sewerage systems in Malaysia according to the established standards
- To encourage and develop the sewerage industry in order that it can be managed efficiently in terms of cost, technology and manpower resources
- To protect the interest of consumers by ensuring the best available sewerage services with a cost that is affordable
- To ensure the national sewerage privatisation project will be implemented successfully and satisfactorily by the concession company without causing problems to the country
- To assist the development of a modern sewerage sector to protect the environment and water resources of the country.

3.2.1.1 Regional and local authorities

The Sewerage Services Department is divided into a two-tier organisation: head office and six other regional offices to cover the whole country. The head office deals mainly with national policy, financial, privatisation and administrative issues, while the regional offices are in charge of the routine functions as regulators and facilitators for the sewerage industry. At present there are 48 major and 96 small Local Authorities in Malaysia, as shown on Table 3.4.

3.2.1.2 Privatisation

Since Independence, the sewerage sector received much lower budgets from the Federal Government compared to the water supply sector. For example, within the period of 1975 to 1995, the Government had spent RM7.785 billion in developing water supply infrastructure, compared to RM328 million for the sewerage sector. This is largely due to the fact that the investment on sewerage facilities, particularly the sewage treatment plants, is hidden in the housing development projects, paid directly by consumers. In addition, owners of private premises in rural and peri-urban areas are responsible for providing individual on-site systems such as septic tanks, themselves, without any subsidy from the Government.

The allocated grants for urban areas were not fully utilised due to a lack of institutional infrastructure to implement many master plans related to centralised sewerage facilities. In the period of 1990-95, for instance, the Federal Government allocated RM550 million to implement centralised sewerage projects in 25 townships. The allocation is a grant towards 50% of the capital cost of these projects. Unfortunately, the projects did not take-off because they required corresponding budgets from the respective State Governments and Local Authorities.

In 1993, the Federal Government awarded a concession of 28 years to Indah Water Konsortium, a private local consortium. The company is responsible for operating and maintaining the existing sewerage facilities and systems, mainly in urban areas, as well as for carrying out capital works with an original total estimated costs of RM6.057 billion within six phases as shown in Table 3.4.

The privatisation of the sewerage sector in Malaysia, is an effort to assist the Sewerage Services Department to implement the following initiatives:

- Operate and maintain all public sewerage systems, including those that will be built by private developers and handed over to the Sewerage Services Department to be declared public systems
- Take over, refurbish and upgrade all public sewerage systems to meet the relevant environmental standards
- Build new systems in existing townships to phase out individual septic tanks and meet set targets for connected services of 84.5% and 29.5%, respectively, in the 49 major Local Authorities and 95 small Local Authorities.

Table 3.4 Sewerage coverage targets to be achieved by the concession company (Sewerage Services Department 1998).

Phase / Year	Category A (48 major towns)		Category B (96 smaller town)	
	Connected	Septic Tanks	Connected	Septic Tanks
Phase 1 (1997)	-	-	-	-
Phase 2 (2002)	63.8%	29.0%	15.8%	50.6%
Phase 3 (2007)	76.2%	19.2%	17.8%	49.5%
Phase 4 (2012)	82.6%	14.3%	19.4%	49.5%
Phase 5 (2017)	84.3%	13.0%	24.0%	47.2%
Phase 6 (2022)	84.3%	15.7%	29.5%	70.5%

3.3 CAPITAL CONTRIBUTION

Capital contribution means the contribution of various parties to finance the capital cost for constructing the sewer network, sewage and sludge treatment plants. The capital cost for the implementation of centralised sewage treatment plants, as shown in Table 3.4 is approximately RM6.057 billion, while the operating cost is estimated to be around RM18 billion.

In 1996, the Cabinet approved a sewerage capital contribution (SCC) policy to partly finance the capital cost of the implementation of the sewerage infrastructures. The SCC rate of up to 1.65% of the selling price or market value was adopted. The SCC is a condition for the approval of all development projects to upgrade the sewerage infrastructure in all new development areas. The payment is to be paid to a Government Trust Account (SSD & MWA 1999). The principles considered in developing the SCC rate are as follows:

- All development projects for residential and commercial areas are required to contribute, except with special exemption from the Department of Sewerage Services
- Discount of SCC is partly allowed based on the level of facilities provided by developers (e.g. sewer lines or pump stations)
- Exemption for development projects that provide the sewerage facilities as shown in Table 3.5

Table 3.5 Permanent works requirements (SSD & MWA 1999).

Criteria	Requirements	
Effluent discharge	To comply with the Environmental Quality (Sewage and Industrial Effluents) 1079	
Design standards	Standard A (mg/l)	Standard B (mg/l)
Biochemical oxygen demand (BOD$_5$)	10	20
Suspended solids	20	40
Buffer zone	In compliance with Department of Environment guidelines for the siting and zoning of industries whereby centralised permanent sewage plant is classified as industry. The buffer requirement is 500 m. However light industries, green areas and infrastructure reserves can be located within the buffer zone. The 30 m clear buffer to building is still maintained	
Environmental impact	To obtain Environmental Impact Assessment approval from the Department of Environment	
Occupational, safety and health	Land area for a minimum of 50,000 projected population equivalent. Modular construction is permitted, however modulation is limited by operational efficiency	
Land status	The sewage treatment plant is to be surrendered to the Government for purpose of Reservation as Federal sewerage reserve	
Tertiary treatment	Provision for adequate land reserve for incorporating additional treatment facilities to meet future effluent discharge standards for nutrients removal	
Sludge treatment	Treatment facilities must meet requirement to dewater sludge greater than 25% dry solids	
Facilities disinfection	Disinfection facilities must be provided to supplement incapability of biological treatment to kill pathogens of waterborne diseases	
Back-up power supply	Two sources of power supply is required. In the absence of one, a generator set shall be provided	
Spare parts	Spare parts for critical equipment and machinery	
Office space	Office space required for operational and maintenance works	
Laboratory	Provide apparatus for taking storing samples and facilities for on-site testing for operational control	

- Exemption for development projects with a housing density less than 25 population equivalent per hectare, providing on-site treatment, such as septic tanks
- All SCCs should be paid prior to the issue of the recommendation for a certificate of fitness
- Off-site connecting sewer is required to connect a new development to the nearest suitable point in a public sewer system.

For the purpose of discounts, the average costs for providing the following sewerage components are determined as shown in Table 3.6. Exemption is also given to low cost houses, community developments, government schools, charitable developments and places of worship. In addition, exemption is also applicable for developments consisting of housing less or equal to 150 PE (SSD & MWA 1999).

Example 3.1 illustrates the calculation for the amount of SCC for a mixed development, comprising housing estate complete with mosque, school and community centres. Example 3.2, in addition, shows the calculation of SCC for a schemed development.

Example 3.1 Determine the total applicable SCC and SCC/PE value for the following residential scheme: 1300 units of houses are to be built in Taman Desa Skudai, Johor Bahru. The developer provides a Government school (500 students), two community centres and a mosque. The houses are divided into four types with different selling prices as shown below.

House Types	Units	PE	Selling price
Type A	50	250	RM500,000
Type B	350	1750	RM200,000
Type C	600	3000	RM100,000
Type D (low cost scheme)	300	1500	RM25,000

Solution: Assume: one unit equals to 5PE.

House Types	Units	PE	Selling price	Sub SCC	SCC
Type A	50	250	RM500,000	RM500,000 x 50 x 1.65%	RM412,500
Type B	350	1750	RM200,000	RM200,000 x 350 x 1.65%	RM1,155,000
Type C	600	3000	RM200,000	RM100,000 x 600 x 1.65%	RM990,000
Type D	300	1500	RM25,000	Exempted	-
School	1	0.2 x 500 =100	-	Exempted	-
Mosque	1	0.2 x 500 = 100	-	Exempted	-
Community centres	2	0.2 x 250 = 50	-	Exempted	-
Total		6750			RM2,557,500

Total applicable SCC is RM2,557500. Total SCC/PE ≈ RM380/PE

Table 3.6 Average costs for the sewerage systems for the purpose of discounts.

Sewerage components	Average costs*
Trunk sewers/pumping stations	RM130/PE
Sewage treatment plant	RM180/PE
Sludge treatment	RM120/PE

*Note: In 2004, the value of USD1 is equivalent to RM3.8.

Example 3.2 Determine the total applicable SCC to be paid by the developers for the following three schemes. Schemes 1 and 3 have started; Scheme 2 will be started in 5 years time. A government school, a mosque and two community centres, are located in Scheme 1, similar to Exercise 3.1.

Scheme 1: PE size = a			
House Types	Units	PE	Selling price
Type A	50	250	RM500,000
Type B	350	1750	RM200,000
Type C	600	3000	RM100,000
Type D (low cost scheme)	300	1500	RM25,000

Scheme 2: PE size = b			
House Types	Units	PE	Selling price
Type A	50	250	RM400,000
Type B	250	1250	RM150,000
Type C	500	2500	RM100,000
Type D (low cost scheme)	200	1000	RM25,000

Scheme 3: PE size = c			
House Types	Units	PE	Selling price
Type A	100	500	RM500,000
Type B	300	1500	RM200,000
Type C	300	1500	RM100,000
Type D (low cost scheme)	300	1500	RM25,000

The sewerage planning is to be centralised with a sewage and sludge treatment plant as shown in the figure (Figure 3.1) below:

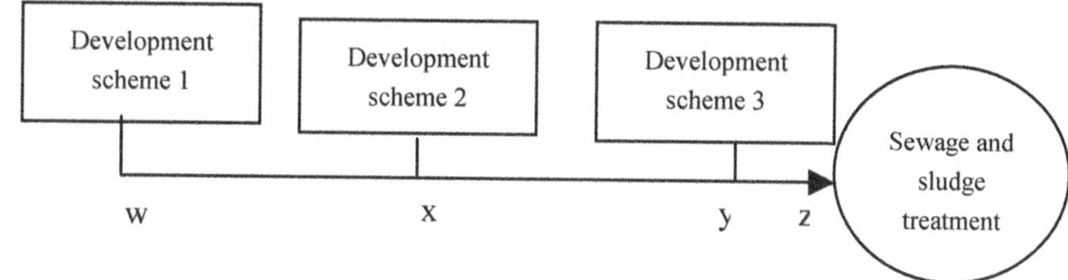

Figure 3.1 Sewage and sludge treatment plant.

Solution: Assume: one unit equals to 5PE. Refer to Table 3.6 for the estimation of capital works:

Cost of sewage & sludge treatment plant =   RM(z)   = RM300 x PE   =
RM4,725,000
Total cost of trunk sewer   = RM130 x PE   =
RM2,047,500

Cost of pipeline yz   =   RM(y)   = RM40 x PE   =
RM630,000
Cost of pipeline xy   =   RM(x)   = RM40 x PE   =
RM630,000
Cost of pipeline wx   =   RM(w)   = RM50 x PE   =
RM787,500

The principle is for the developers to pay according to the principles as calculated in Example 3.1. SCC for a schemed development is as follows:

The developer for Scheme 1 pays:

$$RM(1) = \left[\frac{a}{a+b+c}\right]RM(z) + \left[\frac{a}{a+b+c}\right]RM(y) + \left[\frac{a}{a+b}\right]RM(x) + RM(w)$$

The developer for Scheme 2 pays:

$$RM(2) = \left[\frac{b}{a+b+c}\right]RM(z) + \left[\frac{b}{a+b+c}\right]RM(y) + \left[\frac{b}{a+b}\right]RM(x) + RM(x)$$

The developer for Scheme 3 pays:

$$RM(3) = \left[\frac{c}{a+b+c}\right]RM(z) + \left[\frac{c}{a+b+c}\right]RM(y)$$

The low cost houses in the three schemes are not included as far as payable PE as shown in solution to Exercise 2.1. Therefore the developer for Scheme 1 pays:

$$RM(1) =$$

$$\left[\frac{5000}{5000+4000+3500}\right]RM\,2{,}047{,}500 + \left[\frac{5000}{5000+4000+3500}\right]RM\,630{,}000 + \left[\frac{5000}{5000+4000}\right]RM\,630{,}000 + RM\,787{,}500$$

$$= RM819{,}000 + RM252{,}000 + RM350{,}000 + RM787{,}500$$
$$= RM2{,}208{,}500$$

(This is cheaper compared to local sewage treatment facilities as illustrated in Exercise 3.1)

The developer for Scheme 2 pays:

$$RM(2) =$$

$$\left[\frac{4000}{5000+4000+3500}\right]RM(z) + \left[\frac{4000}{5000+4000+3500}\right]RM(y) + \left[\frac{4000}{5000+4000}\right]RM(x) + RM(x)$$

$$= RM655{,}200 + RM201{,}600 + RM280{,}000 + RM630{,}000$$
$$= RM1{,}766{,}800$$

The developer for Scheme 2 has two options:

- Pay now in which case no further contributions are payable in the future
- Pay in 5 years time when the development proceeds, in which case the payment due is RM(A) escalated at annual long term bank interest rates. For example the annual long-term bank interest rate for the next three years is 7% and 5% for the following two. The payment due: $(1.07)^3\,(1.05)^2\,(1{,}766{,}800) = RM2{,}553{,}690$.

The developer for Scheme 3 pays:

$$RM(3) = \left[\frac{3500}{5000+4000+3500}\right]RM(z) + \left[\frac{3500}{5000+4000+3500}\right]RM(y)$$

$$= RM1{,}323{,}000 + RM176{,}400$$
$$= RM1{,}499{,}400$$

3.4 CATCHMENT STRATEGY

Wastewater contains mainly liquid removed from residential and municipal areas, and industrial premises. It is important to collect the wastewater according to its sources. In Malaysia, sewerage catchment is limited to household and municipal wastewater, mainly from residential dwelling, public toilets, laundries, toilets, kitchens and canteens which are located in commercial, industrial and institutional buildings, and from hospitals and restaurants.

Catchment is broadly defined as a composite area with well demarcated boundaries within which independent self-contained sewerage services can be instituted and managed on an economic perspective whilst meeting regulatory standards for treated effluent quality. There are many procedures and techniques to conduct sewerage catchment strategy. In Malaysia, the procedures are as follows (SSD & MWA 1999):

- Catchment description
- Catchment details and development of Geographical Information Systems (GIS)
- Existing conditions
- Future conditions and development planning
- Sewerage catchment strategy options
- Recommended option
- Description of recommended strategy
- Construction

In principle, the catchment strategy for an area is prepared by the Department of Sewerage Services. However, in practice, the strategy is jointly developed by private developers and the national concession company for new development areas. This is part of the requirements for sewerage approval from the Department of Sewerage Services.

3.4.1 Sewerage system

A typical sewerage system in Malaysia consists of physical facilities for collection, conveyance, treatment plants and sludge disposal. Storm water runoff, on the other hand, is collected and transferred to water bodies using a separate storm water drainage system. The Department of Irrigation and Drainage is responsible for planning, construction and maintenance of storm water management. However allowances are incorporated in sewerage planning to accommodate storm waters that may unintentionally enter the sewerage system through uncovered manholes, leaking pipes or illegal roof drainage connections. In general the physical sewerage facilities can be listed as follows (SSD & MWA 1999):

- Property connection from wastewater generation points
- Reticulation sewers (225 mm to 300 mm in diameter) which collect sewage from a cluster of generation points
- Branch sewer (300 mm to 450 mm in diameter) which connect reticulation sewers to main sewers
- Main sewers (450 mm to 900 in diameter) channel sewage to local small sewage treatment plants, or to trunk sewers
- Trunk sewers (greater than 900 mm in diameter) form the spine of a large catchment and convey sewage to a large centralised sewage treatment plant
- Small local sewage treatment plant
- Centralised sewage treatment plants
- Sludge treatment and disposal facilities

3.4.2 Basic principles

The aim of sewerage catchment planning is to ensure sewage flows, generated within the catchment areas, are properly collected, transferred, treated and disposed. It also identifies the appropriate strategies and implementing facilities. In specific the objectives of sewerage catchment planning are as follows:

- To define the boundaries for a catchment that will optimise the management of sewerage services in that particular area, and support sewerage services in adjoining catchments
- To estimate the generation of current and future loading
- To outline the implementation strategy for sewerage facilities in line with privatisation programmes
- To maximise the utilisation of the existing facilities within the catchment, to serve a particular development, and to serve future development schemes
- To ensure affordable implementation at various stages

3.4.3 Sewerage management alternatives

The procedures as prescribed in the *Guideline for Developers: Sewerage Policy for New Developments* (Vol. 1, SSD & MWA 1999) are the basis for the data gathering exercise within a sewerage catchment. The data is required as background information to identify and assess the potential strategies. In general the data described the following:

- Existing sewerage system

- Proposed changes in sewage coverage in terms of flows
- Issues and constraints related to sewerage management
- Options for plant locations, sewer routes and catchment physiography

Sewerage catchment planning will also provide alternatives for sewerage management for that particular catchment. The alternatives can be summarised as follows:

- **System upgrade, replacement and refurbishment** This is appropriate for areas that are not expected to have significance increases in flow and load. Refurbishment scheme is normally conducted for replacement of broken sewer or pump stations. The upgrading is appropriate for inefficient or ineffective treatment plants.
- **New sewerage works** In most cases new sewerage works are required for new development areas and for areas served by inadequate systems, such as septic tanks or Imhoff tanks. For a new development scheme, developers are mainly responsible for providing the facilities.
- **Plant rationalisation and economics of scale** At present there are more than 6000 sewage treatment plants in Malaysia. Most of them are catering for small catchments, less than 1500 PE. However such catchments are neighbouring to many other catchments serving almost similar sizes. It is not cost effective in terms of economies of scale with the construction and operation of such small decentralised sewage treatment plants. In many cases, a single centralised sewage treatment plant is more cost-effective. The rationalisation of smaller plants or of plants which are no longer viable or require major reinvestment is expected to be an important factor to consider, in evaluating alternative sewerage management options.
- **Sludge management** Sludge treatment and disposal is an expensive exercise. At present only a few sludge facilities are available in Malaysia. Therefore, it is important to strategise the facilities in order to reduce the cost of sewerage handling.
- **Rationalisation of small sewage treatment plants** The aim is (a) to reduce the total number of individual sewage treatment plants serving a catchment (b) to eliminate plants that are of limited capacity to serve future loadings (c) inefficient plants (septic tanks or Imhoff tanks). There are two options i.e. (a) to build a new centralised sewage treatment plant (b) to convert the existing sewage treatment plants into a larger scheme, thus enabling sewage flows from small plants to be channelled to this plant.

3.4.4 Financial analysis and options

The financial analysis and options are normally the key factors in determining the viability of any project, including sewerage facilities, especially in developing countries. The validity of any financial comparison between various options is dependent on pricing of every item in a sewerage system. The SSD and MWA (1999) guidelines proposed the following items to be considered in order to establish a reliable costing procedure:

- Data on similar recent projects
- Information on current construction unit rates
- Preliminary assessment of costs and complexity of specific works
- Weighting factors for unit rates for specific conditions
- Data on costs of specislised equipment and processes
- Information from knowleadgeable personnel on operation and maintenances costs
- Information from local and overseas costing manuals

On top of that it is also important to determine the life of assets and residual value after several years of service. An estimate of the life of sewerage assets, normally used in Malaysia is presented in Table 3.6. It shows that mechanical and electrical components will have zero residual value after 25 years of service. Sewers and outfalls have approximately 25% residual value and civil structures have a small residual value (SSD & MWA 1999).

The other factors which are normally considered in order to determine the financial options, are as follows:

- **Operating and maintenance costs** This is important since sewerage facilities are costly. In the privatisation scheme in Malaysia, Indah Water Konsortium, the estimated capital cost was only RM6.057 billion compared to RM18 billion for operation and maintenance costs.
- **Comparison of costs** Options are evaluated by comparing the capital and operating costs together for each alternative. The basis is either in terms of amortised cost or a net present value basis. Amortised cost is the annual cost that is used to distribute the capital costs over a defined period of time. The alternative is to convert operating and maintenance costs into a capital sum.
- **Rate of return on capital investment**
- **Average incremental cost**

Table 3.7 Expected life of sewerage scheme assets (SSD & MWA 1999).

Sewerage assets	Years*
Sewers	80
Pumping stations – Civil	50
Pumping stations – Mechanical/electrical	25
Treatment plants - Civil	50
Treatment plants – Mechanical/electrical	25
Treatment plants - Control	15
Outfalls	80

* The values are often set by the Government and once final cannot be changed for a particular asset.

3.5 CONCLUSION

Malaysia has experienced an extensive reengineering and modernisation process of its sewerage policy and infrastructures. Without neglecting the importance of on-site individual systems such as septic tanks and ventilated improved pits for rural areas and small townships, there is a massive improvement in small and localised sewerage systems in major cities. The driving force is the formulation of a national framework for sewerage services, particularly after the establishment of the Sewerage Services Department.

Sustainable sanitation in the Malaysian context can be summarised as a sustainable framework, supported by partnerships with the private sector in order to provide various options for sewerage facilities, ranging from individual on-site treatment systems to highly sophisticated centralised sewage and sludge treatment plants. However, the obvious lacking component – which requires more input from scientific, technical and financial aspects - are reuse schemes. Reuse of treated effluent or anything related to wastes is still a taboo for most Malaysians. Probably this is due to the fact that people are not yet concerned about the shortage of water resources, since Malaysia is still considered a water-rich country.

REFERENCES

Khoo, K. K. (1996) *Kuala Lumpur: The Formative Years*. Berita Publishing, Kuala Lumpur.
Malaysia 2002 Yearbook. Berita Publishing, Kuala Lumpur.

Sewerage Services Department (1998) *Sewerage Services Report 1994-97*. Ministry of Housing and Local Council, Malaysia, Kuala Lumpur.

SSD & MWA (1999) *Guidelines for Developers Vol. 1: Sewerage Policy for New Developments*. Second Edition. Sewerage Services Department and Malaysian Water Association, Kuala Lumpur.

Ujang, Z. and Buckley, C.A. (2002) Water and wastewater in developing countries: Present reality and strategy for the future. *Wat.Sci.Tech.* **46**, 1-9.

Ujang, Z. (2003) Hazardous waste management in Malaysia. *UNESCO Encylopedia of Life Support System for Hazardous Waste Management in Developing Countries*. UNESCO, Paris (in press).

4

Wastewater treatment technology for developing countries

Mogens Henze, Zaini Ujang and Eddy Soedjono

4.1 INTRODUCTION

The types and facilities for municipal wastewater treatment technologies adopted in developing countries are closely related to many non-technical factors and issues, such as regulatory requirements, financial capability and mechanism, and sustainability in terms on operation and maintenance. In general many developing countries have adopted matured treatment technologies, such as constructed wetlands, conventional activated sludge and waste stabilisation ponds from developed countries, particularly those which require a large investment from external funding. This is also obvious for small and decentralised facilities by adopting package plants, similar to the conventional technologies available in Japan (better known as *Jokasou*), Germany or the United States.

However, there is also progress to improve the technologies in developing countries, through local activities in research and development such as in the case of South Africa, Malaysia, Thailand, India, Indonesia, Mexico and China. As a result, many local technologies, including advanced processes and systems, have been developed and adopted based on local conditions, and financial capability and mechanism. This is also an indication of the progress in human resource development programmes and collaborative efforts between engineers or researchers in those countries and their counterparts from developed countries.

In general, the treatment technologies appropriate for developing countries are almost similar to those in developed countries, provided detailed cost-benefit analysis should be given to the sustainability aspect, in terms of capital financing and plant operation and maintenance. In addition, the technologies should be designed, based on local climate, operator expertise and legal requirements, as well as the appropriateness with government policy and other public utilities. For example, sewage from major cities in China and Malaysia has been mostly sewered and treated using large plants of the conventional activated sludge system. The technologies and facilities are designed by both local and foreign engineers, and locally assembled, built and maintained. The other facilities are served by adopting on-site, small and decentralised sewerage systems, such as communal and individual septic tanks.

It is interesting to note that this development has been a recent phenomenon. In the past sewage treatment plants were mainly waste stabilisation pond systems and Imhoff tanks, in the case of small and centralised facilities. For individual on-site facilities, septic tanks and ventilated improved pits have been common and considered the most appropriate low cost option. Other systems were also available such as pit latrines and pour flush toilets.

In order to guide professionals and policy makers to choose an appropriate sewage treatment system, it is important to understand the overall perspective of wastewater treatment systems available as follows:

- Biofilm system
- Activated sludge system
- Hybrid system

In Malaysia, a piece of software for a decision support system to choose an appropriate wastewater treatment system has been developed by the Universiti Teknologi Malaysia, and used by the Sewerage Services Department and the Department of Environment. The software is known as WASDA (Wastewater Treatment Plant Design Advisor) and has been useful to assist managers and professionals to conduct process design using various types of wastewater treatment systems. It is also useful for the authorities to check the design of wastewater treatment plants submitted by consultants and contractors for approval.

The contents of this chapter are based on Henze et al, 1997.

4.2 BIOFILM SYSTEM

Biofilm systems are also widely known as fixed film or attached growth systems. Biofilm refers to a biological growth developed on the surface of filter media, known as a biofilter, in biological wastewater processes. Biofilters are characterized by bacteria being attached to a solid surface in the form of a biofilm. Biofilm is a dense layer of bacteria characterised by their ability to adhere to a solid medium and form a fixed film of polymers in which the bacteria are protected against sloughing off.

The disadvantage of biofilters can be the low efficiency of the biomass. The reason is that the substances must be carried through the biofilm to be removed by the bacteria. This transport takes place by molecular diffusion, which is a slow process. In practice it proves that the general rule is that the removal is limited by diffusion. This phenomenon must be understood in order to understand the functioning of biofilters.

4.2.1 Mass balances for biofilters

Normally, it is not necessary to recycle sludge over a biofilter plant as the sludge concentration becomes sufficiently large by means of the biofilm and a sufficiently large specific solid surface. As recycle of sludge is not necessary, we can, in certain cases, leave out a secondary settling tank. Normally, however, the sludge present in the water, which has passed the biofilter, needs a subsequent settling. The sludge comes from sloughed off biofilm and from the content of suspended solids in the influent water.

4.2.1.1 Biofilters without recycle

A mass balance for material over the biofilter itself is shown in Figure 4.1.

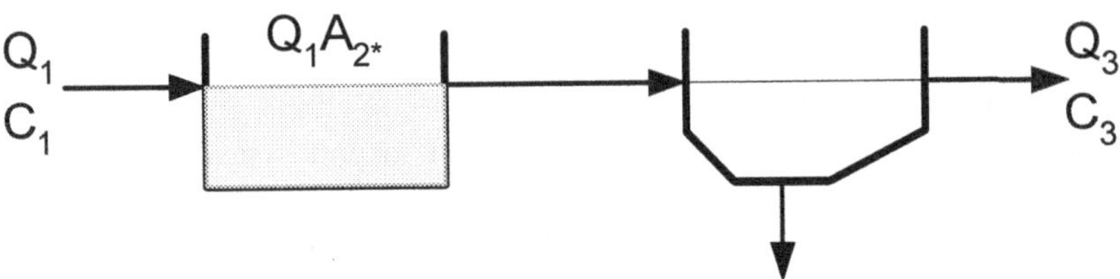

Figure 4.1 Biofilters without recycle.

The mass balance is shown as follows:

$$Q_1 . C_1 - r_{v,s} . V_2 = Q_3 . C_3 \qquad (4.1)$$

where Q_1 = flow to the filter
           C_1 = concentration in the influent
           $r_{V,S}$ = volumetric removal rate of the filter
           V_2 = volume of aeration tank
           Q_3 = flow leaving the settling tank
           C_3 = effluent concentration of the settling tank

The removal can also be expressed by means of the sludge concentration, X_2, as follows:

$$r_{V,S} \cdot V_2 = r_{X,S} \cdot V_2 \cdot X_2 \tag{4.2}$$

As X_2 should only include the active biomass, which is usually unknown, it is frequently common practice to relate the removal to the volume or the surface of the surface. If the removal is expressed per unit area of surface, Equation (4.2) is transformed to:

$$Q_1 \cdot C_1 - r_{A,S} \cdot A_{2*} = Q_3 \cdot C_3 \tag{4.3}$$

where

$r_{A,S}$ = the removal rate per unit area of surface (unit: kg COD/m^2.d)

A_{2*} = the overall area of the carrier (unit: m^2)

4.2.1.2 Biofilters with recycle

Most biofilter plants have recycle which is normally made by recycling directly over the filter as shown in Figure 4.2.

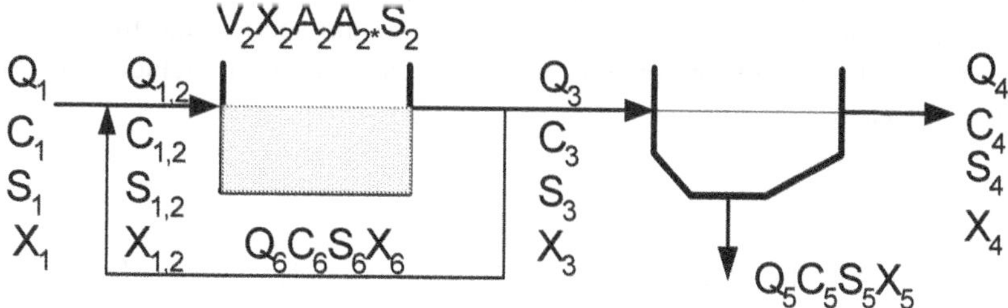

Figure 4.2 Biofilters with recycle.

The recycle ratio R, is defined as:

$$R = \frac{Q_6}{Q_1} \tag{4.4}$$

The purpose of recycling is to ensure a sufficient flow rate of wastewater and biomass through the biofilter. This is important for most types of biofilter plants. Also, the recycle lowers the influent concentration to the biofilter and can therefore affect the process, both through the order of reaction and as to whether oxygen is limiting or not. The equation for the removal of substrate will be the same as for biofilters without recycle, but the reaction term will be influenced by the recycle.

4.2.2 Concept and definitions for biofilters

Treatment efficiency, based on Figure 4.2, is defined as follows:

$$E = \frac{C_1 - C_4}{C_1} \tag{4.5}$$

The treatment efficiency for practise includes the effect of the secondary settling. Recycle ratio is defined as shown in Equation 4.4. Volumetric loading rate is defined as follows:

$$B_V = \frac{Q_1 . C_1}{V_2} \tag{4.6}$$

The mass flow of material added per unit volume per unit time, $Q_1 . C_1$, is calculated at a position prior to a possible recycle. The volumetric loading rate cannot be increased by an internal recycle in the plant. Biofilm surface area has two meanings, the area of the filter tank in a horizontal section and the surface area of the surface. Biofilm area is normally calculated to include specific surface of the carrier, as follows:

$$A_2 = \omega . V_2 \tag{4.7}$$

where ω is the specific surface of the carrier (for example m^2/m^3). The surface area in a horizontal section, A_2, is used for the calculation of the hydraulic surface loading rate.

Organic surface loading rate is an expression of the mass flow to the surface of the carrier in the filter as follows:

$$B_{A,C} = \frac{Q_1.C_1}{A_{2*}}$$ (4.8)

This is the formula, which comes closest to the kinetic conditions in the filter. In addition, the hydraulic surface loading rate (or nominal filter rate) is defined by the symbols in Figure 4.2 as:

$$B_{A,V} = \frac{Q_{1,2}}{A_2} = \frac{Q_1 + Q_6}{A_2}$$ (4.9)

This parameter is a simple expression of the flow passing the biofilm and hence an expression of the hydraulic erosion, which is considered to be crucial for the control of the thickness of the biofilm through sloughing.

Sludge production, F_{SP}, is the mass of sludge leaving the plant per unit time. Based on Figure 4.2, the sludge production can be calculated as follows (if measurements are available)

$$F_{SP} = Q_5 . X_5 + Q_4 . X_4$$ (4.10)

If sludge yield is used, the sludge production can be estimated as:

$$F_{SP} = Y_{obs} (C_1 - C_3) . Q_1$$ (4.11)

Surplus sludge production is the part of the sludge production to be further treated in the plant that is $Q_5 . X_5$ as shown in Figure 4.2.

4.2.3 Design of biofilters

Based on experiences from many years operation of conventional biofilters, a set of design criteria has been established, which are generally used for the practical design of these plants. These criteria are based on the idea that a given type of plant can be loaded hydraulically, or with organic matter up to a limit, which can be determined for ordinary use in the design phase. This may sometimes be combined with requirements for the retention time in the system. These design criteria are not based on any special understanding of the phenomena on which the treatment is based, but they are exclusively based on a systematic approach and experiences from an endless number of thoroughly measured plants. The criteria are very simple, just like it is remarkable that in spite of this simplicity they have actually been most useful for decades. Likewise, there is no doubt that they have given rise to design errors because the circumstances change from one plant to another to a greater extent than it can be expressed in such simple relationships, as it will be illustrated for trickling filters and for rotating disks.

4.2.3.1 Design of trickling filters

According to the conventional design criteria for trickling filters, the treatment results exclusively depend on the organic volumetric loading rate and the hydraulic surface loading rate. Based on these parameters and many years of experience in Germany, a design criteria for trickling filters is shown in Table 4.1.

Table 4.1 Criteria design for trickling filters used in Germany.

Load category	Low	Moderate	Normal	High	Unit
Organic volumetric loading rate	200	200 - 450	450 - 750	> 750	g BOD/m^3.d
Hydraulic surface loading rate	approx. 0.2	0.4 – 0.8	0.6 – 1.2	> 1.2	m/h
Estimated treatment efficiency	92 ± 10	88 ± 12	83 ± 15	75 ± 20	%
Estimated effluent concentration	< 20	< 25	20 - 40	30 - 80	g BOD/m^3

The German criteria (Treibel, 1975) states the following treatment which is to be expected:

$$E = 93 - 0.017\,B_V \tag{4.12}$$

for $B_V < 1{,}000$ g BOD/m^3.d (unit: %)

There is a high uncertainty as to the observation. For the settled effluent from an ordinary trickling filter loaded with domestic wastewater where $B_V = 400$ g BOD/m^3.d, a 24 hours flow weight composite sample will show concentrations in the range $5 - 20$ g BOD/m^3; for $B_V = 1{,}000$ g BOD/m^3.d in the range $20 - 50$ g BOD/m^3.

It is recommended to carry out the design on the basis of $B_V = 400$ g BOD/m^3.d, 0.5 m/h $< B_{A,V} < 1.0$ m/h and $R < 1.0$. For trickling filters with plastic media, it is recommended that $B_A = 4$ g BOD/m^3.d, $0.8 < B_{A,V} < 1.8$ m/h in the range $100 < \omega < 200$ m^2/m^3 and $R < 1$. For these loadings, no nitrification will take place.

The values in Table 4.1 only apply to ordinary municipal wastewater, which is considered to have a relatively homogenous composition. All the data of treatment results (expressed as treatment efficiency, or as concentration in the effluent), which can be found in high numbers in the literature, are characterized by a very large spread. This indicates that these simple loading rules cannot take into account numerous special circumstances, which can occur in practice. For example, filter medium, surface area, influent concentration and the special property of the organic matter have not been taken into consideration. A design engineer has to be creative and critical using the above loading values, although this approach has been the basis of the successful design of numerous existing full-scale trickling filters.

Example 4.1
A trickling filter with a diameter of 10 m and a height of 2 m is loaded with 235 m^3/d of mixed domestic and industrial wastewaters. The concentration of BOD is 500 g/m^3. Find the volumetric loading rate and the necessary recycle.

Solution:
Using Equation 4.6, the volumetric loading rate can be calculated as follows:

$$B_V = Q_1 . C_1/V_2$$
$$= Q_1 . C_1/(\pi . r^2 . h)$$

$$B_V = 235 \text{ } m^3/d . (500 \text{ g BOD}/m^3)/(\pi . 5^2 m^2 . 2m)$$
$$= 748 \text{ g BOD}/m^3.d$$

Hence the filter is interfacing normal and high loading. The necessary recycle is controlled by the requirement of the hydraulic surface loading rate which, based on Table 4.1 in this case, is estimated at 1.2 m/h.

From Equation 4.9, the necessary recycle, Q_6, can be calculated as:

$$B_{A,V} = (Q_1 + Q_6)/A_2$$
$$1.2 \text{ m/h} . 24 \text{ hr/d} = (235 \text{ } m^3/d + Q_6)/(\pi . 5^2 m^2)$$
$$Q_6 = 2{,}027 \text{ } m^3/d$$

$$R = Q_6/Q_1$$
$$= (2{,}027 \text{ } m^3/d)/(235 \text{ } m^3/d)$$
$$= 8.6$$

This is a high recycle ratio value. It appears from Table 4.1 that the requirement of the hydraulic surface loading rate, and hence the recycle, increases by increasing the organic volumetric loading rate.

4.2.3.2 Design of rotating biological contactors or discs

Some of the rules for rotating biological contactors or discs are very good examples to show how different the loading values could be considered. The disks and their design rules were developed in the seventies. Table 4.2 lists North American and European design values and recommended loadings. It clearly appears that there is no agreement and that the variation is quite significant: from 5 to 26 g BOD/m^2.d. It also appears that the more optimistic values are those given by consulting companies. It also can be noticed that the allowable load is stated per m^2 surface of the discs, which is a theoretical, much more relevant size to relate the treatment to, than to m^3 of the tank in which the discs rotate.

Table 4.2 Recommended loading rate for discs.

Recommended loading g BOD/m^2.d	Year	References
14	1974	Steels (1974)
8.5	1978	Murphy and Wilson (1978)
26	1976	Autotrol Corp. (1976)
13	1977	Envirodisc Corp. (1977)
4.9	1974	Ontario Ministry of Environment (1974)
8 - 10	1980	Abwassertechnische Vereinigung (1989)

The biofilm growth on rotating discs is regulated by the rotational speed, which is usually stated as a requirement of a peripheral speed on the discs of not less than 0.3 m/s. To this should be added a minimum requirement of the distance between the discs, which is usually stated at 1.5 – 2.5 cm.

4.2.3.3 Other type of filters

None of the submerged filter types have gained much common practical use that real recommendations are available in respect of loading. The biggest volumetric loading has been achieved with fluidized filters, which could be loaded with 10 kg BOD/m^3.d.

4.2.4   Technical conditions concerning biofilters

A number of important process conditions will be briefly discussed below. The process or technical conditions that are important for the optimum function of biofilter plants include:

- aeration
- growth and sloughing off of the biofilm

4.2.4.1   Aeration of biofilters

The supply of oxygen to biofilters is of greater importance than for other biological plants because the oxygen supply is not just stoichiometrically decisive, but frequently it is also governs the reaction rate. The following types of aeration can be used:

- separate aeration of the flow of recycle
- aeration in the filter itself
- aeration of the biofilm

4.2.4.2 Growth and sloughing off the biofilm

Biofilters are based on the ability of the bacteria to attach and develop on a solid medium. Technically, it is desirable that a stable state with equilibrium between growth and sloughing off is developed. No matter how fundamental these conditions might be it is, however, a fact that too little is known about these conditions, apart from the very technical empiricism.

Adhesion is the mechanism in which micro organisms, particularly bacteria adhere to any surface or filter media. In fact, a partiality for adhesion occurs, irrespective of the type of surface (except where counter-measures have been taken, for example in marine coating). The start of a biofilm is simply that the medium concerned is in contact with water, in this case with the wastewater. In practice, the development of a functional biofilm takes approximately 14 days under aerobic conditions. The building up is commenced faster on surfaces with earlier growth than on completely clean surfaces. The building up process is selective and the bacteria that are not attached will simply be washed out of the plant. For special substances and processes, starting up may be a big problem.

Types of biofilm indicate that the selected bacteria can grow in certain environmental or process conditions. No study at the moment has been conducted dealing with the relationship between microbial determination of species and measured biofilm characteristics. It can only be concluded that two types of biofilms generally occur: (a) the dense film and (b) the filamentous film. The dense film is a coherent, immobilized biomass. In its ideal form, it has a plane, smooth surface. For this type of biofilm there is clear experimental proof of the good and theoretical description of the removal. The filamentous biofilm is dominated by filamentous bacteria that grow in continuation of each other and form filaments attached to the filter medium, typically the Chlamydobacteriales *Sphaerotilus natans*, also known from highly polluted streams under the name of sewage fungus. In its extreme form, this type of film forms a mat whose filaments are waving in the liquid flow. This causes turbulent movements around the filaments for which reason the transport of substrate into the biofilm is increased compared with molecular diffusion. These films will therefore remove material according to a zero order reaction.

The bacterial density in the film is decisive for the removal rate. The mass of the bacteria, which carry out the process is difficult to determine quantitatively by measurement, but it can be calculated as a fraction of the total mass in the same way as for the bacteria in the sludge in an activated sludge plant. The total mass may vary within wide ranges: $10 - 100$ kg VSS/m^3; but data primarily fall in the range $40 - 60$ kg VSS/m^3.

Sloughing off the biofilm is due to the unbalanced condition between organic removals and continuous growing of the bacteria. Sloughing off may cause media clogging. The following conditions are important for the sloughing off and are used technologically, intentionally or unintentionally, as part of the control of the biofilm:

- Hydraulic erosion acts continually on the surface of the biofilm and leads to a steady sloughing off on the outer side. In its extreme version it is the total sloughing off of the biofilm on the sand in the recycle flow of a fluidized filter in a tank where a stirrer creates a very high turbulence level. The requirements of the hydraulic loading of trickling filters and of the rotational speed of rotating discs are empirically dictated from a desire to cause sloughing off, but the flow of water is not so strong that it can by itself cause the sloughing off. Other factors release the film so that it can be sloughed off hydraulically.
- Degradation of starved bacteria in the bottom of biofilms may cause a weakening of the adhesion. Biofilms will in practice have a tendency to grow to a thickness where they are only partially penetrated with substrate. In the aerobic filter, whose reaction rate is controlled by the oxygen, anaerobic conditions will occur in the bottom of the film. This will degrade the bacteria in the bottom and destroy the adhesion. Sufficiently weakened,

the film will be sloughed off completely over a smaller area by hydraulic erosion.

- Super saturation and bubble formation in the bottom of the biofilm may destroy the adhesion – examples are methane production under anaerobic conditions and the production of pure nitrogen by denitrification.

In the last two cases the film is sloughed off completely, and a naked area is left on the filter medium where new growth starts. The filter is thus continuously in a state of sloughing off and regrowing. Therefore the biofilm never has a well-defined thickness in practice, which applies to the whole film.

A special phenomenon is the grazing of higher animals on the bacterial mass in the filter. Very little is known about this, but it is undoubtedly of importance and is for example reflected in seasonal variations in the biomass in a trickling filter and in sudden changes in the biofilm in for example nitrifying plants. Occasionally all biofilms disappear. A conventional trickling filter will contain most biomass in the early spring which makes the risk of clogging correspondingly higher.

4.3 ACTIVATED SLUDGE TREATMENT SYSTEM

Activated sludge treatment plants, also known as dispersed growth systems, have been designed since early 1900s, primarily to remove organic matters and recently nutrient removal from wastewater. To date, many types of activated sludge plants have been successfully built to achieve its design objectives, also in developing countries. The principle in an activated sludge plant is that a mass of "active" sludge is kept moving in wastewater by stirring or aeration. Apart from the living biomass, the suspended solids in activated sludge contain inorganic matters, as well as organic particles. Some of the organic particles can be degraded by a hydrolysis process, whereas others are non-degradable or inert.

The amount of suspended solids in the treatment plant is regulated through recycle system of the suspended solids, and by removing the so-called excess sludge from the system. Organic matter that enters an activated sludge process has three major outlets, i.e.: carbon dioxide gas, excess sludge or the effluents.

4.3.1 Mass balance in activated sludge plant

The schematic diagram for an activated sludge plant is shown in Figure 4.3. The purpose of recycle is to increase the "active" sludge concentration in the aeration tank. As a result, sludge age (or also known as solid retention time) is increased and the value of hydraulic retention time can be separated from the sludge age.   Having the recirculation, the biomass can be accumulated consisting of rapid- and/or slow-growing micro-organisms. The flocculated biomass will determine the settling characteristics and facilities. The settled concentrated in the settling tank is either recycled or disposed.   Higher recirculation rates increase the sludge concentration (thus increase the sludge age) in the aeration tank to indicate a high concentration of active biomass, which leads to a smaller size of aeration tank.

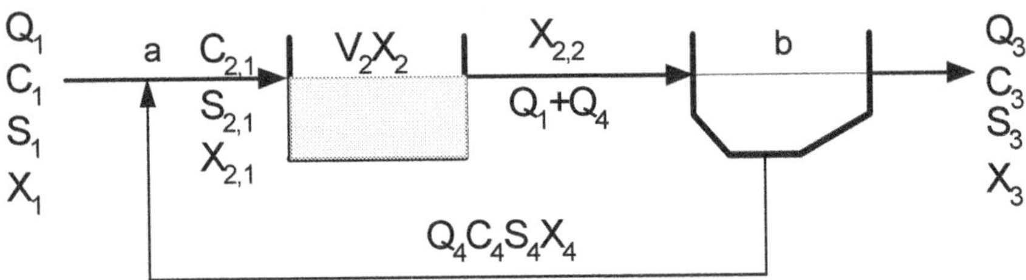

Figure 4.3 Schematic diagram of conventional activated sludge plant.

A mass balance can be set up at points *a* and *b*. At point *a*, the mass balance of the sludge to be discharged to the aeration tank ($X_{2.1}$) can be calculated as:

$$Q_1 . X_{B,1} + Q_4 . X_{B,4} = (Q_1 + Q_2) . X_{B,2.2} \qquad (4.13)$$

An equivalent mass balance at point *b* may for example be used to calculate the flow rate for return sludge, Q_4, if the sludge concentrations are known which flow from the top and bottom of the settling tank ($C_{B,3}$ and $X_{B,4}$) as well as the sludge concentration in the effluent of the aeration tank ($X_{B,2.2}$).

$$(Q_1 + Q_4) . X_{B,2.2} = Q_3 . X_{B,3} + Q_4 . X_{B,4} \qquad (4.14)$$

$$Q_4 = \frac{Q_1.X_{B,2.2} - Q_3.X_{B,3}}{X_{B,4} - X_{B,2.2}} \qquad (4.15)$$

The concentration of sludge increases while passing through the aeration tank. A sludge mass balance around the aeration tank itself, as shown in Figure 4.1, may be used to calculate the effluent concentration, $X_{B,2.2}$ as follows:

$$(Q_1 + Q_4) . X_{B,2.1} + (Q_1 + Q_4) . (C_{2.1} - C_{2.2}) . Y_{obs} = (Q_1 + Q_4) . X_{B,2.2} \qquad (4.16)$$

or

$$X_{B,2.2} = X_{B,2.1} + (C_{2.1} - C_{2.2}) \cdot Y_{obs} \tag{4.17}$$

Example 4.2 illustrates the method to calculate the effluent concentration from the aeration tank. The amount of sludge concentration in the aeration tank itself, $X_{B,2}$, depends on the hydraulics. In an ideal mixed aeration tank such as an activated sludge process, $X_{B,2} = X_{B,2.2}$.

Example 4.2
Calculate the effluent concentration, $X_{B,2.2}$, of activated sludge from the aeration tank of an activated sludge plant. The following data is known:

$X_{B,2.1} = 4.5$ kg SS/m^3
$C_1 = 0.4$ kg BOD/m^3
$C_3 = 0.015$ kg BOD/m^3
$Y_{obs} = 0.9$ kg SS/kg BOD
$Q_1 = 2,500$ m^3/d
$Q_4 = 2,000$ m^3/d

Solution:
$X_{B,2.2}$ is calculated using Equation 4.17:

$$X_{B,2.2} = X_{B,2.1} + (C_{2.1} - C_{2.2}) \cdot Y_{obs}$$

$C_{2.1}$ and $C_{2.2}$ are not known. It is normally assumed that no reactions take place in the settling tank, that is $C_{2.2} = C_3$.

$C_{2.1}$ can be found by means of a BOD-balance at point a as shown in Figure 4.3:

$$Q_1 \cdot C_1 + Q_4 \cdot C_4 = (Q_1 + Q_4) \cdot C_{2.1}$$

Assuming that removals only take place in the aeration tank, $C_4 = C_3 = C_{2.2}$, substitution of these values gives:

$$C_{2.1} = (Q_1 \cdot C_1 + Q_4 \cdot C_3) / (Q_1 + Q_4)$$

Substitution of these values by known data or values gives:

$$
\begin{aligned}
C_{2.1} &= (2,500 \cdot 0.4 + 2,000 \cdot 0.015) / (2,500 + 2,000) \\
&= 0.229 \text{ kg BOD/m}^3
\end{aligned}
$$

In addition by substitution of these calculated and known values in Equation 4.6 gives:

$$X_{B,2.2} = 4.5 + (0.229 - 0.015) \cdot 0.9 = 4.69 \text{ kg SS/m}^3.$$

Example 4.2 shows that the variation from the influent entering the aeration tank to the effluent leaving the tank is, in this case, only an increase of the sludge concentration of 4 – 5 %. Considering the uncertainties in the determination of the incoming quantities, the variation may often be ignored and the sludge concentration may be assumed to be constant throughout the tank.

NOTE. The variation **cannot** be ignored during the process of calculating the sludge production, which takes place in the plant because the variation constitutes the sludge production. Therefore, after all designed calculations have been conducted, the design engineer should rationalise the practicality of the results.

4.3.2 Concept and definitions of the activated sludge process

There are many concepts, jargons and definitions related to activated sludge plants. Some of the concepts and definitions are not very applicable and logical, but it is useful to know what their meanings are when other designers use them. Some of the concepts are briefly discussed below.

Treatment efficiency is normally used to describe the overall efficiency of the treatment process, that is both the aeration and settling tanks. The efficiency is defined as follows, with typical E value around 80 to 95 %:

$$E = \frac{(C_1 - C_3)}{C_1} \tag{4.18}$$

Recycle rate is defined as the ratio between the flow rates of return sludge and raw wastewater, with typical R-value around 50 to 100 % as follows:

$$R = \frac{Q_4}{Q_1} \tag{4.19}$$

Volumetric loading in an activated sludge is defined as inlet flow rate and concentration within a volume of bioreactor. The unit of the loading is kg BOD/m^3.d and was used in the early design of activated sludge process.

$$B_V = \frac{Q_1.C_1}{V_2} \tag{4.20}$$

Sludge concentration (or biomass concentration), X, of wastewater in the aeration tank, or in the flow of recycle sludge is often given in the unit of suspended solids, X_{SS}, volatile solids, X_{VSS}, or as COD, X_{COD}. Calculations of the activated sludge processes can be made in any of these required units. For example, if SS unit is used, all quantities must be calculated in this unit (for example: yield constant, sludge mass, sludge production, removal rate, etc.). It should be noted that only the part of the sludge concentration measured, is considered as active biomass (living bacteria), which can be shown as follows:

$$X_{COD} > X_{B,H} + X_{B,A} \tag{4.21}$$

where $X_{B,H}$ = the concentration of heterotrophic biomass

$X_{B,A}$ = the concentration of autotrophic biomass (nitrifying bacteria)

Sludge mass, M_X, in an activated sludge plant comprises any sludge in the part of the treatment plant handling activated sludge, including the sludge, which is not currently being aerated in the activated sludge tank. Sometimes large sludge masses may be hidden in secondary settling tanks. The sludge mass is defined as:

$$M_X = \Sigma \, V . X \tag{4.22}$$

Hydrolysis is carried out in the whole sludge mass, irrespective of the location of the sludge, however, at various rates depending on the process conditions in the individual tanks.

Sludge loading states the amount of organic matter applied to the sludge per day:

$$B_X = \frac{Q_1.C_1}{V_2.X_2} \tag{4.23}$$

The sludge concentration is often expressed as kg SS/m^3 and the concentration of organic matters, C_1 as kg BOD/m^3. Hence the sludge loading is expressed as kg BOD/kg SS.d. The sludge loading is often used as a design parameter. When using the unit SS, attention should be paid to the wastewater conditions (for example, the suspended solids) or to the operation of the plant (for example, the addition of precipitants) which may modify the ratio between VSS and SS in the activated sludge. A more reliable unit for sludge loading for design purposes is kg BOD/kg VSS.d.

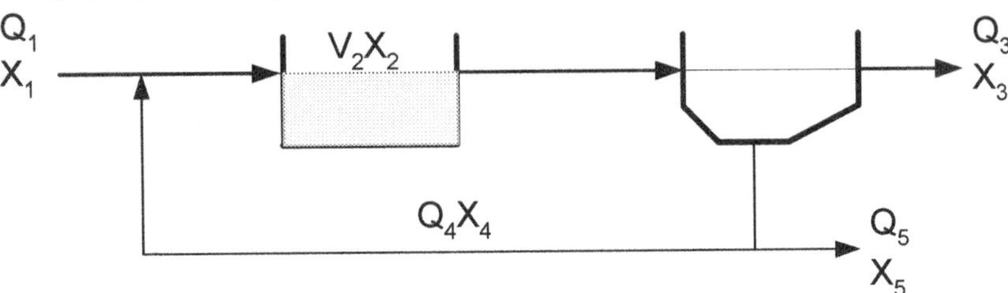

Figure 4.4 Activated sludge plant with sludge recycle and excess sludge withdrawal.

Sludge production, F_{SP}, states the quantity of sludge leaving the plant per unit time. Based on Figure 4.4, sludge production can be calculated as follows:

$$F_{SP} = Q_3 \cdot X_3 + Q_5 \cdot X_5 \tag{4.24}$$

The sludge in the influent ($Q_1.X_1$) is included in the sludge production and should therefore not be deducted in Equation 4.24. The sludge production can be measured in many ways and hence be given in different units like kg SS/d, kg VSS/d and kg COD/d. Table 4.3 shows that components must be included in different measuring methods for sludge and sludge production.

Table 4.3 Measuring method for sludge and sludge production.

Component / Sludge production	Inorganic SS, raw wastewater	Organic SS, raw wastewater	Biological growth, aeration tank	Chemical precipitation
kg SS/d	+	+	+	+
kg VSS/d		+	+	(+)
kg COD/d		+	+	

Different units for the sludge production are needed, depending on the purpose of applications. For sludge dewatering purposes, the unit kg SS/d is most relevant. In case of an anaerobic or an aerobic sludge stabilisation, the unit kg VSS/d or kg COD/d is more appropriate. The volume of the sludge production thus depends both on the influent of wastewater to the plant and on the process in the plant.

The sludge production can be **measured** in an existing plant and is calculated from Equation 4.24. The sludge production can be **estimated** in a design situation using Equation 4.25.

$$F_{SP} = Y_{obs} (C_1 - C_3) \cdot Q_1 \tag{4.25}$$

In order to be able to estimate the sludge production using Equation 4.25, Y_{obs} values must be estimated. The values vary as seen in Figure 4.5. Example 4.3 illustrates the calculation of the sludge production.

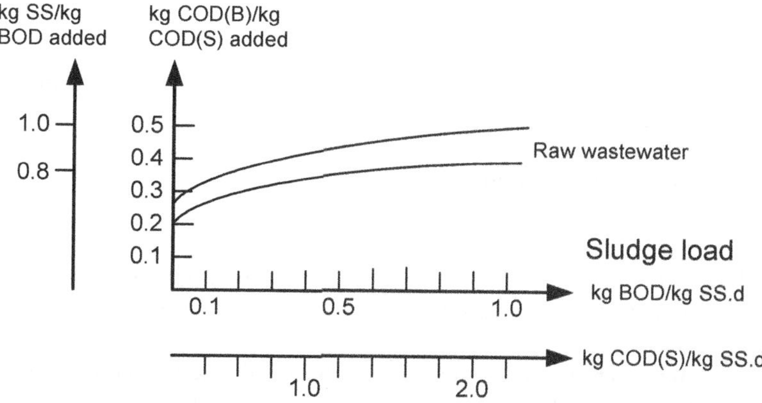

Figure 4.5 Y_{obs} values for the activated sludge process.

Example 4.3
Calculate sludge loading and sludge production for an activated sludge plant for the treatment of municipal wastewater in a city of Johor Bahru, Malaysia. The following data is provided by the project owner:

$Q_1 = 11,000 \text{ m}^3/\text{d}$

$C_1 = 0.4$ kg BOD/m^3
$C_3 = 0.04$ kg BOD/m^3

The concentration of activated sludge in the aeration tank is set at 5.2 kg SS/m^3. The volume, V_2, of the aeration tank is calculated at 5,600 m^3.

Solution:
The sludge loading is calculated using Equation 4.23:

$$B_X = Q_1 . C_1/(V_2 . X_2)$$

Substitution of these value gives:

$$B_X = 11,000 \times 0.4/(5,600 \times 5.2)$$
$$= 0.15 \text{ kg BOD/kg SS.d}$$

The sludge production is estimated from Equation 4.25:

$$F_{SP} = Y_{obs} . (C_1 - C_3) . Q_1$$

Y_{obs} value is estimated using Figure 4.5 at 0.8 kg SS/kg BOD. Substitution of these value gives:

$$F_{SP} = 0.8 (0.4 - 0.04). 11,000$$
$$= 3,170 \text{ kg SS/d}$$

According to Figure 4.4, a small amount of sludge is leaving the plant with the treated wastewater. It corresponds to the quantity of $Q_3.X_3$.

Excess sludge production, F_{ESP}, is the portion of the sludge production, which is actively withdrawn and further, treated at the sludge treatment plant. Using the symbols from Figure 4.1, the quantity is $Q_5.X_5$. The excess sludge production can be withdrawn from the recycle flow as shown in Figure 4.4, together with the primary sludge by leading it to the influent of the sludge treatment plant if there is mechanical treatment. If sludge age control is required in the plant, direct excess sludge withdrawal from the aeration tank is the best solution.

Sludge age is the mean cell residence time of sludge (biomass) in the plant. It can be estimated on the basis of measurements on an existing plant as shown in the following equation:

$$\theta_X = \frac{M_X}{F_{SP}} \tag{4.26}$$

Based on Figure 4.3:

$$M_X = V_2 . X_2 \tag{4.27}$$

$$F_{SP} = Q_3 . X_3 \tag{4.28}$$

and, $X_2 = X_3$ for a completely mixed reactor,

$$\theta_X = \frac{V_2 . X_2}{Q_3 . X_3} = \frac{V_2}{Q_3} = \theta \tag{4.29}$$

where θ_X = sludge age
θ = hydraulic retention time

In the case of no sludge recycle, the retention time for the biomass and liquid are identical. It is, however, not the case with recirculation. The main idea of recycling is to increase sludge age over the hydraulic retention time.

For a plant scheme as shown in Figure 4.4:

$$M_X = V_2 . X_2 \tag{4.30}$$

(the sludge mass in the settling tank being set at 0 is not quite correct!)

$$F_{SP} = Q_3 . X_3 + Q_5 . X_5$$
$$= Y_{obs} . (C_1 - C_3) . Q_1 \tag{4.31}$$

Substitution into Equation 4.29 gives:

$$\theta_X = \frac{V_2 . X_2}{Y_{obs}.(C_1 - C_3).Q_1} = \frac{\theta . X_2}{Y_{obs}.(C_1 - C_3)} \tag{4.32}$$

The calculation of sludge age is shown in Example 4.4.

Example 4.4

Calculate the hydraulic retention time in the aeration tank of an ideal mixed system, and its sludge age for the following plant:

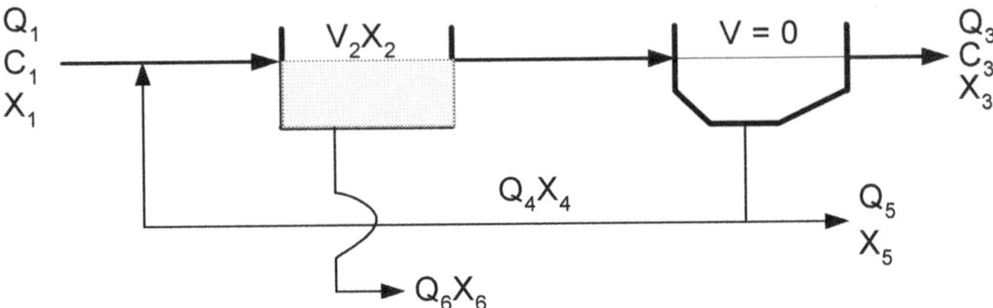

Figure 4.6 Ideal mixed system.

The following data is known, provided by the project owner:

$V_2 = 15{,}000 \text{ m}^3$
$X_2 = 4.0 \text{ kg COD/m}^3$
$Q_1 = 1{,}500 \text{ m}^3/\text{h}$
$Q_4 = 750 \text{ m}^3/\text{h}$
$Q_5 = 100 \text{ m}^3/\text{d}$
$Q_6 = 1{,}000 \text{ m}^3/\text{d}$
$X_3 = 0.05 \text{ kg COD/m}^3$
$X_5 = 11.0 \text{ kg COD/m}^3$
$Y_{obs} = 0.4 \text{ kg COD(B)/kg COD(S)}$

Solution:

The hydraulic retention time is:

$$\theta = V_2/Q_1$$
$$= 15{,}000/1{,}500 = 10 \text{ h}$$

However, it is important to note that the hydraulic retention time is **not**:

$$\theta = V_2/(Q_1 + Q_4)$$
$$= 15{,}000/(1{,}500 + 750) = 6.7 \text{ h},$$

(as it is the average time for passage through the tank for a water particle).

The sludge age or solid retention time, θ_X, is calculated from Equation 4.26:

θ_X = M_X/F_{SP}
M_X = $V_2 . X_2$ (the volume of the settling tank = 0, so that no sludge is present)
F_{SP} = $(Q_3 . X_3) + (Q_5 . X_5) + (Q_6 . X_6)$

Substitution of this value gives:

θ_X = $V_2 . X_2/(Q_3 . X_3 + Q_5 . X_5 + Q_6 . X_6)$

Here all quantities are known with the exception of X_6 and Q_3. As the Q_6 water flow is withdrawn from the aeration tank, $X_6 = X_2$. Q_3 is found from a water balance for the whole plant.

Q_1 = $Q_3 + Q_5 + Q_6$
Q_3 = $Q_1 - Q_5 - Q_6$
      = $1,500 . 24 - 100 - 1,000$
      = $45,900 \ m^3/d$

Substitution of Q_3 and X_6 and other known quantities give:

θ_X = $15,000 . 4.0/(34,900 . 0.05 + 100 . 11.0 + 1,000 . 4.0)$
      = 8.8 d

NOTE. This example gives more information than needed to come up with an adequate solution (redundant information). Very often design engineers have problems with insufficient information, and such systems are therefore likely to be subjected to engineering estimates.

Aerobic sludge age is important for nitrification processes and for processes where in particular slowly degradable, e.g. xenobiotics compounds must be removed biologically. The aerobic sludge age states the average time during which a sludge particle (for example, nitrifying bacteria) stays in the plant under aerobic conditions. The aerobic sludge age is usually shorter than the sludge age (the total sludge age). However the definition of aerobic sludge age is parallel to the sludge age:

$$\theta_{X, aerobic} = \frac{M_{X, aerobic}}{F_{SP}} \tag{4.33}$$

In a tank with aerobic conditions, half of the time (and in the other half for example anoxic conditions), the aerobic sludge mass will be 50 % of the total sludge mass, that is 0.5 . V. X.

4.3.3 Design of the activated sludge processes

The design of an activated sludge plant can use one of the following methods, but only the first two methods will be discussed in detail:
- volumetric loading
- sludge loading
- Computer-aided process design.

Computer-aided process design requires a special treatment, and is not covered in this book. For information on models and their use, see e.g. Henze *et al.*, 2000.

4.3.4 Design using volumetric loading

The design is based on the BOD-volumetric loading as follows:

$$B_{V,BOD} = \frac{Q_1.C_1}{V_2} \tag{4.34}$$

This design method corresponds historically to the first one to be used. If the treatment plants receive wastewater of a uniform composition, and if they are operated in such a way that the sludge concentration in the aeration tank is the same from plant to plant, reasonable results can be obtained with this simple procedure. In reality, however, it is seldom that conditions are so uniform from plant to plant.

Table 4.4 states simple design quantities for activated sludge plants for ordinary municipal wastewaters commonly used in Europe at the average temperature 10°C with influent concentration of 0.2 to 0.4 kg BOD/m^3 and 0.4 to 0.8 kg COD/m^3. At lower temperatures, the operation of the plants will be less efficient, and vice versa for higher temperatures.

The necessary aeration tank volume, V_2, can be found from the following equation:

$$V_2 = \frac{Q_1.C_1}{B_{V,BOD}} \tag{4.35}$$

Table 4.4 Characteristics of activated sludge plants treating domestic wastewater in Europe (Henze *et al.*, 1997).

Parameters	Symbols	Units	Sludge loading, kg BOD/kgSS.d		
			0.05 - 0.10	0.20 - 0.30	0.50 - 0.60
Sludge concentration	X_{ss}	kg SS/m^3	4.0 - 7.0	3.0 - 5.0	3.0 – 5.0
Sludge concentration	X_{vss}	kg VSS/m^3	2.5 – 4.5	2.0 – 3.5	2.0 – 3.5
Sludge concentration	X_{COD}	kg COD/m^3	3.5 – 6.5	3.0 – 5.0	3.0 – 5.0
VSS of SS in sludge	-	%	65 - 70	70 - 75	70 - 75
Volumetric loading	$B_{V,BOD}$	kg BOD/(m^3.d)	0.2 – 0.6	0.6 – 1.5	1.5 – 3.0
Sludge loading	$B_{X,BOD}$	kg BOD/(kg VSS.d)	0.08 – 0.15	0.3 – 0.45	0.7 – 0.85
Yield constant	Y_{BOD}	kg SS/kgBOD	0.6 – 0.9	0.9 – 1.1	0.9 – 1.2
Yield constant	Y_{COD}	kg COD/kg COD	0.3 – 0.5	0.4 – 0.6	0.4 - .6
Treatment efficiency	E_{BOD}	%	90 - 95	85 – 90	80 - 90
Treatment efficiency	E_{COD}	%	75 - 85	70 – 80	65 - 80
Sludge age	θ_X	d	15 - 20	3 - 6	1 - 3

The volumetric loading is a secondary (indirect) design parameter and should therefore be used with caution. It cannot be used for the design of more complicated processes. Calculation example is shown in Example 4.5.

Example 4.5

On the basis of the volumetric loading, design an activated sludge plant to treat wastewater with the given values shown in Table 4.5 to an effluent concentration of 0.020 kg BOD/m^3.

The dry-weather and wet-weather treatment efficiency for BOD, respectively, will be:

$E_{BOD} = (240 - 20)/240 = 92\ \%$ for dry-weather
$E_{BOD} = (160 - 20)/160 = 88\ \%$ for wet-weather

Table 4.4 shows that the volumetric loading, $B_{V,BOD}$, must be in the range of $0.2 - 0.6$ kg BOD/m^3.d in dry weather to obtain a $90 - 95\ \%$ BOD removal. In wet-weather the volumetric loading can be $0.6 - 1.5$ kg BOD/m^3.d. It means that the design must follow the dry-weather condition of loading about $B_{V,BOD} = 0.4$ kg BOD/m^3.d.

Table 4.5 Given wastewater characteristics for a town on dry- and wet-weather conditions (Henze *et al.*, 1997).

Parameters	Units	Dry-weather	Wet-weather
Q_1	m^3/d	5000	9000
	m^3/h (max)	420	750
C_1	kg BOD / m^3	0.240	0.160
	kg COD / m^3	0.510	0.360
	kg TN / m^3	0.040	0.028
	kg TP / m^3	0.012	0.007
$Q_1.C_1$	kg BOD / d	1,200	1,440
	kg COD / d	2550	3240
	kg TN / d	200	252
	kg TP / d	60	63

Solution:
Based on Equation 4.35, the volume is:
$$V_2 = Q_1 . C_1/B_{V,BOD}$$
$$= (1{,}200\ kg\ BOD/d)/0.4\ kg\ BOD/m^3.d$$
$$= 3{,}000\ m^3.$$

Based on Figures in Table 4.5, the population equivalent represented by the quantity of the wastewater is:

$$PE_{water} = Q_1 / 0.2$$
$$= (5{,}000\ m^3/d) / (0.2\ m^3/(PE.d)$$
$$= 25{,}000\ PE$$

$$PE_{BOD} \quad = Q_1 . C_1 / 0.06$$
$$= (1,200 \text{ kg BOD/d}) / (0.06 \text{ kg BOD}) / PE.d$$
$$= 20,000 \text{ PE}$$

NOTE. The two calculations of the population equivalent will usually deviate from one to another. This is due to the influence of industrial discharges, infiltration, exfiltration, etc.

4.3.5 The design using sludge loading or sludge age

Using sludge loading or sludge age, a design engineer can carry out a more sophisticated design exercise. Sludge loading can be used for ordinary biological removals, whereas the sludge age should be used in connection with nitrification processes and other processes utilizing slowly growing bacteria for the removal of special pollutants in the wastewater, such as phenol and cyanide. When designing using BOD-sludge loading, the following equation is the basis:

$$B_{X,BOD} = \frac{Q_1.C_{BOD,1}}{X_2.V_2} \tag{4.36}$$

The volume, V_2, of the aeration tank can be calculated using the following equation:

$$V_2 = \frac{Q_1.C_{BOD,1}}{X_2.B_{X,BOD}} \tag{4.37}$$

Typical treatment performances, as a function of the sludge loading, have been shown in Table 4.4.

Sludge loading has been preferred by design engineers for many years, as the main design parameter for biological treatment plans. However, with the introduction of advanced processes such as biological phosphorus removal, simultaneous precipitation, nitrification and denitrification, it will in many cases be difficult or quite impossible to use sludge loading as the design basis.

If the sludge age is to be used as the design basis, the aerobic sludge age will in most cases be incorporated as the design basis. Equation 4.26 can be transformed to:

$$\theta_X = \frac{M_X}{F_{SP}} = \frac{V_2.X_2}{F_{SP}} \qquad\qquad (4.38)$$

$$V_2 = \frac{\theta_X.F_{SP}}{X_2} \qquad\qquad (4.39)$$

In this case it is assumed that $M_X = (V_2 . X_2)$. This is normally a good assumption if the aerobic sludge age is used as the design basis; and hence is included in the equation. Example 4.6 shows the calculation using the sludge age.

Example 4.6

Using the sludge loading, design an activated sludge plant to treat wastewater in Table 4.5 to an effluent concentration of 0.020 kg BOD/m^3. The sludge concentration in the activated sludge plant is in dry weather 4.5 kg SS/m^3 and in wet weather 4.0 kg SS/m3.

Solution:
From Example 4.5:

E_{BOD} = 92 % (dry weather)
E_{BOD} = 88 % (wet weather)

From Equation 4.21, is found:
$V_{2,dry\ weather}$ = (1,200 kg BOD/d) / (4.5 kg SS/m^3) . $B_{X,BOD}$

Table 4.4 shows that a sludge loading of 0.05 – 0.1 kg BOD/kg SS.d gives a treatment efficiency for BOD in the range of 90 – 95 %. As a treatment efficiency of 92 % is needed, the necessary $B_{X,BOD}$ ~ 0.07 (dry weather) is estimated. For wet weather $B_{X,BOD}$ ~ 0.2 is assumed.

$V_{2,dry\ weather}$ = 1,200/(4.5 . 0.007) = 3,810 m^3
$V_{2,wet\ weather}$ = 1,440/(4.0 . 0.20) = 1,800 m^3

The size of the aeration tank must be approximately 3,800 m^3, as the dry-weather loading is taken as the design basis. The size here is greater than 3,000 m^3, which was calculated in Example 4.5. The reason is that the sludge concentration in the aeration tank was lower. As design by means of volumetric loading does not take sludge concentration into consideration, the necessary volume is always underestimated. The result may be, at worst, that the plant built does not meet the effluent legal requirements.

4.4 HYBRID TECHNOLOGY

Hybrid system is a new technology, incorporating both biofilm and activated sludge systems. The state-of-the-art of this technology is still at the initial stage, mainly for packaged plants for industrial wastewater treatment. The common reason for using hybrid systems has been to enhance treatment efficiency where individual processes such as biofilm or activated system has not been effective.

The most common hybrid system is the application of floc enhancement filter media in activated sludge process. The objective is to maintain stable and big floc size to enhance biodegradation of organics in bioreactors and to improve settling performance in secondary clarifiers. However, such systems requireadditional operation and maintenance costs to deal with the filter media.

4.5 CONCLUSION

Technology for municipal wastewater treatment has been developed and matured in developed countries. This know-how is applicable to developing countries in the form of design guidelines and instrumentations. A few technical adjustments are urgently needed in relation to climatic conditions, waste characteristics and operational parameters. These adjustments can be made by conducting research and development activities in developing countries. The parameters developed in many developed countries can be used as a guideline to move forward, and should be adjusted to suit local conditions, in terms of technical, economical and climatic conditions.

REFERENCES

Henze M., Harremoes P., la Cour Jansen J. and Arvin E. (1997) *Wastewater Treatment* (Second edition). Springer, Berlin.

Henze, M., Gujer, W., Mino, T., & van Loosdrecht, M.C.M. (2000). Activated sludge models ASM1, ASM2, ASM2D and ASM3. *IWA Scientific and Technical Report No.9*, IWA Scientific and Technical Reports, IWA, London. ISBN 1025-0913

Malaysia 2002 Yearbook. Berita Publishing, Kuala Lumpur.

Sewerage Services Department (1998) *Sewerage Services Report 1994-97*. Ministry of Housing and Local Council, Malaysia, Kuala Lumpur.

SSD & MWA (1999) *Guidelines for Developers Vol. 1: Sewerage Policy for New Developments*. Second Edition. Sewerage Services Department and Malaysian Water Association, Kuala Lumpur.

Treibel (ed.) (1975) *Lehr- und Handbuch der Abwassertechnik, Bd II* (Textbook for Wastewater Engineering, Volume II). Verlag von Wilhelm Ernst & Sohn, Berlin.

Ujang, Z. and Buckley, C.A. (2002) Water and wastewater in developing countries: present reality and strategy for the future. *Wat.Sci.Tech.* **46**, 1-9.

Ujang, Z. (2003) Hazardous waste management in Malaysia. *UNESCO Encyclopaedia of Life Support System for Hazardous Waste Management in Developing Countries*. UNESCO, Paris (in press).

5

Collection systems
- dry and wet weather performance

Jes Vollertsen and Thorkild Hvitved-Jacobsen

5.1 INTRODUCTION

A prerequisite for a modern society is the supply of potable water. Abundant quantities of good quality water are essential. However, when the water has been mixed with substances from households, institutions, industries and farms the water becomes waste. Not only used, potable water causes nuisances in urban areas, but also rainwater falling on impermeable surfaces gives rise to numerous problems.

The management of wastewater and stormwater runoff in urban areas – in short: urban drainage – covers the initial pollution of the waters, its collection into the conveyance system, the conveyance itself, the treatment of the polluted water and the subsequent effects on the receiving environment. The dry weather flow, i.e. the wastewater, is associated with quite different properties compared to the stormwater runoff. In this case, particularly, quality, quantity and time scale of the flow rates. Consequently, the methods applied to manage the two types of flow are quite different.

Historically, the first engineering within urban wastewater management was the construction of the collection system. In the Iron Age, and probably already before that, covered drains inside houses were build. With increasing urbanization, organized ways to manage wastewaters outside housing developed. A well-known example hereof is the Cloaca Maxima that was built in Rome under the regime of Tarquinius Priscus (616-579 B.C.) to drain Forum Romanum, but also in other roman cities, wastewater drains were built. Furthermore, stormwater drains are known from ancient cities all over the world.

In ancient times, most urbanized areas had, however, none or only ineffective drainage systems. Those systems often had insufficient capacity and could not satisfy the need for removing pollutants from the catchment. These circumstances resulted in poor hygienic conditions, which again resulted in epidemics; the large cholera epidemics of the European Middle Ages being a well-known example hereof.

During the 19th century, the cholera epidemics in Europe initiated the construction of sewers. It began in the middle of the 19th century in the large, industrialized areas of the United Kingdom, Germany and France. Towards the end of the 19th century and at the beginning of the 20th century, the rest of the European cities followed, as did other cities around the world.

Today, every city in the developed world is serviced by adequate sewer systems. Also many cities in the developing countries have sewers, however, many a city of the undeveloped world lacks appropriate sanitation. Without proper sewer systems, the same epidemics that harassed medieval Europe today plague the undeveloped world.

The collection and conveyance system is the backbone of the urban drainage system. It stands for by far the largest capital investments – typically about 90% – compared to the other parts of the urban wastewater system, e.g. the treatment facilities. However, the collections system is not the most visible part of the system. The upstream and downstream boundaries of the urban drainage system are in general more visible to the public. i.e. the need to manage the sources of wastewater, its treatment and subsequently its impact on the receiving waters are more readily perceived by the public than the need to manage the wastewater conveyance. Typically, the general public only perceives the collection system when it fails; i.e. when it interacts with its surroundings in undesired ways.

It is consequently common, that the environmental engineer perceives the need of investments in the wastewater collection infrastructure quite differently from how it is perceived by the public. The dissonance between on the one hand a need for large investments and on the other hand a low public visibility, makes it a challenge to focus stockholder awareness on the fact that a good sewer system is a prerequisite for a satisfactory operation of the other parts of the urban drainage system.

5.2 TYPES OF COLLECTION SYSTEMS

Collection systems fall into two groups: separate collection systems and combined collection systems. In separate systems, sanitary wastewaters from housings, industries, institutions, business areas and etceteras are kept separate from stormwater (Figure 5.1). Consequently, separate systems are constructed with two parallel conveyance channels: one for wastewater and one for stormwater runoff particularly originating from impervious urban surfaces and roads. Combined collection systems, on the other hand, mix the two types of waters in one channel, conveying both wastewater and stormwater (Figure 5.1).

In separate collection systems, the wastewater is taken into treatment, while the stormwater runoff is discharged into adjacent receiving waters. Treatment of separate stormwater runoff from urban areas and highways has – especially in the USA and Canada – become frequent over the last decades.

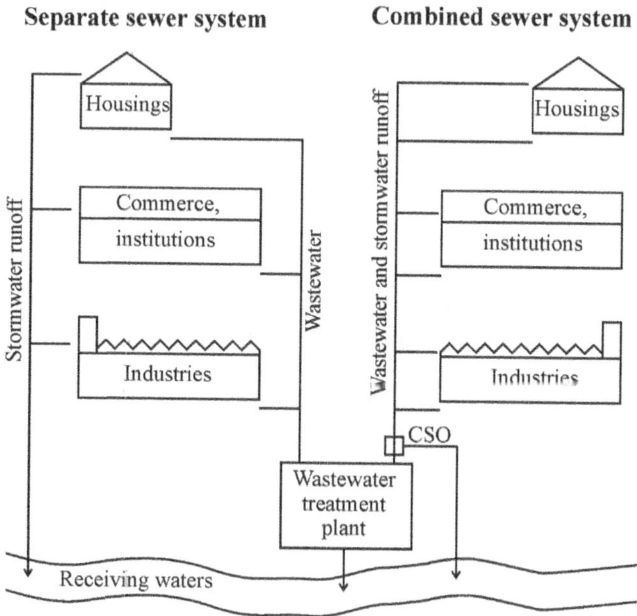

Figure 5.1 Separate and combined sewer systems.

In a combined collection systems, all dry weather flow from the catchment is taken into treatment. During rain, part of the stormwater is taken into treatment together with the wastewater, however, when the combined volumes of wastewater and stormwater runoff exceeds a design value, the excess mixed water flows over into the receiving waters. Combined sewer overflow (CSO) structures are situated within the catchment wherever convenient receiving waters are close by.

Every separately sewered catchment has, unfortunately, a number of drains connected to the wrong conveyance system. Some connector pipes carry wastewater but drain to the storm water system and other connector pipes carry stormwater but drain to the wastewater system. These connections are termed false connections. Depending on the number of false connections, a system originally designed as a separate system, can end up behaving more or less like a combined system.

It is crucial that the environmental engineer takes false connections into account, as they can cause significant pollution loads to be discharged untreated into the receiving waters. Two possibilities exist to manage false connections: Firstly, all false connections can be identified and properly reconnected. This is the optimal solution, but it is also a rather expensive solution. It involves political and legal difficulties related to the fact that work has to be carried out on private properties and probably even paid for by the property owners. The second solution, the more pragmatic one, is to treat separate systems containing large numbers of false connections as if they were combined systems. i.e. to treat waters from both systems during dry weather conditions and to allow overflow into receiving waters during storm events.

A number of factors determine if a collection system originally was constructed separately or combined. One main factor is the historical context in which the system was developed. In Western Europe, the collection systems of the major cities were established as combined systems some 100-150 years ago. There were at this time no reasons to keep stormwater and wastewater separated, as receiving water quality and consequently the treatment of wastewaters were not an issue. The systems were established to fight health problems associated with the large cholera epidemics that had ravaged Europe for many centuries. The focus on receiving water quality has first risen in the last decades of the last century. Compared with several areas in the world, storm intensities in Europe are relatively moderate, making it a pragmatic engineering solution to convey both stormwater and wastewater in the same channel.

Today the cities of the developed world use much effort to bring down the pollution load on the urban receiving environments. Urban watercourses – which in the last century have been covered and hidden away – are today 'day-lighted' to serve as recreational areas within the cities, and urban lakes and coastal regions are desired to be of bathing water quality. The concept of combined sewered catchments is detrimental to this modern way of viewing the urban water environment. Much effort is consequently put into mitigation of the combined sewer overflows and – where economically and practically possible – into separation of catchments.

The engineers of the past have – for reasons that in their time were good and sound – mixed stormwater and wastewater. Unfortunately, they hereby removed an important degree of freedom for the future development of the urban drainage system. The above example illustrates that flexibility should be emphasized when establishing new sewer systems. We cannot today predict the demands of tomorrow. However, by maximizing system flexibility we can increase the possibility for a long-term sustainable evolution and operation of urban drainage systems.

5.3 SOURCES AND QUANTITIES FOR DRY WEATHER WASTEWATER

Dry weather flow from urban catchments can conveniently be divided into waters from household, institutions, business areas, industries, drainage of buildings and infiltration. For all wastewater types, a number of general characteristics related to the wastewater flow pattern and wastewater pollutant content can be identified.

5.3.1 Wastewater from households

The water consumption pattern of people in their residences does not change much from year to year or from place to place. The pattern of wastewater production and the wastewater constituents from residential areas are, consequently, rather predictable.

As an example, Figure 5.2 shows a typical weekday of dry weather wastewater flow from the residential area of Frejlev, Denmark. The town has a population of approximately 2,000 inhabitants and covers an area of 85 ha. During weekdays, most people rise between 6 and 7 in the morning, resulting in a sharp increase in the sanitary wastewater flow. Visual observation of this wastewater reveals a significant change in colour from greyish to brownish due to faecal matter. During the later morning hours, the flow slowly increases towards a maximum around noon, accompanied by a colour change back to greyish. Late in the afternoon the wastewater production increases again as people get home from work. In the early evening hours a second peak, caused by household activities, is seen. A short pumping main within the catchment causes the numerous small flow peaks.

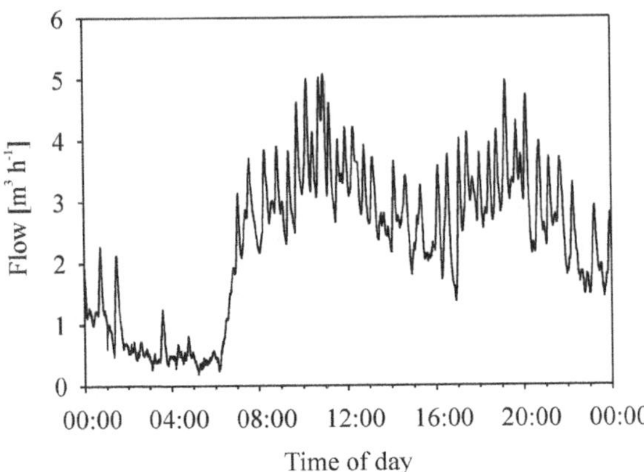

Figure 5.2 Flow pattern from a purely residential catchment, cf. text.

In this example, the transport time for most wastewater contributions to the point of flow measurement is less than 30 minutes. In catchments covering larger areas, the diurnal variation becomes less clear and easy to interpret due to longer transport times. However, here the pattern of the wastewater contribution from households will tend to be constant.

The amount of wastewater produced by households can in most cases be equalled to the water consumption, as most potable water used in households ends up being discharged to the drainage system. The magnitude of household water consumption depends on a number of factors, of which water availability, cultural differences and the pricing of water are some of the most important elements.

If water is difficult to obtain or expensive, less will be used. An estimate of average daily water consumptions based on the availability of the source of water is shown in Table 5.1. The pricing of water is a very important factor. For example, in Denmark the water consumption per capita decreased from 64 m^3 year^{-1} capita^{-1} in 1989 to 53 m^3 year^{-1} capita^{-1} in 1995 (Danish EPA, 1997). This decrease was caused by a change in public behaviour as water prices increased many times in the 1980's and 1990's.

When estimating the future wastewater production, the future water consumption must be estimated, taking into account not only the population development but also aspects like the availability of water and the economic and political development of the society as such (Campos and von Sperling, 1996).

Sanitary installations, cloth washing and the kitchen are the main sources of household wastewaters and its pollutants. The amount and composition of pollutants discharged from these sources depends on the water consumption and the number of inhabitants. However, cultural aspects related to e.g. the use of water or paper for anal cleansing play an important role. Another example is the widespread American custom to use kitchen grinders to dispose of kitchen waste into the drain, whereas in Europe, it is common to dispose of kitchen waste into the solid waste collection system.

On a regional level significant difference are found in terms of volumes, pollutant contents and flow patterns. Such changes are based on the fact that people within different communities behave differently. It is consequently not possible to simply copy default values from e.g. the US to South-East Asia. For each region, data must be collected systematically to obtain the information needed.

Table 5.1 The dependency of the average daily water consumption of households on the available water supply (IRC/WHO, 1981).

Type of water supply	Average daily water consumption [liters capita^{-1} day^{-1}]
Distance to communal water source > 1000 m	5-10
Distance to communal water source 500-1000 m	10-15
Distance to village well > 250 m	15-25
Distance to public standpost > 250 m	20-50
Courtyard connection	20-80
House connection, one tap	30-60
House connection, several taps	70-250

5.3.2 Institutions, business areas and industries

The constituents of wastewater from business areas are largely identical to those found in domestic wastewater, as the main part of the wastewater from business areas is wastewater from sanitary installations. However, the flow pattern from business areas is quite different from the flow pattern from households, and follows the activities in question.

The same is the case for wastewater from most institutions, e.g. schools and kindergartens. However, the wastewater from e.g. hospitals is quite different – not so much with respect to the flow pattern as with respect to the wastewater constituents. Many problematic wastewater components may be found in hospital wastewaters. The concentrations of pathogens can be significantly higher than for household wastewater, as can the concentration of medicine residuals. Furthermore, hospitals use numerous chemicals that are not used in households.

The wastewater flows from industries are strongly dependent on the type of industry. In general, estimates of wastewater productions from industries should be based on measured water consumptions and pollutant contents. Where such detailed information is not available, water flows and pollutants must be estimated based on the size and type of the industry. The prediction of industrial wastewater flows and qualities is significantly more uncertain than the prediction of flows and qualities from e.g. households.

When the type and size of industries is only approximately known – e.g. when a catchment is still in the planning state – the wastewater production must be determined from the size of the area and the planed type of land-use. Experiences from developed countries on average daily wastewater production of such areas are shown in Table 5.2. However, it must be emphasized that certain industries – e.g. food processing industries – can have significantly different water consumptions than those given in Table 5.2. Typical rates of water uses for a broad number of commercial facilities, industries and institutions can be found in e.g. Tchobanoglous *et al.* (2003).

Not only private water consumption depends on the price of water and wastewater. For instance, in the period from 1989 to 1995, increased water prices caused the water consumption of Danish industries, business areas and institutions to drop by 15-20% (Danish EPA, 1997).

Table 5.2 Commercial and industrial water consumption (ATV – A 118E, 1977).

	Average daily water consumption [litres ha^{-1} s^{-1}]
Business and industries with low water consumption	0.5
Business and industries with medium water consumption	1.0
Business and industries with high water consumption	1.5
If no information on water consumption is available	1.0

5.3.3 Infiltration and drainage of buildings

In areas where the groundwater table is above the sewer invert, infiltration of groundwater into the sewer system may occur. The extent of the infiltration varies much, and in severe cases the infiltrating groundwater amounts to several times the sanitary wastewater flow.

The extent of infiltration should be assessed by measurements and inspections. For whole catchments infiltration can be roughly estimated based on the difference between measured wastewater production and measured potable water consumption. This method is, however, often limited by lack of data, especially as sewer catchments and water distribution areas are not always identical. Another method is to use tracers. Hereby, a rather accurate estimate of the infiltration into individual pipe stretches can be made. However, to cover a whole catchment, a large number of tracer measurements must be performed. A cruder method is to estimate the wastewater flow in the early morning hours, assuming that no or only little wastewater is produced by housing and industries. This method is simple and effective, however, it can only be used on small to medium sized catchments, where wastewater transport times are limited. Figure 5.3 shows the flow pattern at a Danish treatment plant subject to severe infiltration. The catchment covers about 10,000 person equivalents and the transport time is in the magnitude of some hours. The minimum water flow during the early morning hours allows the infiltration to be estimated to be around 40 l s^{-1}, corresponding to roughly 80% of the total wastewater load on the treatment plant.

When buildings have basements, and the groundwater table is above the basement floor, it is common to prevent intruding moisture by installing circumferential drains to lower the groundwater table around the cellar. The drained groundwater is subsequently discharged into the drainage system. The significance of this contribution is of course strongly dependent on catchment characteristics, i.e. on the location of the groundwater table and if houses have basements or not. Infiltration can also occur when a leaking potable water pipe or stormwater pipe is placed above a leaky wastewater pipe. However, this type of infiltration is typically of minor importance compared to infiltration of groundwater and drainage of buildings.

Water from circumferential drains or infiltrating groundwater holds few pollutants. Its main effect is to dilute the wastewater and to reduce the wastewater and stormwater conveyance capacity of the system.

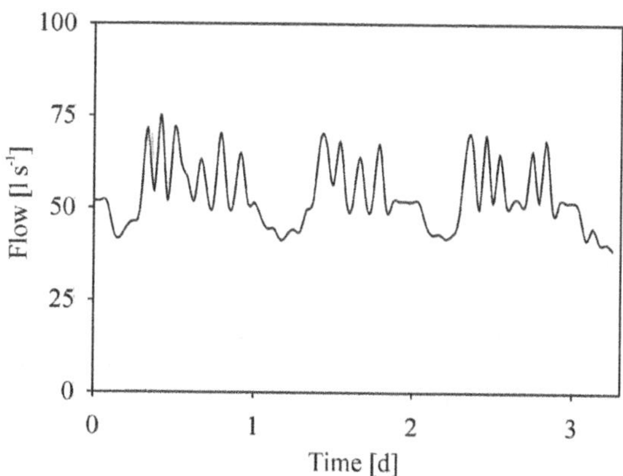

Figure 5.3 Flow pattern at a treatment plant receiving severe infiltration.

5.4 STORMWATER QUANTITIES

The first step in stormwater runoff analysis is the analysis of the catchment. Such analysis includes information on:

- Precipitation. Ideally local, historical rainfall series should be available.
- Catchment surfaces in terms of pervious and impervious areas and the connection of such areas to the drainage system.
- Location, dimensions, slopes, materials, and etceteras of the existing drainage system.
- Structures within the system, e.g. stormwater outfalls, combined sewer overflow structures and basins.
- Type, capacity and environmental classification of receiving waters.

It is often advantageous to divide large catchments into sub-catchments with more or less homogeneous characteristics. The division of catchments into sub-catchments is generally based on drainage patterns, surface slopes, and land-use patterns.

Urbanization of rural areas tends to modify the hydrological characteristics of the catchments so that stormwater is collected and removed more rapidly from an urbanized catchment than from a rural catchment. In rural catchments, infiltration, vegetation, ponds or wetlands tend to detain and retain stormwater.

Urban development results in vegetation being removed, surface permeability being decreased and natural storages being diminished. Shorter stormwater detention times and increased runoff peak flows are consequences hereof. Furthermore, total runoff volumes increase due to decreased evaporation and reduced infiltration. Together, this leads to an increased risk for exceeding the hydraulic capacities of receiving waters and to flooding of river floodplains.

Discharge of stormwater to receiving waters reduces the available water resource in the catchment. Reduced infiltration gives rise to less groundwater formation, a consequence of which is less groundwater being available for potable water uses. Wetlands and streams are also depending on high groundwater tables, and are at risk of drying out when the groundwater table becomes too low. Where wetlands and streams dry out during part of the year , the diversity of the ecosystem will decrease.

To amend these negative effects of urbanization on receiving water quality, ground water quality and the diversity of the ecosystems, the hydrological balance of a developed urban catchment should – as far as possible – resemble the hydrological balance of the corresponding undeveloped catchment.

On the other hand, communities want a certain level of protection against flooding, and the costs tend to be high when fulfilling this goal together with a semi-natural hydrology within the urban environment. Planning stormwater management systems, the costs of a higher protection level must be weighted against the overall community benefit. When only limited resources are available it might be appropriate to choose less sophisticated but cheaper stormwater management solutions and leave the limited community resources for other crucial purposes. This means that poor communities must accept lower protection levels, poorer ecosystems, poorer groundwater quality, and so on, compared to rich communities. The engineer designing the stormwater management system must advise the community of costs, benefits, and consequences resulting from different stormwater management solutions so that the most feasible stormwater protection level can be found. It must be the community and not the designing engineer who decides whether a lower risk of damages due to flooding and surcharging as well as better environmental qualities are more appropriate than e.g. improved health care or infrastructure.

5.4.1 Precipitation and design storms

A widely used engineering practice for design of stormwater runoff systems is the use of design storms, i.e. to define artificial rainfall events as a design criterion. The use of design storms for designing stormwater runoff systems requires a minimum of resources. This and the fact that other design methods demand detailed regional knowledge on long term series of historical rains, has contributed to the popularity of this approach. A return frequency is assigned to such a design storm and design is carried out assuming that the return frequency of the rainfall event equals the return frequency of the runoff event. The return frequency (f) is the inverse of the return period (T), i.e. the statement that a certain storm event occurs with a frequency of 0.5 year^{-1} is equivalent to stating that it has a return period of 2 years. In other words, this rain will on average occur once in two years. Although design storms must reflect required levels of protection, the local climate, and catchment conditions, they need not be scientifically rigorous (ASCE and WEF, 1992).

During a real storm event, the rain intensity is rather variable (Figure 5.4). To construct a synthetic storm from such a storm event, the rain is enveloped in a simple geometric form. The most commonly used form is a rectangular envelope with a predefined duration. Storm durations shorter than 5-10 minutes are normally not considered, as the shortest concentration times of interest are of the same order. Furthermore, storage on surfaces during concentration will flatten the surface runoff hydrograph and lover the peak runoff of very short and high intensive rains.

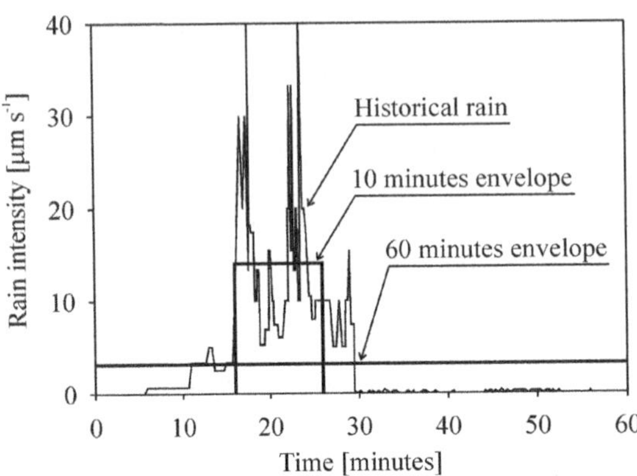

Figure 5.4 Historical rainfall from August 12, 1998, Hasseris, Denmark. The solid lines envelop the most intensive 10 and 60 minutes intervals.

In Figure 5.4 two synthetic rainfall events have been constructed from a historical rain by finding the most intensive 10 minutes and 60 minutes intervals. For Danish conditions, the measured rain was quite intensive, having a 10 minutes rain intensity of 14 $\mu m\ s^{-1}$ corresponding to a return frequency of 0.5 years^{-1}.

After collecting and compiling long series of historical rainfall, the derived synthetic rainfalls can be ranked with respect to their intensities. Hereafter, relations between the return frequency of the design storm and the corresponding rain intensity can be found. Plotting rain intensities against rain durations for a given return period allows rainfall intensity-duration-frequency curves to be drawn. Figure 5.5 gives an example of such curves, showing curves for Miami in USA and Flanders in Belgium.

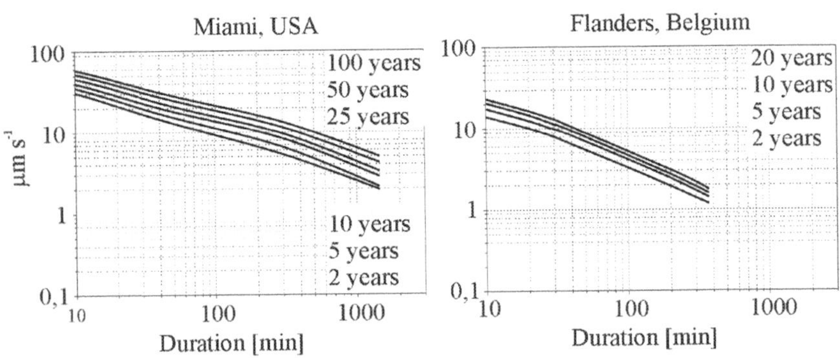

Figure 5.5 Rainfall intensity-duration-frequency curves.

To be appropriate as the data behind synthetic rainfalls, rainfall must be registered with a high resolution, e.g. the data depicted in Figure 5.4 has a resolution of 0.2 mm of rainfall. Rainfall registrations on a fixed time-basis – e.g. an hourly basis – are not well suited for construction of synthetic rainfalls of shorter durations.

Each rainfall intensity-duration-frequency curves can be approximated by mathematical expressions of e.g. one of the following forms:

$$i = \frac{a}{\left(t_d + c\right)^b} \qquad i = \frac{d}{t_d^e + g} \qquad i = ht_d^j$$

where i is the average intensity for the rainfall duration t_d and a, b, c, d, e, g, h and j are constants depending on the return frequency and local characteristics. Rainfall intensity-duration-frequency curves and their mathematical approximations are site-specific and must not be applied outside the area and the range of rainfall data to which they are calibrated. However, if no detailed knowledge of rainfall is available rain intensity-duration-frequency curves may be constructed as a 'best-guess' based on information available.

The total rainfall depth for a certain rainfall duration and frequency is a function of the climate and may vary much even over short distances. For instance, significant differences in precipitation patterns can be found between the foot and the top of a hill. Spatial variability must be taken into account as uncertainty on the rainfall data, as knowledge on the rainfall pattern of a catchment seldom reaches such details.

Rainfall intensity-duration-frequency curves and their mathematical equivalents are available in many developed countries. However, in developing countries such data are often missing. Rainfall depth, intensity and duration for design storms must in these cases be estimated from what information can be collected. This includes rainfall data from regions with similar characteristics and short-term rainfall measurements performed on-site.

5.4.2 Choosing return frequency and storm duration

Based on a desired – or affordable – level of protection, design storm return periods are selected. Choosing a higher level of protection results in larger or more intense design storms and hence stormwater management systems with higher capacities.

Return frequencies (or return periods) and hence the desired protection levels should be based on the potential damage from flooding and surcharging. For example, for surcharge of an infiltration basin resulting in overflow into a nearby stream, a design storm return period of 0.25-0.5 years might be reasonable. Thus, it is acceptable that the structure is surcharged 2-4 times every year. On the other hand, for flooding of major storm sewer elements in central parts of town, return periods of 5-20 years would be more appropriate.

It is widely accepted to design surcharge of interceptor drains in tropical cities for a return period of 5 years. However, for a number of occasions, lower return periods probably would be more appropriate, taking into account both the seriousness of the damage inflicted by the flooding and the costs of the stormwater system construction. In Mombasa, for example, a return period of 1 year has been used for most stormwater sewers and in Calcutta some drains have been designed for a return period of only 2 months (WHO, 1991).

The duration of the design storm depends on both the hydrologic time-response characteristics of the catchment and the typical durations of intense storms in the region. Storm durations for stormwater transport in pipes or channels are typically in the range from 10 minutes for small catchments to a couple of hours for large catchments. However, for design of storage and infiltration facilities, storm durations are typically in the order of hours and for design of stormwater treatment ponds, appropriate storm durations are in the order of days.

5.4.3 Impervious surfaces and runoff coefficients

A key characteristic of an urban catchment is the size of the impervious surface area. Impervious surfaces typically account for 10-30% of the total surface area in low-density residential districts, 30-60% in high-density residential districts, and 80-100% in central business districts. The impervious surfaces within a catchment are further divided into two categories: connected and non-connected impervious surfaces. Connected impervious surfaces drain directly to storm sewers or drainage channels. Non-connected impervious surfaces drain onto pervious surfaces, e.g. roofs that drain onto lawns are non-connected impervious surfaces.

Maximum discharges from urban catchments are largely determined by the runoff from connected impervious surfaces. It is therefore essential that the connected impervious areas be estimated with good accuracy. When analysing existing drainage systems, the engineer should visually inspect the impervious surfaces to determine their size and how they are connected to the sewer system.

Not all precipitation that falls on an impervious surface ends up as stormwater runoff. Some is stored in surface depressions and some evaporates. Furthermore, a surface can be only partly impervious so that some of the precipitation infiltrates and some runs of. The runoff coefficient (φ) accounts for the integrated effects of rainfall interception, infiltration, and evaporation on the stormwater runoff volume (Table 5.3). The initial loss (i) accounts for depression storage and surface wetting. Both φ and i depend on rainfall intensity and duration, catchment characteristics, and length of the inter-event dry period. The bigger the rainfall depth for a given rainfall duration, the lesser the relative effect of rainfall abstractions on the runoff volume, and correspondingly the runoff coefficient increases. Both the initial loss and the runoff coefficient must be expected to increase with rainfall intensity because marginal areas – typically with higher surface permeability – will contribute to the runoff. Typically initial losses are around 0.5-1.5 mm of rainfall.

The runoff coefficient for green areas is set to zero where it can be ascertained that stormwater will infiltrate without runoff. Unpaved soils with very low permeability and without good vegetation cover – e.g. rock or heavy clay – may have runoff coefficients similar to pavements. Furthermore, the steeper a surface is, the higher φ will become.

Table 5.3 Normal range of runoff coefficients* (ASCE and WEF, 1992).

Character of surface	Runoff coefficients (φ)
Pavement	
Asphalt and concrete	0.70 – 0.95
Brick	0.70 – 0.85
Roofs	0.75 – 0.95
Lawns, sandy soil	
Flat (slope less than 2%)	0.05 – 0.10
Average (slope from 2% to 7%)	0.10 – 0.15
Steep (slope more than 7%)	0.15 – 0.20
Lawns, heavy soil	
Flat (slope less than 2%)	0.13 – 0.17
Average (slope from 2% to 7%)	0.18 – 0.22
Steep (slope more than 7%)	0.25 – 0.35

* The ranges of φ presented are typical for return periods of 2-10 years. Higher values are appropriate for larger design storms.

5.4.4 Runoff hydrographs

Converting rain hyetographs into runoff hydrographs, a linear relation between runoff volume and rainfall volume for storm events is found to describe runoff satisfactorily (Figure 5.6). By field measurements, the initial amount of water that is not transferred into the drainage system – the initial loss – can be found as the intersection with the rainfall volume axis. The slope of the linear relationship is interpreted as φ (Figure 5.6). In the design situation – and most often also when existing systems are analysed – no measured relation is available. In this case values for initial loss and φ must be estimated. Initial loss has to be estimated from experience and φ from analysis of the imperviousness of the catchment, Table 5.3.

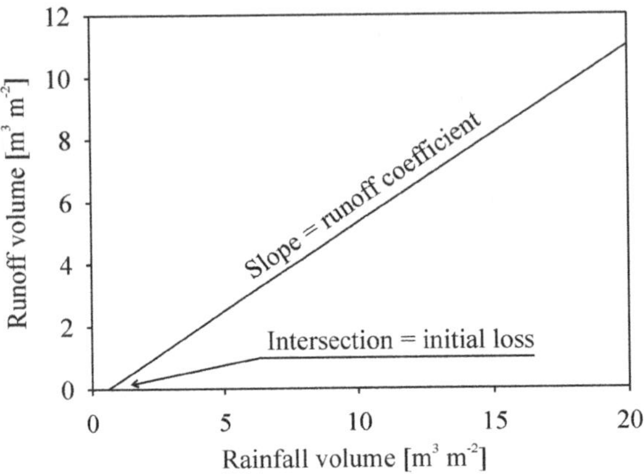

Figure 5.6 Runoff coefficient and initial loss for a catchment.

5.5 ROOTING OF DRY AND WET WEATHER FLOW

After identification of the flows to be conveyed, the system must be designed to perform the conveyance at minimum maintenance requirements. For this purpose, two fundamental properties must be fulfilled: the pipe must have sufficient conveyance capacity and the pipe must be self-cleansing in order to avoid blockages to build up.

5.5.1 Dry weather flow

The capacity of a sewer conveying dry weather flow should never be used fully. There are several reasons for this, for example to have an excess capacity for later, upstream development and the fact that full flowing sewers increase the risk for backwaters that may flood private properties. As a rule of thumb, not more than half of the conveyance capacity should be used during the diurnal peak flow.

Self-cleansing of separate wastewater sewers can be ensured by requiring a certain flow velocity or bottom shear stress to occur at the daily peak flow. At the upper part of the sewer, where the flow pattern is erratic, a minimum pipe slope is more appropriate. What flow velocities, shear stresses or slopes to be chosen is the point of many discussions and studies. Generally, the higher the minimum values chosen, the higher the safety level with respect to avoiding dry weather blockages. Typical minimum values for shear stresses ensuring self-cleansing in separate wastewater sewers are 0.5-1.5 N m^{-2}.

5.5.2 Wet weather flow

After determination of design rainfall and hydrologic properties of catchments the necessary hydraulic transport capacity of the stormwater management system must be found. In the simplest design approach, the rational method, the storm duration is set to the travel time (concentration time) from the remotest point of the catchment to the point under design. However, it is never set to less than a minimum time of e.g. 5 or 10 minutes.

As an example, a catchment of 45 ha with an average degree of imperviousness of 0.30 is located in Miami, Florida. It has a concentration time of 15 minutes and must be designed for a return period of 5 years. The design flow based on the rational method can now be calculated: From Figure 5.5 it is seen that the intensity is 30 μm s^{-1}. The downstream part of the conveyance system must hence be designed for a flow of: 30 10^{-6} m s^{-1} * 45 10^4 m^2 * 0.30 = 4.1 m^3 s^{-1}.

The time-area method is an improvement of the rational method, relating storm durations to the actual area that contributes to the runoff. In this method, all storm durations – in the interval from the minimum storm duration to the concentration time of the catchment – are rooted through the stormwater system in order to obtain the necessary hydraulic capacity.

Typically, one of these methods is used for design of small and simple catchments – e.g. new developments. The methods give good estimates of the flow coming from such catchments, however, when the catchments include storage and overflows, it becomes problematic to use these approaches. Furthermore, the methods only give the design flow but no information on pipe surcharge levels. Consequently, they can be used where the design criterion is a return period of pipe surcharging but not where the design criterion is a return period of surface flooding.

Where the design criterion is a return frequency of surface flooding, computer models solving non-stationary hydraulics for both surface and pipe flow must be used. These models are furthermore necessary when analysing catchments with storage and overflows. Such hydrodynamic models have been shown to yield excellent simulation results if they are well calibrated to the catchment. However, if the models are not calibrated they cannot be expected to yield proper results. They furthermore need excellent information on catchment structure, catchment surfaces and precipitation. Thus, the model does not give better answers than the quality of the input data allows for. On the other hand, even with poor quality input data the models do yield important information on where in the drainage system problems are most likely to occur, even though the actual return periods of flooding or surcharge are not accurate.

Today computational power and advanced runoff models makes it possible to relate large series of historic rainfall hyetographs to their runoff hydrographs, producing reliable, local runoff statistics. The drawback of this method is that it requires valid, historical rainfall series with a time resolution of 5-10 minutes. The length of such rainfall series must be several times the maximum return period envisioned in order to produce reliable runoff statistics. As return frequencies for surcharge of sewer systems often are in the order of years or decades, historical rainfall covering at least several decades must be used. Such series are not always available in developed countries and seldom available in developing countries. Therefore, design storms based on educated estimations often will have to suffice, disregarding the limitations of this approach.

Self-cleansing of separate stormwater sewers can pragmatically be assured by choosing a minimum flow velocity for the full flowing sewer, setting it so high that the sewer is also kept reasonably free of sediments in-between large storm events. Alternatively, a flow occurring e.g. on average once a month can be chosen to fulfil the self-cleaning criteria.

Combined sewers should be designed for self-cleansing in a similar way to separate wastewater sewers, however, demanding higher flow velocities and shear stresses. The rational behind the need for a higher self-cleansing criterion is that inorganic particles from the stormwater runoff become mixed with organic particles from the dry weather flow, yielding a material that is more difficult to convey.

5.6 WASTEWATER QUALITY

Potable water is used as a convenient carrier of undesired substances. By this process the potable water changes its nature and becomes wastewater. Wastewater is water thrown out because it has become useless or even harmful. At a first look, the term 'wastewater quality' consequently seems to contradict the very nature of wastewater. How can something that is undesired by everybody be associated with good or poor quality? The rational behind introducing the term 'wastewater quality' is a method for distinguishing between its properties related to impacts on the subsequent wastewater treatment, the receiving environment and processes in the sewer itself.

5.6.1 Types and concentrations of wastewater quality parameters

Wastewater consists of water into which a multitude of different chemical and microbial components have been mixed. It is impractical or even impossible to measure most of these components individually. Consequently, a number of lumped parameters are traditionally used for assessing wastewater quality. The quality measures most commonly used in urban wastewater management can be organized into groups:

- Measures for organic matter (Table 5.4)
- Measures for inorganic matter (Table 5.5)
- Measures for physical parameters (Table 5.6)
- Measures for toxic compounds and pathogens (Table 5.7)

All the methods described in the tables are standardized methods, and defined in *Standard Methods for the Examination of Water and Wastewater* (APHA *et al.*, 1998).

The measures for wastewater quality shown in the Tables 5.4-5.7 give some of the information needed when approaching the design of a subsequent treatment plant or when managing problems related to in-sewer transformations of the wastewater. However, when going into detailed analysis of transformation processes in sewers or treatment plants, more information on the wastewater is necessary (see section 5.9).

Table 5.4 Measures for organic matter.

Chemical oxygen demand (COD)	COD is a measure for the total organic matter. By this method, organic matter is chemically oxidized into inorganic compounds. The equivalent oxygen mass needed for the oxidation is used as the measure for the organic matter content. i.e. the unit of COD ($gCOD\ m^{-3}$) is identical to the unit $gO_2\ m^{-3}$.
Soluble chemical oxygen demand (COD_{sol})	COD measured on the liquid permeating through a glass fiber filter of about 1 μm pore size. i.e. the definition between "soluble" and "particulate" is made by the filter pore size used.
Biological oxygen demand (BOD)	BOD is a measure for the biodegradable organic matter in a sample. It does, however, not measure all biodegradable organic matter. The mass of oxygen needed for this partial breakdown of the organic matter is used as a measure of the organic matter. i.e. as for COD, the unit of BOD ($gBOD\ m^{-3}$) is identical to the unit $gO_2\ m^{-3}$.

Volatile total solids (VS)	VS is a measure for the total organic matter. The sample is dried and the organic matter is pyrochemically oxidized at 550°C. The matter lost by the oxidation is assumed to be all and nothing but organic matter. The unit is gVS m^{-3}.
Volatile suspended solids (VSS)	VSS measures the organic matter content of suspended solids (TSS, cf. Table 5.6). The sample is dried and the organic matter contained in the suspended solids is pyrochemically oxidized at 550°C. The matter lost by the oxidation is assumed to be all and nothing but organic matter. The unit is gVSS m^{-3}.

Table 5.5 Measures for inorganic matter.

Total N	Total N measures the sum of all nitrogen compounds in the sample.
Total Kjeldahl N (TKN)	Total Kjeldahl N measures the organic bound nitrogen and ammonia/ammonium.
Ammonia/ ammonium	The term 'ammonia' is for simplicity used as a label for the sum of ammonia and ammonium ($NH_3+NH_4^+$). Ammonia is a reduced, inorganic nitrogen compound with an oxidation level of the nitrogen atom of -3.
Nitrate	Nitrate (NO_3^-) is an oxidized, inorganic nitrogen compound with an oxidation level of the nitrogen atom of $+5$.
Nitrite	Nitrite (NO_2^-) is an oxidized, inorganic nitrogen compound with an oxidation level of the nitrogen atom of $+3$.
Nitrate+Nitrite	Often nitrate and nitrate are measured lumped into one fraction.
Total P	Total P measures the sum of all phosphorus compounds.
Ortho P	Ortho P measures the inorganic, soluble phosphorus compounds, often written as PO_4.

Table 5.6 Measures for solids and physical parameters.

Total solids (TS)	A measure for all the matter in a sample. The sample is dried at temperatures a bit above the boiling point of water. The residual matter is assumed to be the total matter in the sample. TS consequently also includes salts.
Total suspended solids (TSS)	A measure for the content of particles in a sample. The sample is filtered through a glass fiber filter of about 1 μm pore size. The matter retained on the filter paper is viewed as the particle content of the sample. i.e. the definition of a 'suspended solid' is made by the filter pore size used.

Temperature (T)	Temperature measured in degrees Celsius or Kelvin
Alkalinity	Alkalinity is a measure for the acid neutralization capacity of a sample. It measures the amount of acid needed to lower the pH to a chosen value. Its unit is $gCaCO_3 \, l^{-1}$ or $m\text{-}eq \, l^{-1}$. The first unit tells how much $CaCO_3$ the neutralization capacity is equal to. The second unit tells how many mole of H^+ are used to lower the pH to the chosen value.
pH	pH is a measure for the acidity/alkalinity of a sample. pH is the negative logarithm to the hydrogen ion concentration.

Table 5.7 Measures for toxic compounds and pathogens.

Heavy metals	A number of heavy metals occur in significant concentrations in stormwater and wastewater. e.g. arsenic, cadmium, copper, chromium, lead, mercury and zinc.
Organic micro-pollutants	A considerable number of organic micropollutants can be found in urban wastewaters and runoff waters. Due to the large number of different pollutants, a few are typically chosen as indicators and analyzed for.
Pathogens	As for micropollutants, a huge number of pathogens exist in especially wastewaters. Consequently, a few are chosen as indicators. e.g. one for bacteria and one for viruses.

The concentrations of wastewater pollutants differ much between wastewater types. However, it is generally true that wastewater can be described as water with a small fraction of pollutants. As an example, a typical COD concentration in wastewater corresponds to the addition of less than a teaspoonful of sugar to one litre of clean water.

Household wastewaters are typically rather well defined, but concentrations can still easily vary a factor 5 between catchments of the same region. Also variations between regions are to be expected, depending on the behaviour of the population and the water availability. Typical pollutant concentrations in household wastewaters can be found in e.g. Tchobanoglous *et al.* (2003).

Industrial wastewaters can contain a multitude of pollutants within a broad spectrum of concentrations. Large regional differences in pollutants contained in wastewaters from identical types of industries must be expected. For example, in some countries, toxic waste is frequently dumped into the drain – disregarding that it might be illegal to do so. In other countries, strict control of wastewater discharges into the urban drainage system or requirements for pre-treatment by the wastewater producing industries, largely prevents this. It is crucial to know to which extent toxic wastes are dumped into the sewers as such wastewaters will be more difficult to treat, and will have a more severe impact on the receiving environment than wastewaters that are free from toxic components.

5.6.2 Characterization of wastewater organic matter

Characterization of wastewater organic matter by means of lumped parameters like COD and BOD gives a rough idea of the organic matter content. Relating BOD to COD as well as to COD_{sol}, allows an estimate of the wastewater biodegradability. For example, the higher the ratio of BOD to COD, the better is the biodegradability. In the same way, the ratio of COD_{sol} to COD_{total} gives some indication towards good or poor wastewater biodegradability.

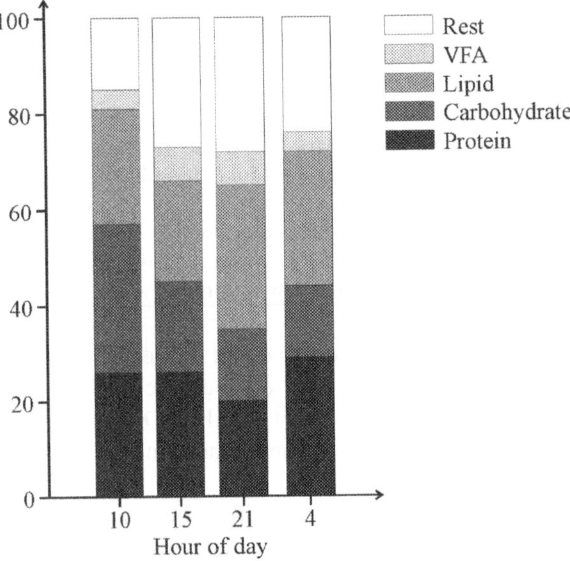

Figure 5.7 Composition of wastewater in the inlet to Aalborg West wastewater treatment plant, June 23-24, 1992 (after Raunkjær, 1993).

More details of wastewater composition can be obtained when analysing the wastewater for different chemical constituents like carbohydrates, lipids, volatile fatty acids, protein, and etceteras (Nielsen *et al.*, 1992; Raunkjær *et al.*, 1994). Figure 5.7 shows an example of such a fractionation of the organic matter in wastewater into different chemical compounds. However, when simulating microbial transformation of wastewater constituents in treatment plants, the sewer itself, and in the receiving environment, a different approach for wastewater organic matter fractionation is needed. The reasoning behind this is that the fractionation of the organic matter must be in agreement with the model concepts used for the respective simulations and use the same component definitions as the applied models.

For activated sludge process simulation, the activated sludge models published by the International Water Association (IWA, former IAWQ, former IAWPRC) are the most widely used model concepts (IAWQ, 2000). In these models, the organic matter is measured in the unit of COD. The total COD of the wastewater is divided into the components representing biomasses and its substrates. As an example, when characterizing wastewater for modelling by means of the Activated Sludge Model no. 3 (ASM3), the total wastewater COD must be divided into the components shown in Table 5.8. Nitrifying biomass is not present in wastewater at any significant concentration and can hence be omitted. In this case, the total COD of a wastewater sample equals the sum of S_S, S_I, X_I, X_S, X_{STO} and X_H.

The concept of aerobic degradation of organic matter in ASM3 is illustrated in Figure 5.8. Here it is assumed that the main part of the slowly biodegradable substrate (X_S) is hydrolysed into readily biodegradable substrate (S_S) and that a minor part is turned into inert, soluble organic matter (S_I). The S_S is transformed and stored inside the biomass as storage products (X_{STO}). The biomass subsequently consumes X_{STO} for growth purposes. Oxygen uptakes are related to four different processes: storage of S_S, growth of X_H, endogen respiration of X_H, and respiration of X_{STO}.

Table 5.8 Organic matter components in raw wastewater according to the Activated Sludge Model no. 3 (IAWQ, 2000).

Component	
S_S	Readily biodegradable substrate
S_I	Inert, soluble organic matter
X_I	Inert, particulate organic matter
X_S	Slowly biodegradable substrate
X_H	Heterotrophic biomass
X_{STO}	Cell internal storage products
X_{AUT}	Autotrophic, nitrifying biomass (negligible in wastewater)

None of the wastewater COD components can be analytically determined, and the components must hence be determined indirectly. A versatile tool for this purpose is measurement of the respiration activity in terms of the oxygen uptake rate of the biomass.

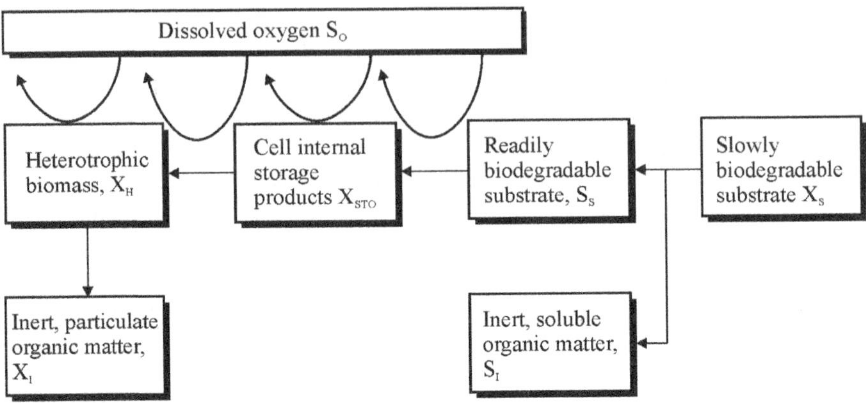

Figure 5.8 Aerobic degradation of organic matter in the Activated Sludge Model No. 3 (IAWQ, 2000).

Table 5.9 Organic matter components in raw wastewater according to the model for wastewater aerobic/anaerobic transformations in sewers (Hvitved-Jacobsen, 2002).

Component	
S_S	Readily biodegradable substrate
$X_{S,fast}$	Fast hydrolysable substrate
$X_{S,slow}$	Slowly hydrolysable substrate
X_{Bw}	Heterotrophic biomass

Using the ASM3 concept on the interpretation of oxygen uptake rate measurements, the fractions X_H, X_{STO}, S_S and X_S can be determined, while the determination of X_I and S_I must be based on a mass balance with the total and the soluble COD. Note that the COD components are not analytical components but virtual components, defined by the model of which they are part. When another concept is used, they will consequently differ in definition and size. For example, when using other versions of the activated sludge models published by IWA (1, 2, 2d, or 3) on the same oxygen uptake rate measurement, components with identical names may have slightly different definitions and consequently give slightly different results.

When simulating activated sludge processes with ASM3, the fractionation into the COD compounds shown in Table 5.8 by means of the ASM3 concept (Figure 5.8) is the right thing to do. However, simulation of a different system calls for another concept and consequently for other definitions of the components – or maybe even for other components.

An example hereof is the model used for simulation of in-sewer transformation of organic matter and sulphur compounds, the WATS model (wastewater aerobic/anaerobic transformations in sewers). Conditions in sewer systems differ significantly from conditions in activated sludge treatment plants, and the processes and components included in the model therefore differ from those included in the activated sludge model concepts. The organic matter components applied in the WATS model are shown in Table 5.9. When characterizing wastewater for modelling by the WATS model, only the aerobic part of the model is needed (Figure 5.9). Two fractions of hydrolysable substrate ($X_{S,fast}$, $X_{S,slow}$) are hydrolysed into readily biodegradable substrate (S_S) that is used by the heterotrophic biomass (X_{Bw}). Oxygen uptake is related to the processes for growth of X_{Bw} and maintenance energy requirements of X_{Bw}.

Also for the WATS model it is true that the components are virtual components and consequently must be determined indirectly from measurements of respiration activity by means of the aerobic part of the WATS model (Figure 5.9). Hereby the components X_{Bw}, S_S, and $X_{S,fast}$ can be determined while the component $X_{S,slow}$ must be found from the mass balance, noting that total COD equal the sum of the 4 model components.

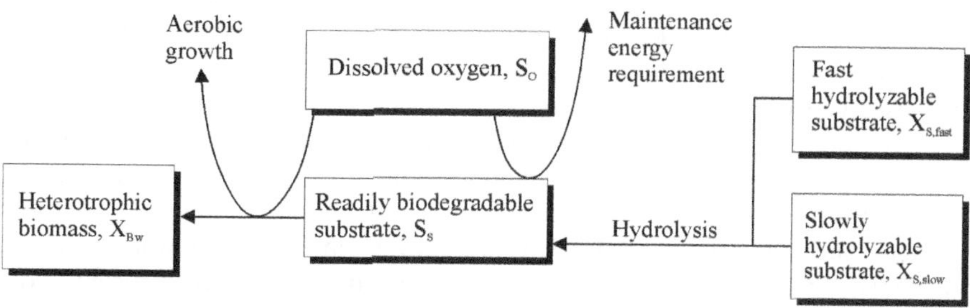

Figure 5.9 Aerobic degradation of organic matter in the WATS model (Hvitved-Jacobsen, 2002).

5.6.3 Variability in wastewater composition

Looking at the producers of wastewater, it becomes clear that the wastewater entering the sewer system mostly does so as discrete loads. For example, the flushing of a toilet is an event that occurs a certain number of times per day, it does not occur continuously. When a toilet flush enters the wastewater stream, it tends to be conveyed as a plug of pollutants, only dissipating slowly. When a sufficient number of such events come together in terms of the total flow from a catchment, the discrete loads tend to equal out.

Over the day, not only the quantity of wastewater is subject to changes but also wastewater quality. These quality changes are related to the variation of the wastewater producing activities within the catchment. Consequently the wastewater quality at any location in the sewer is subject to diurnal variations that are overlaid with short-term variations.

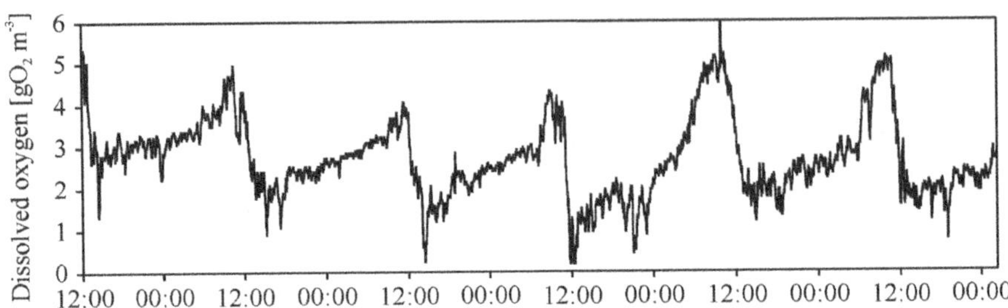

Figure 5.10 Variation in dissolved oxygen concentrations over 6 days in a sewer intercepting wastewater from app. 3,000 person equivalents (after Gudjonsson *et al.*, 2002).

Figure 5.10 shows an example that illustrates the diurnal wastewater composition variability overlaid with a short-term variability. Actually, the figure shows the variability of dissolved oxygen in the bulk water. However, as the dissolved oxygen concentration is governed by changes in the biological activity and wastewater biodegradability, the dissolved oxygen concentration mirrors the changes in wastewater composition. Two important lessons can be learned from Figure 5.10. Firstly the curve is not smooth but ragged. The rapid changes are caused by short-term variability in the wastewater quality. Secondly, there is a clear diurnal variation in the oxygen concentration, reflecting a diurnal variation in the quality of the wastewater produced within the catchment.

5.7 STORMWATER QUALITY

The quality of stormwater depends on the pollution washed of from the catchment surfaces, the pollutants resuspended in the conveyance system itself, and the retaining of pollutants prior to discharge. The two types of conveyance systems – combined and separate (section 5.2) – result in quite different qualities of stormwater to be discharged.

5.7.1 Separate systems

Assuming that there are no false connections to the separate storm sewer, the only pollution input to the stormwater runoff (SWR) are the pollutants adsorbed in the raindrops and the pollutants carried by the particles washed of the catchment surfaces. Catchment surfaces become increasingly polluted as a consequence of urbanization. During stormwater runoff, pollutants are eroded from the catchment surfaces and subsequently conveyed by the runoff water. The types and concentrations of the pollutants trapped in the stormwater depend on the land use and activities within the catchment. Table 5.10 shows an example of pollutants from highways with high traffic loads in urbanized areas compared to rural highways with less traffic.

The environmental impacts of discharging untreated stormwater runoff can be severe, as the amounts of water discharged often are large. It is consequently typically appropriate to implement stormwater mitigation technologies in order to protect the receiving waters (section 5.8).

Table 5.9 Mean pollutant concentrations (g m^{-3}) in runoff from urban and rural highways (Driscoll *et al.*, 1990).

Pollutant (g m^{-3})	Urban[1]	Rural[2]
Total Suspended Solids (TSS)	142	41
Volatile Suspended Solids (VSS)	39	12
Chemical Oxygen Demand (COD)	114	49
Nitrate+Nitrite (NO_3+NO_2)	0.76	0.57
Total Kjeldal Nitrogen (TKN)	1.83	0.87
Dissolved Phosphorous (PO_4)	0.40	0.16
Total Copper (Cu)	0.054	0.022
Total Lead (Pb)	0.40	0.080
Total Zinc (Zn)	0.329	0.080

[1] Average daily traffic > 30,000 vehicles
[2] Average daily traffic < 30,000 vehicles

5.7.2 Combined systems

Catchment surfaces also contribute to the pollution in combined stormwater runoff. However, in combined systems, pollutants with the origin in the wastewater flow and the sewer network must also be taken into account. The contributions to the pollutants of combined stormwater runoff that must be taken into account are:

- Pollutants washed off from catchment surfaces
- Sediments and biofilms formed in the sewer network during dry weather conditions
- The dry weather flow

Where pipe slopes are low, sediments will deposit during dry weather and may build up to significant depths. The sediments are a mixture of inorganic and organic particles. The main part of the inorganic particles originates from the catchment surfaces and is washed into the drains during consecutive storms. The main part of the organic particles originates from the dry weather periods where it become deposited and mixed into the inorganic particles. During storm runoff, flow velocities and corresponding shear stresses increase, and the deposits become resuspended into the flow.

The dry weather flow in the pipe at the onset of the storm as well as the wastewater running into the pipe during the storm becomes mixed into the stormwater runoff. How large this contribution to the pollution content is, depends on the ratio between the dry weather flow and the stormwater flow.

In contrast to stormwater runoff from separate sewers (SWR), not all stormwater from combined sewers is discharged to the receiving waters. Part of the mixed water goes to the treatment plant and first after exceeding the capacity of the treatment plant, surplus waters are discharged into the receiving waters (Figure 5.1). Depending on the hydraulic characteristics of the combined sewer overflow (CSO) structure, a large fraction of the pollutants can be retained in the system and conveyed to the treatment plant. To the contrary, what is discharged from CSOs is generally more polluted than SWR.

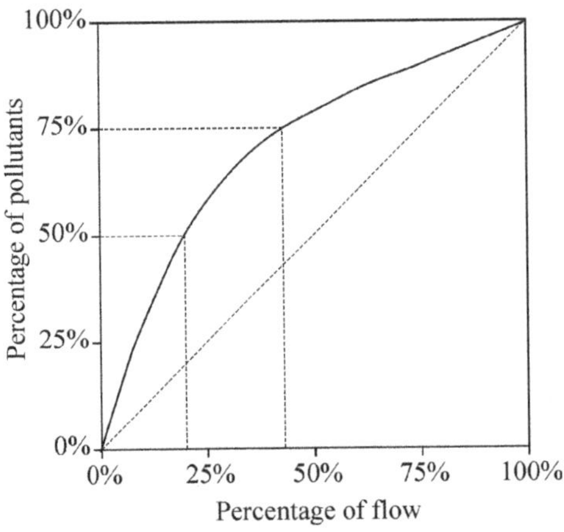

Figure 5.11 First flush in stormwater runoff, cf. text.

5.7.3 Pollutants variability

Pollutant concentrations in stormwater runoff are subject to large variations, both between events and during events. The term 'site mean concentration' (SMC) is often used to describe the average concentration of a pollutant discharged at a site, e.g. a stormwater outlet. When characterizing an event, the term 'event mean concentration' (EMC) of a pollutant is used. Variability's between the site mean concentrations of different sites and the event mean concentrations of a single site are of similar magnitude.

During a storm event, the pollutant concentrations in runoff water from both combined and separate systems are highly variable. In most cases, concentrations in the beginning of a runoff event are significantly higher than at the end of an event. This phenomenon is often called 'first flush'. For most catchments, first flushes must be expected. A convenient approach to illustrate first flushes is shown in Figure 5.11. The higher above the diagonal the curve is located, the more pronounced is the first flush effect. In the example, 50% of the pollutants are contained in the first 20% of the runoff volume, and 75% of the pollutants are contained in the first 43% of the runoff volume. Knowledge on the temporal and spatial variations of pollutants in SWR and CSOs is crucial when designing management systems for stormwater impact mitigation as well as analysing receiving water impacts.

5.8 STORMWATER IMPACT MITIGATION

Pollution loads discharged from developed areas increase with decreasing stormwater detention times. Consequently, receiving water impacts from catchment with short detention times are more pronounced than impacts from catchment with long detention times. When protecting receiving waters against pollution from stormwater runoff, a number of management systems are introduced as integrated parts of the conveyance system (Young *et al.*, 1996). The common denominator of these systems is that they combine detention or retention with different treatment technologies.

- *Extended detention ponds* are depressions in the ground with a controlled outlet, draining all water from the pond between storms. They control the maximum runoff level to protect downstream watercourses and channels against erosion. They also enhance removal of particulates by sedimentation.
- *Wet ponds* have a permanent pool. They fulfil the same goals as extended detention ponds, but in addition to sedimentation, microbial degradation processes and physico-chemical processes in the pond remove pollutants between runoff events.

- *Infiltration trenches* are excavated trenches lined with a filter fabric and filled with stones or plastic materials. The cavities of the filling material serve as storage. Water infiltrates into the surrounding soil through the sides and bottom of the trench.
- *Infiltration basins* follow a similar concept. However, they contain no filling materials but have a visible water surface that slowly dissipates after a storm event.
- *Sand filters* are conceptually similar to infiltration trenches and infiltration basins. However, drains beneath the structure collect the infiltrating water and discharge it as surface water.
- *Water quality inlets* are oil and grit separators that treat the water before it enters the stormwater drainage system. Water quality inlets are used where large amounts of particles, oil, and etceteras must be expected.
- *Grassed swales* are open stormwater conveyance systems that combine conveyance with pollutant removal. Pollutants are removed by infiltration and filtering through the vegetation.
- *Filter strips* are vegetated strips of land, conveying stormwater as overland flow to the stormwater conveyance system or directly into the receiving waters. The vegetation can be of any form, and pollutant removal processes are similar to grassed swales.
- *Constructed wetlands* remove pollutants by mechanisms similar to wet ponds. However, water depths are low and vary between dry land and pools of water.
- *Porous pavement* is any pavement that allows stormwater to infiltrate. Such pavement is typically of asphalt or concrete.

Choosing stormwater management strategies, a number of considerations must be made. Not only environmental benefits must be taken into account, but also costs of construction and maintenance are important issues. Some stormwater management strategies with high pollutant removal efficiencies, e.g. wet ponds, occupy much space, and have permanent water pools that call for e.g. mosquito control. Others are expensive to construct and maintain, e.g. sand filters. The choice of strategy must be based on an evaluation of all community added costs and values.

5.9 CHEMICAL, BIOLOGICAL AND PHYSICAL PROCESSES IN SEWERS

The initial motivation for removing stormwaters and wastewaters by means of a conveyance system was to reduce offending impacts on the urban environment from urban pollutants. Especially as the pollutants in wastewater cause nuisances because they become microbial degraded, resulting in a number of offensive degradation products. By removing the pollutants from the catchment surface, the problem of the urban environment is largely solved. However, the microbial degradation of pollutants continues within the conveyance system, causing impacts on the conveyance system itself, the urban environment, the treatment plant, and the receiving environment.

Processes of a physical, chemical and biological nature take place within the conveyance system. Diurnal build up and erosion of sewer sediments is an important physical process in many sewers. Gas transfers across phase boundaries are physico-chemical processes taking place where there is a free water surface, i.e. in gravity sewers. Biological processes are the transformations of wastewater compounds associated with growth and maintenance of biomass.

Transformations of wastewater in sewer networks are related to processes in the wastewater phase as well as the slime layer (biofilm). These biological transformations change the quality of the wastewater and are important for sewer corrosion, odour problems, wastewater treatment and combined sewer overflows. The transformations going on in sewers are strongly interlinked, and the importance of a process cannot be predicted without knowing the importance of all the other processes. As an example, to the question if hydrogen sulphide will be formed in a certain gravity sewer, it must first be determined if conditions are anaerobic (absence of both oxygen and nitrate). To do so, the complete mass balance for oxygen must be established, i.e. reaeration as well as oxygen consumptions in bulk water and biofilm must be known. In order to predict reaeration, the sewer geometry, temperature and flow conditions must be known, and in order to determine the oxygen consumptions in bulk water and biofilms, the quality of the wastewater COD must be defined (Hvitved-Jacobsen, 2002).

5.9.1 Why simulate sewer processes?

The need for simulating sewer processes is not a new one. Already in the first part of the last century, attempts were made to simulate processes related to corrosion of the conveyance system. Today, further reasons for sewer process simulation have been named:
(1) Corrosion of pipes and structures
(2) Odour problems
(3) Treatment plant concerns
(4) Receiving water impacts

5.9.2 Corrosion and odours

The first two issues, corrosion and odour, are related to anaerobic conditions, i.e. those cases where no oxygen and no nitrate is present. Such conditions allow sulphate reduction and fermentation to proceed, the products of which cause corrosion and odour. Typically such problems are observed at one of the following conditions:

- *After transport in pressure mains.* When wastewater is transported in pressure mains, anaerobic conditions   will develop rapidly as no reaeration takes place. At typical transport times and organic matter concentrations, significant amounts of hydrogen sulphide and odour will be produced even at low temperatures. Downstream of the pressure main outlet – especially at points of high turbulence – hydrogen sulphide and other volatile, odorous compounds are stripped from the water phase. Once present in the sewer atmosphere, they cause corrosion and odour problems.
- *In gravity sewers.* Odour and corrosion are seen in some gravity sewer systems. Conditions that increase the risk of such problems are: High temperatures, low flow velocities, high water depths, large biofilm to bulk water ratios, high organic matter concentrations, sediment deposits, and stagnant wastewater in e.g. septic tanks. Furthermore, if such wastewater becomes subject to high turbulence, stripping of hydrogen sulphide and other odorous compounds take place, potentially resulting in corrosion and odour problems.

5.9.3 Treatment plant impacts

The third issue, impacts on treatment plants, can be caused by hydrogen sulphide and odorous substances formed in the sewer system. This is especially the case in the inlet structures and grit chambers of treatment plants, where these substances are easily stripped from the water phase.

Another treatment plant impact is associated with the quality of the wastewater organic matter. Treatment plants with nitrogen removal and biological phosphorous removal need easily biodegradable organic matter in the removal processes. However, where wastewater is conveyed aerobically, easily biodegradable organic matter is rapidly consumed. On the other hand, where wastewater is conveyed anaerobically, easily biodegradable organic matter is conserved and even produced. Consequently, anaerobic sewers yield wastewater well suited for biological nutrient removal but of much odour and containing hydrogen sulphide. On the other hand, aerobic sewers yield wastewater with little odour and hydrogen sulphide but with low biodegradability.

5.9.4 Receiving water impacts

The last issue, receiving water impacts from combined sewer networks (CSO's), is related to storm events and consequently to the resuspension of sediments in the networks. At low flow velocities, wastewater particulates become deposited and form sediments. Depending on the flow conditions and inputs to the system, the sediments formed contain more or less organic matter (section 5.7). During dry weather, the organic matters in the sediments become partly broken down into more easily degradable components. When resuspended, the degradation products affect the degradability of the discharged stormwater.

5.9.5 Integrated urban wastewater management

Wastewater treatment plants are traditionally designed after the collection and conveyance systems are in place. Even intercepting sewers conveying wastewaters from whole catchments to a centralized treatment site are typically only designed for capacity and not included in the design of the biological wastewater treatment processes. Today, the state of the art allows a more sustainable solution, namely the integrated design of sewers and treatment plants.

Today, the wastewater conveyance system can be engineered to optimise the wastewater quality for subsequent treatment. Firstly, the optimal wastewater quality must be defined with respect to the treatment processes. Generally speaking, mechanical/chemical treatment plants produce the best effluent with wastewater of low biodegradability and a high content of particulates. Biological nutrient removal plants, on the other hand, produce the best effluent at the lowest cost with easily biodegradable wastewater. Secondly, the sewers must be designed to yield these qualities without operational problems and at low operational costs.

It is the challenge of the designing engineer to conceive sewer networks that can be operated to yield the desired wastewater quality at a minimum of environmental problems. Impacts on and operation of the sewer itself, the urban environment, the treatment plant and the receiving waters must be planned integrated and optimised to yield the best overall environment (Hvitved-Jacobsen, 2002; Vollertsen et al., 2002).

Assessing receiving water impacts, it is important to include all pollution loads. Stormwater discharges and combined sewer overflows must be included, as must discharges from treatment plants and other point sources. Furthermore, non-point pollutions from e.g. agriculture must be identified. All pollutant loads together with the transport and transformation processes in receiving waters ultimately allows an assessment of the receiving water quality.

Assessing environmental impacts of specific pollutants, their effect on the environment must be known. Some pollutants have acute impacts on the aquatic ecosystems and some have accumulated effects. Pollutants with acute toxic effects are e.g. ammonia and oxygen consuming organic matter, but also high water velocities causing erosion must be counted as an acute impact. Pollutants with accumulated toxic effects are e.g. nutrients (nitrogen and phosphorous) and heavy metals.

Sediments and bulk water residence times vary significantly between streams, lakes, coastal regions and the ocean. This fact in combination with the time scale of pollutant effects results in different responses to the pollutants from different types of receiving waters. For example, the effect of nitrogen and phosphorous on a receiving water is eutrophication, which causes extensive algae growth and consequently lowers the penetration depth of sunlight. Furthermore, after some time, the algae die off and cause oxygen depletion. This blooming of algae needs significantly more time than typical residence times in running waters allow for. Eutrophication is consequently not an issue in running waters. Generally speaking, the assessment of impacts from urban wastewater and stormwater runoff demands the study of processes in the sewerage system, the locale receiving water characteristics and the effects of the pollutants discharged.

5.10   CONCLUDING REMARKS

Collection and conveyance systems do more than collect and convey stormwater and wastewater. The characteristics of the stormwater collection system influences the pollutants collected, and stormwater treatment can be integrated already at the top end of the system. The design of the wastewater collection system influences the wastewater quality that is discharged to the treatment facility. This illustrates that design of the conveyance system should be integrated with the design of stormwater and wastewater collection and with the design of the treatment facilities. A more sustainable sewerage system is obtained by having all the properties of the whole system in mind when designing its parts.

REFERENCES

APHA, AWWA, WEF (1998) *Standard Methods for the Examination of Water and Wastewater.* APHA, AWWA, WEF, 20th Edition, Washington 1998.

ASCE and WEF (1992) *Design and Construction of urban stormwater management systems.* ASCE Manuals and Reports of Engineering practice No. 77; WEF Manual of Practice FD-20, USA.

ATV – A 118E (1977) *Standards for the Hydraulic Calculation of Wastewater, Stormwater and Combined Wastewater Sewers*. Abwassertechnische Vereinigung e.V., Germany, 1977.

Campos, M. and von Sperling, M. (1996) Estimation of domestic wastewater characteristics in a developing country based on socio-economic variables. *Water Science and Technology*, **34**(3-4), 71-77.

Danish EPA (1997) Udviklingen i den danske vandforsyningsstruktur (The development in the Danish water supply structure). *Arbejdsrapport til Miljøstyrelsen* nr 62, 1997.

Driscoll, E., Shelley, P.E. and Strecker, E.W. (1990) *Pollutant loadings and impacts from highway stormwater runoff, Volumes I.IV*, FHWA/RD-88-006-9, Federal Highway Administration, Woodward-Clyde Consultants, Oakland, CA, USA.

Gudjonsson, G., Vollertsen, J. and Hvitved-Jacobsen, T. (2002) Dissolved oxygen in gravity sewers - measurement and simulation. *Water Science & Technology*, **45**(3), 35-44.

Hvitved-Jacobsen, T. (2002) *Sewer processes: microbial and chemical process engineering of sewer networks*. CRC Press, Boca Raton, Florida, USA.

IAWQ (2000) *The Activated Sludge Models (1,2, 2d and 3)*. IWA Scientific and Technical Report, by the IWA Task Group on Mathematical Modelling for Design and Operation of Biological Wastewater Treatment. Edited by M. Henze.

IRC/WHO (1981) *Small Community Water Supplies, Technology of Water Supply Systems in Developing Countries*. Technical Paper Series 18, Den Haag Aug. 1981.

Nielsen, P.H., Raunkjær, K., Norsker, N.H., Jensen, N.Aa. and Hvitved-Jacobsen, T. (1992) Transformation of wastewater in sewer systems - a review, *Water Science & Technology* **25** (6), 17-31.

Raunkjær, K. (1993) *Characterization and transformation of wastewater organic matter in sewer systems*. Ph.D. dissertation, Aalborg University, Environmental Engineering Laboratory

Raunkjær, K., Hvitved-Jacobsen, T. and Nielsen, P.H. (1994) Measurement of pools of protein, carbohydrate and lipid in domestic wastewater, *Water Research* **28** (2), 251-262.

Tchobanoglous, G., Burton, F.L. and Stensel, H.D. (2003) *Wastewater Engineering – Treatment and Reuse*. Fourth edition. Metcalf & Eddy, Inc. Published by McGraw-Hill, Inc.

Vollertsen, J., Hvitved-Jacobsen, T., Ujang, Z. and Abdul Talib, S. (2002) Integrated design of sewers and wastewater treatment plants. *Water Science & Technology* **46** (9), 11-20.

WHO (1991) *Surface water drainage for low-income communities*. World Health Organization (WHO).

Young, G.K., Stein, S., Cole, P., Kammer, T., Graziano, F. and Bank, F. (1996) *Evaluation and management of highway runoff water quality*. FHWA-PD-96-032, Federal Highway Administration, GKY and Associates, Inc, Springfield, VA, USA.

6

Conventional small and decentralised wastewater systems

Robert Hughes, Goen Ho and Kuruvilla Matthew

6.1 INTRODUCTION

This chapter will detail the types of conventional small and decentralised wastewater systems that are used in developing nations and will inform the reader on the application of these systems in urban, peri-urban and rural areas, highlighting their strengths and weaknesses.

Current methodologies used to identify appropriate technologies and treatment processes for a specific situation, including social participation and acceptance, will be identified and explained to the reader. In particular, there will be a focus on the use of conventional small and decentralised wastewater systems to provide sustainable sanitation.

Sanitation is often a secondary factor when infrastructure is initially developed. It is commonplace for electricity and water to be primary objectives when implementing infrastructure, as these development aspects have the most apparent beneficial outcome (Keraita *et al.*, 2003). However it has long been pointed out that sanitation is not of secondary importance, due to its ability to reduce disease transfer and ensure the maintenance of public and environmental health (UNEP, 2004).

Sanitation and water are duly linked in their ability to provide an equitable and beneficial resource. For example, without sanitation it is common for water supplies, especially those from bores and wells, to be contaminated with disease transmitting organisms and some xenobiotic compounds as well as high levels of nutrients. The public health impacts of contaminated water supplies can include losses in production and expensive medication and immunisation programs. Currently about 2.4 billion people lack basic sanitation, 80% of these people live within Asia (UNDP, 2002).

6.1.1 Current practices in developing countries

The traditional approaches of centralised wastewater management provided in many developed nations have been inadequate for most developing nations. Although these approaches are often viewed as the most appropriate method for centralised management, they are associated with high development costs, especially infrastructure costs (e.g. deep sewerage costs), and high maintenance and operation costs (Ali, 2002). Generally, the higher socio-economic groups within a population can only maintain these costs because the high price of infrastructure is associated with higher land prices. Hence centralised approaches are not applicable for the whole population within a developing nation (Parkinson and Tayler, 2003).

In many of the urban centres of developing countries, smaller centralised and decentralised systems coupled with sewerage infrastructure may provide sanitation for the medium to higher socio-economic people (UNEP, 2004). The lower socio-economic people may have some provision of sanitation through these smaller systems. Commonly however in urban areas, the majority of sanitation is provided by small decentralised and onsite systems, such as pit latrines or water closets (Keraita *et al.*, 2003). These systems are often poorly managed in terms of operational performance and can lead to wastewater overflows or discharge of poorly treated wastewater (Bapat and Agarwal, 2003). The percentage of onsite systems in operation in many developing nations is likely to grow in the future, because the gradual development approach allowed by small systems is supported by non-government organisations (NGOs) and the World Bank (Wilderer and Schreff, 2000).

In a majority of poorly developed urban centres untreated wastewater may be discharged into water bodies leading to poor public and aquatic health (Senzia *et al.*, 2003). In other urban centres untreated wastewater may be piped to open ground and vacant plots away from the source, leading to pollutant dispersal into other urban areas and imminent health risks (Parkinson and Tayler, 2003). In these cases often the people causing the problem are unaware of the effect their sewage is causing to people down pipe in the disposal area. This can lead to an unwillingness to implement more effective sanitation (UNEP, 2004). In some of the poorest areas, sewage may be discharged locally, near the source into pathways, water courses and road verges (Keraita *et al.*, 2003). In these areas people may be aware of the direct effect of inadequate sanitation (UNEP, 2004).

In peri-urban areas often there is a lack of decentralised sanitation services, especially where rapid urbanisation occurs and infrastructure development is impeded by financial constraints. In these areas, sewage is largely treated by the available onsite technologies. The technologies are often inadequate for the quantity of wastewater produced by the majority. Wastewater in peri-urban areas like urban centres may be discharged away from, or at the source, and is commonly discharged without treatment onto areas for food production e.g. agriculture and aquaculture (Keraita *et al.*, 2003).

6.1.2 Conventional and decentralised wastewater systems

Decentralised and onsite wastewater treatment has gained importance across the world to treat wastewater from single sources and cluster scale housing. Onsite wastewater treatment is currently employed to treat wastewater for approximately 25% of the US population (Dawes and Goonetilleke, 2003). The use of decentralised systems in development situations is often more flexible than centralised approaches, allowing the design to fit into a number of development locations and scenarios (Wilderer and Schreff, 2000; Randall, 2003).

Traditionally, were sanitation capital was available, decentralised approaches were used in rural and remote areas (Dawes and Goonetilleke, 2003). Currently they are used within integrated urban water management (IUWM) and/or water sensitive urban design, to allow reuse of wastewater in an urban setting (Fane *et al.*, 2002). The treated wastewater can be used onto public open spaces, private garden/lawn areas, wetlands and other aesthetic landscape features (Foley *et al.*, 2004). In some situations the decentralised technologies can be used for urban food production, where nutrients and water are recycled onsite within the urban/peri-urban setting (UNEP, 2002).

There are obvious risks when using many decentralised systems and many uncertainties in respect to system performance and maintenance. These types of factors can be dealt with effectively through good design, following best management practices and understanding the role of decentralised technology within the nutrient cycle (Fane *et al.*, 2002). The nutrient cycle approach takes into account the effects to surrounding landscapes and land use practices (Ho, 2003).

This approach can be site-specific in terms of nutrient loadings or can follow a regionally accepted long-term irrigation rate. The maintenance of decentralised systems in many developed nations, such as the US, are now following certification programs, to ensure that appropriately trained personnel maintain these systems.

Centralised management of decentralised systems offers the most appropriate approach for ensuring that they are maintained for performance and discharge criteria (Parkinson and Tayler, 2003). Centralised management of small and onsite systems is often inadequate in most countries. Presently in developed nations, centralised authorities will assess the systems performance but the maintenance of the system is typically borne by the owner. Such an approach reduces the need for subsidisation of wastewater systems, which typically occurs with the use of centralised systems in developed nations (Hoehn and Krieger, 2000). However, when considering the adoption of ownership of a wastewater system to a developing nation, the cost of the system in terms of maintenance and operation should be within the capacity of the owner, including the possibility of operation and maintenance by the owner. If the owner is unable to maintain their own system, then appropriate maintenance personnel for a particular system should be locally available. The approach may involve high or low-tech systems, depending on the available expertise and money (See section 6.6 Selection of small and on-site systems for further details on technology selection).

6.2.   SMALL SYTEMS AND SUSTAINABILITY

Small systems are relied upon by a majority of developing countries across the world to provide sanitation for a substantial percentage of their population (UNEP, 2002). This is because they have been identified as one of the most appropriate means for sustainable sanitation in many developing nations, in terms of their implementability (Wilderer and Schreff, 2000; UNEP, 2004). Small systems have been further identified for uses within ecological sanitation and offer sustainability outcomes such as their ability to reuse wastewater and nutrients onto peri-urban and urban lands (Ho, 2003).

6.2.1 Relationship between small systems and sustainability

Small systems can be applied in urban and peri-urban areas as a method to increase sustainability, because they are more cost-effective at recycling water and nutrients than traditional centralised wastewater systems (Fane *et al.*, 2002; Ho, 2003; Mulder, 2003). The use of small wastewater systems has been put forth as one of the best means to recycle nutrients within the anthropogenic nutrient loop (UNEP, 2002). Small systems can be located near the source of nutrient and water use (e.g. agriculture), allowing the nutrients and water to be reused back at the source of application, after adequate treatment (Jonsson, 2002; Maurer *et al.*, 2003). This cycle process allows the recycling of nutrients within the natural biogeochemistry protecting aquatic water bodies, whilst also ensuring that primary industries such as agriculture have available nutrients and water for production (Figure 6.1) (Ho, 2003). In addition the added advantage of recycling non-renewable nutrients such as phosphorus, in a cost-effective manner (e.g. close to the source of application), can enhance long-term sustainability of resources (Jonsson, 2002).

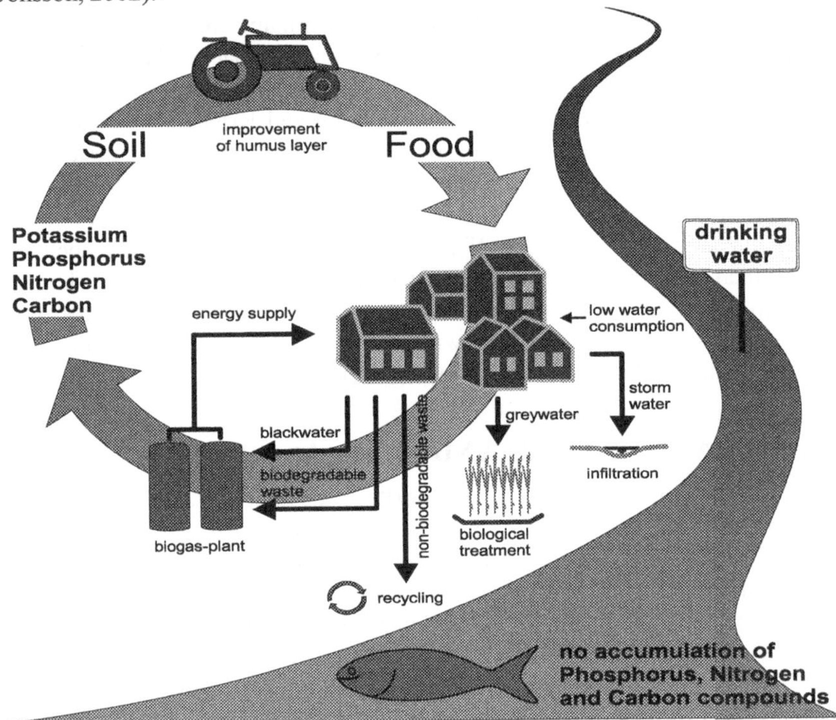

Figure 6.1. The recycling of nutrients using small systems and the protection of natural water bodies (Lange and Otterpohl, 1997).

Centralised treatment systems often do not ensure adequate recycling of nutrients and water and have in many cases increased nutrient production (e.g. fertiliser production) for primary industry use, because they are unable to recycle the nutrients and water onto the areas whence they came (Figure 6.2) (Fane *et al.*, 2002; Ho, 2003; Maurer *et al.*, 2003). Centralised approaches divert wastewater to a location a long distance from the source, which reduces the ability of the nutrients and water to be effectively (e.g. costs > benefits) recycled back to the source. Further to this, the concentration and mixing of the wastewater, especially with industrial discharges, can make treatment problematic and may reduce the reuse potential of the wastewater (and biosolids) further. Centralised methods can in many cases increase the nutrients within the local biogeochemistry with effects such as water pollution e.g. eutrophication (Ho, 2003).

Besides, the ability of small systems to recycle nutrients and water, non-flush technologies such as pit latrines, and vermicomposting and composting toilets, can reduce water demand for sanitation. In the vermicomposting and composting toilet example, recycling of nutrients through the production of vermicompost and compost is also maintained. These approaches are not always preferred however, due to poor public acceptance of non-flush technologies, from problems such as slight odours, sight of degrading wastes and the view of the technologies being inferior to flush technologies.

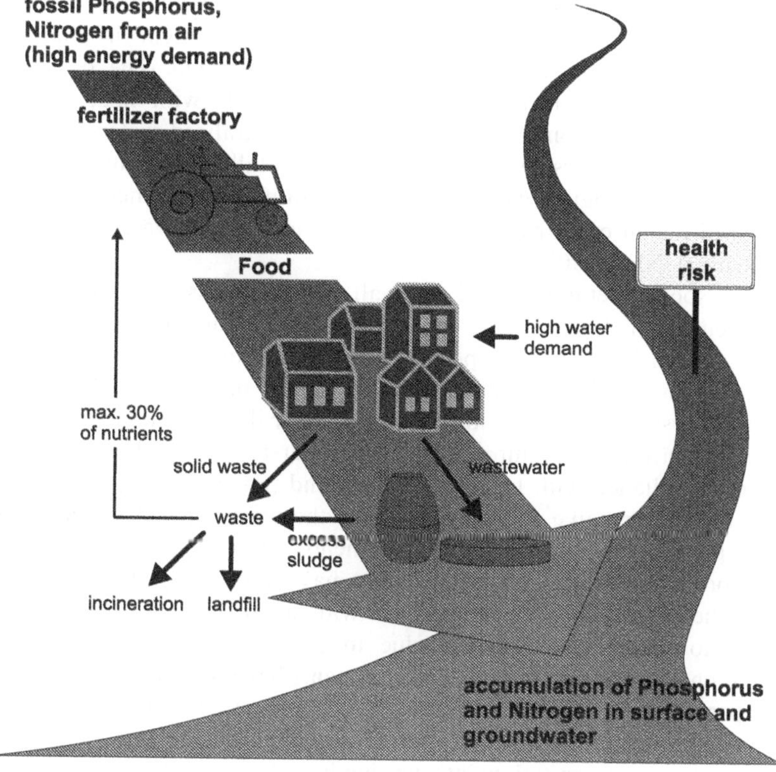

Figure 6.2 The use of centralised approaches to wastewater production and pollution of natural water bodies (Lange and Otterpohl, 1997).

6.2.2 Economic, social and cultural implications of small systems

Small systems have implications beyond environmental sustainability. The lower cost of small systems, allows them to be designed for the available regional costs. The small scale reduces the capital cost involved in the design of these systems, and the simple technology involved in most small systems can ensure that operation and maintenance costs are within specified local budgets. The small scale further ensures that as demand increases for sanitation, so also can the number of units in an area be increased, thus allowing a gradual implementation to be developed that is catered to immediate demands. Gradual development reduces the need for large scale developments in one period of time, that can be exorbitant to newly developed local economies (Wilderer and Schreff, 2000). Gradual development also ensures that the units within an area are working at optimal capacity e.g. 70 - 100% long-term acceptance rate.

The simple technological approaches, such as pit latrines and composting toilets, can also ensure that community involvement in the design and implementation of the small systems is maintained. The desired demand-driven approach can be utilised more effectively with small systems, when compared to the complexity of larger scale centralised systems. However, it is important to understand the underlying downfalls of the small scale approach as bad technological choice may cause pollution of immediate environments e.g. diffusion of nutrients and pathogens into groundwater from a pour flush latrine (UNEP, 2002).

A social factor involved in small scale systems is the ownership of the system itself. Localised or onsite wastewater treatment ensures the owner of the system is partially responsible for its maintenance and operation. Hence, the owner will in many cases gather a greater understanding and appreciation of the system, when it is located within their daily living area. Further to this, the owner may be required to have an understanding of the system to ensure it has compliance with local regulations and policy. Finally, in some cases an owner may receive an appreciation of the effect of bad system performance via pollution of surrounding environments, although this is an unadvisable and unfortunate scenario where the owner may learn through bad experience.

The opportunity driven approach with small scale systems is likely to be a key to their future success, due to their ability to treat wastewater for appropriate reuse scenarios. Small systems have the added advantage of being a part of the agricultural or aquacultural process and may offer an important means to increase production. Thus, their value may go beyond the importance of sanitation into production. As a greater variety of high tech small systems become available in the growing market, they will be invaluable tools for IUWM, making sure water is applied to specific uses depending on its quality (e.g. BOD, SS and nutrients), thus reducing potable water demand (Randall, 2003).

6.3 SEWERAGE SYSTEMS

Conventional sewerage systems were designed to remove wastewater and odours from the contact zone of humans e.g. water supplies, living and working areas. The improvement in public health from sewerage was notable from these early systems, and deep sewerage infrastructure was put in place in many western countries with associated high costs. These costs however were thought to offer long term investments to public health, productivity and economic outcomes and were in some cases accepted out of necessity e.g. public desire and demand for increased health and disease prevention.

In many developing nations, the high costs of conventional sewerage may not be within the means of regional or national governments. As noted above the high costs of this type of infrastructure also ensures high land prices, which may not be accessible by the majority. Additionally, alternative measures may be essential where future population increases are imminent or where the population fluctuates considerably. Further to this, community participation and sustainability within conventional sewerage may be questioned and other methods of sewerage may be more pertinent to the communities needs e.g. simplified sewerage (Mara, 2001).

6.3.1 Settled sewerage (small bore sewerage)

Settled sewerage is sewerage that is designed for primary treated wastewater e.g. primary treatment in a septic tank (Figure 6.3). This type of sewerage is used for two main reasons. The first reason is due to the soil infiltration capacity, where the soil has a low permeability and is incapable of treating the wastewater through percolation. The second reason is to remove the solids in the wastewater that negates the need for high velocity self cleansing pipelines. In this way settled sewerage can reduce the cost in pipeline infrastructure making it a viable option where there are large distances between homes, and where traditional onsite disposal is failing e.g. sodicity of clay soils.

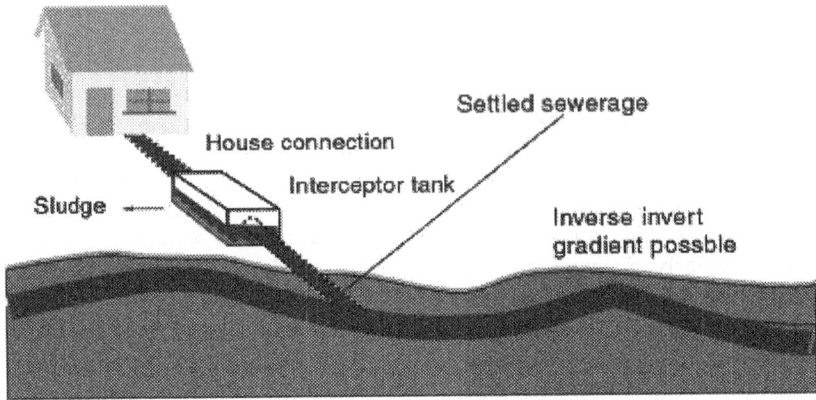

Figure 6.3 The depiction of a settled sewerage system that uses an inceptor tank for solids removal (UNEP, 2002).

6.3.2 Simplified sewerage (shallow sewerage, including condominial sewerage)

Simplified sewerage, also known as 'shallow sewerage', is a type of sewerage system developed to reduce the infrastructure and maintenance costs when compared to traditional deep sewerage. Simplified sewerage is designed on a similar hydraulic theory as conventional sewerage infrastructure but involves smaller diameter pipes and pipe depths as low as 0.2m. Inspection chambers in simplified sewerage such as manholes are typically inspection cleanouts.

The use of shallow sewerage is designed for quicker installation of sewerage infrastructure and is designed for up to 20 years when compared to the 30 years plus for deep sewerage. Shallow sewerage is utilised where uncertain population increases in urban centres are occurring. Mara (2001) noted that in developing countries simplified sewerage is a pragmatic method to offer sanitation to rapidly urbanising centres at a low cost.

Condominium sewerage is an example of shallow sewerage which involves sewerage piping passing through housing lots instead of on the side of the street (UNEP, 2002). Figure 6.4 shows a generic example of infrastructure built on the principles of condominium sewerage. The figure below also shows the use of the condominial approach to sewerage where the individual unit is an urban block of land rather than an individual house. This approach can increase community participation.

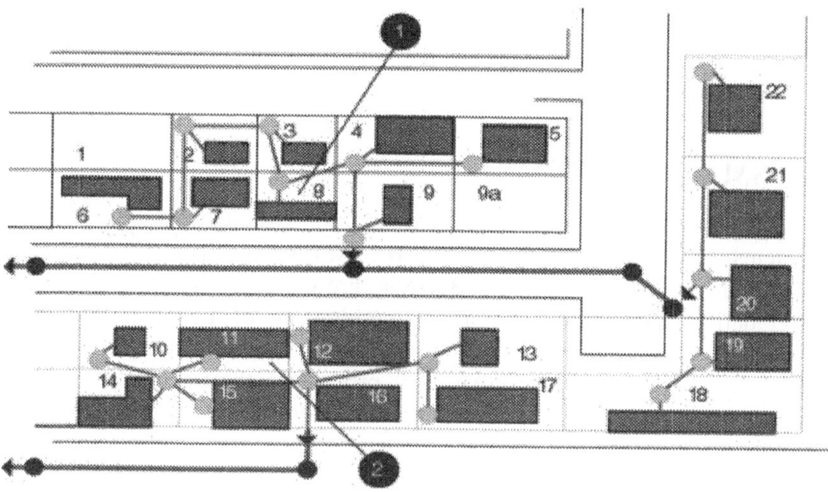

Figure 6.4 An example of condominium sewerage crossing numerous household lots (UNEP, 2002).

6.3.3 Low cost sewerage and community involvement

Simplified sewerage is more cost-effective than conventional sewerage due to its ability to allow community participation. Simplified sewerage has been estimated to cost between 30 to 50% less than conventional sewerage (UNEP, 2002). Many of the construction costs in conventional sewerage are due to the use of highly specialised machinery. The machinery is utilised to dig deep trenches, carry the pipes and other infrastructure into place and cover the trenches.

In simplified sewerage, such as condominial sewerage, the sewage connection to the main system may occur from a whole urban block and the block may become the unit assessed by the local regulatory authority. The regulatory authority may agree with the community to connect to the local sewerage system and/or treatment plant when appropriate measures and regulations have been met. In this case, the community must work together to ensure they have an adequate sewerage system. However due to the low depth of piping in the soil profile and the small diameter and length of the pipe it can allow a 'hands on' approach, where the community can dig the trenches and place the pipework in place under appropriate supervision. The community, with appropriate hygiene practices and management in place, may also undertake maintenance of the clean out chambers. The use of condominial sewerage in Brazil, its place of origin, has substantially increased sanitation services in this country and it is expected that this type of sewerage will be used extensively in the developing world.

6.4 SMALL SYSTEMS

Small systems have been developed for small scale wastewater flows. The systems typically rely on biological processes, such as bacteria and uptake via plants for removal of nutrients. Bacteria through consumption and biochemical degradation often remove the organic components of the wastewater. Many of the small systems employed across the world are cost effective because they utilise ecological processes (e.g. soil processes) that typically occur in the natural environment, rather than mechanical or chemical dependent treatment processes.

6.4.1 Ponds and lagoons

The use of ponds and lagoons in wastewater treatment is effective at removing BOD and SS as would occur in mechanical treatment systems such as activated sludge treatment. Ponding or lagooning has a longer residence time (e.g. order of days) than other mechanical treatment methods (e.g. hours for activated sludge) so many of the pathogenic bacteria, viruses, helminth eggs and cysts will be inactivated through natural die off to a very high standard.

Lagoons are made through a shallow excavation of around 1 to 2m and are generally unlined, allowing for soil percolation. After a period of time, sediment on the base of the pond will often form an impermeable barrier. Impermeable barriers may be placed on the pond base with a plastic liner or clay barrier to prevent any unwanted leaching into the soil, where high water tables are present. Developing an annual water balance is often important when developing a lagoon, as precipitation and evaporative loss can affect performance and sizing parameters. For more information on lagoons, please refer to Chapter 7.

6.4.2 Constructed Wetlands

Constructed wetlands are used for wastewater treatment and utilise similar processes to lagoons and land based treatment. Wetlands rely upon bacteria for the degradation of organic substances and uptake of nutrients by plants. Adsorption by the media used to construct the wetland (e.g. zeolite, red mud or bentonite) is also an important process and is a method for removal of nutrients, heavy metals and trace organic compounds.

Wetland like lagoons can be designed with the use of a water balance, but are often lined to ensure wastewater passes through the system not through the soil. The majority of a wetland bed is constructed with the use of coarse sand or aggregate. Other materials are also added in smaller quantities such as those with high adsorption properties or which are locally available (Dallas and Ho, 2004). Subsurface flow wetlands have been identified for use where air borne vectors of disease such as mosquitoes are present (Stephenson, 2001). For more information on wetlands, please refer to Chapter 8.

6.4.3 Land based treatment systems

Land based treatment of wastewater has been principally applied as a treatment mechanism due to the ability of bacteria to biodegrade organics within the wastewater. The properties of soil, particularly the clay mineral content, also play an important role in adsorption of heavy metals, trace organic compounds (e.g. surfactants) and nutrients. Often the adsorptive properties of the underlying soil, such as its cation exchange capacity and phosphorus retention index, will determine its ability to remove heavy metals, trace organics and phosphorus before infiltration into the groundwater.

Soil aquifer treatment is an approach that relies on the soil within an unlined basin to treat the wastewater to an appropriate quality before it reaches the underlying water table (Figure 6.5). The application of wastewater occurs in a cycle depending on the reduction of soil infiltration from organic clogging. In a typical situation, the cycle will involve one week of wastewater flooding where infiltration is reduced from organic build up, and one week of drying where bacteria consume the organics and soil drying takes place. After the drying period, soil infiltration is commonly acceptable once again for wastewater application.

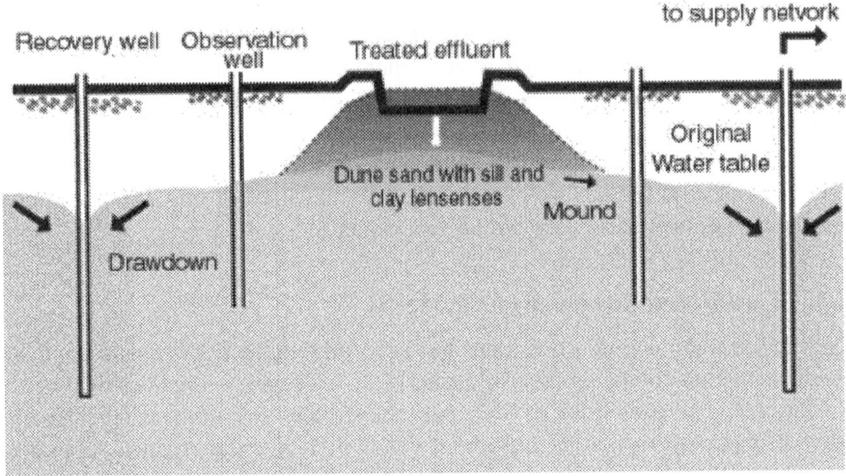

Figure 6.5 A typical soil aquifer treatment system showing the ground water table and observation wells (UNEP, 2002).

In slow-rate land application systems wastewater is disposed into a distribution channel upslope (Figure 6.6). There is commonly a collection channel downslope of the distribution channel. Between the channels, the land has a gradient that ensures wastewater can percolate through the soil whilst dispersing over the land area between the two channels. Vegetation, such as grasses, covers the area between the two channels and is used as food for livestock grazing either through direct feeding on the dried application area, or through 'cut and carry' approaches. The season, soil infiltration and types of vegetation covering the area between the two channels, will be used to define the rate of wastewater application over the land area.

The final type of land application system involves surface treatment through grass filtration or overland flow. This method is commonly employed during the rainy season when soil infiltration is reduced and wastewater treatment through percolation is ineffective. This method involves two channels as described above for slow rate application systems. The vegetation between the channels is relied upon to treat the wastewater through organic biodegradation from surface bacteria and vegetation uptake of nutrients. Typically this method involves numerous health risks and untreated wastewater should only be applied in small quantities (UNEP, 2002).

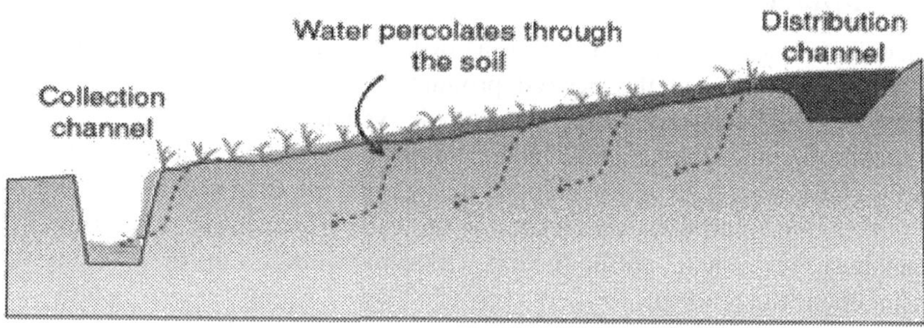

Figure 6.6 The typical design of a slow-rate land application system including upslope and down slope channels (UNEP, 2002).

6.4.4 Reuse

Wastewater reuse is practiced for agriculture, aquaculture, horticulture (e.g. flowers and forestry), natural flow allocation, groundwater recharge and industrial processes e.g. cooling systems. The reuse of wastewater should adhere to standard values, such as those described in Chapter 2, to minimise public and environmental health impacts. Hence, wastewater should be treated to an appropriate level (e.g. standard or site specific criteria) for the reuse purpose.

Small systems offer one of the most appropriate methods of wastewater treatment for reuse. Land based treatment systems are in effect reusing the wastewater where grazing of the treatment area or cut and carry procedures are undertaken. In many areas of the world land based treatment has been used for growing of food crops, although this is highly hazardous when considering disease transfer through direct uptake or disease vectors (Parkinson and Tayler, 2003). The use of ponds and wetlands before land application are an excellent approach for reuse of wastewater and can negate many problems from land application of raw wastewater.

Pond and wetland systems are simple cost effective approaches to treat wastewater to desired treatment level. Normally, the desired treatment level using these systems can be attained using a specific residence time to ensure nutrient uptake and pathogen inactivation. Furthermore, these systems can store quantities of wastewater for reuse and can be designed in parallel to allow irrigation from a final holding pond that has the desired standard of treated wastewater.

6.4.5 Aquaculture systems

Aquaculture systems can be designed as deep and shallow ponds, and artificial wetlands depending on the desired product that is to be produced from the wastewater e.g. fish, duckweed or aquatic vegetables. In many cases, the product of one aquaculture pond such as duckweed can be the food for the next pond such as grass carp, although both species may be used in the one pond. As with other reuse purposes, the desired wastewater treatment level is critical for aquaculture to reduce public health risks to workers in the system and consumers of the product.

Wastewater applied into aquaculture systems may be delivered in a variety of ways e.g. overhung latrine and pumped. In larger scale systems the aquaculture ponds are applied with treated wastewater (BOD 10- 30kg/ha/d) so that the higher nutrient wastewater can drive the system by providing nutrients for plant growth and subsequent fish growth. In some developing nations such as Indonesia and Vietnam contaminated surface waters may act as a source of these nutrients (UNEP, 2002).

6.4.6 Sludge management

The production of sludge occurs in all wastewater treatment systems, including decentralised and centralised systems. Sludge is produced because the primary aim of wastewater treatment is to remove solid organic and inorganic wastes from the wastewater. The level of sludge decomposition will depend on the treatment provided but in most processes, the solid organic material is converted to bacterial cells.

The level of sludge contamination will often reflect the level of wastewater pollution, as many treatment processes biodegrade organic hazardous substances, some of which are converted into hydrophobic forms, allowing adsorption to the sludge. Many of the inorganic substances, such as heavy metals are also removed with the sludge. To reduce the problems of highly hazardous sludge production, industrial wastewater should be pre-treated, to remove toxic organic compounds and heavy metals. Preferably, methods of reducing hazardous chemicals in both industrial and domestic wastewater should be used, such as the use of inventories described in Chapter 2.

There are numerous options for the treatment of sludge. Many of these provide useful by-products. These include, but are not limited to, stabilisation, thickening, dewatering, drying and incineration. Incineration is often expensive because fuel is needed for combustion and air pollution control requires extensive treatment of combustion gases. Stabilisation typically only occurs after the sludge has been thickened or dewatered. Stabilisation of sludge through composting or vermicomposting can produce biosolids useful for land application or soil remediation. The level of sludge contamination will often determine its usefulness as a by-product (Benitez *et al.*, 1999; Bajsa *et al.*, 2003; Maboeta and van Rensburg, 2003; Vinneras *et al.*, 2003).

6.5 ONSITE SYSTEMS

Onsite systems are systems contained within the household lot or within the close vicinity of the source of wastewater production. They typically have very little sewerage piping. The low cost and simplification of onsite systems has increased their application in developing nations and when applied correctly they can provide substantial sanitation. The new high tech onsite systems can offer wastewater treated appropriately for reuse applications (Randall, 2003).

6.5.1 Ventilated improved pit (VIP) latrine

The ventilated improved pit latrine (VIP) is an improvement on the traditional pit latrine. While the common pit latrine may have a single vault, the VIP consists of a shallow pit divided into two vaults, each with a volume of about $1.5m^3$. The pit covers are made of pre-cast concrete and are removable. Emptying can be done manually or mechanically in a 3-4 year cycle. Unlike the single pit latrine the VIP has the advantage of being permanent due to cycling of pit use.

The VIP, similar to the normal pit latrine, has ventilation provided by a vent pipe. The vent pipe has a fly screen on the outer opening, to avoid insect infestation. The vent pipe is often placed well above the pit, painted black and faced towards the sun to increase convection and airflow out of the pit. Pit latrines may be lined to prevent groundwater pollution.

6.5.2 Vermicompost toilets

Although vermicomposting has been applied to solid organic waste (e.g. sewerage sludge), its application to directly treat raw wastewater is a recent development. There are a number of designs of vermicompost toilets that include dry systems, batch systems and wet systems.

The dry vermicomposting systems separate excess liquids by urinal separation, or leachate separation within the system e.g. leachate pipe at the bottom of the tank. The systems that separate excess liquid are designed principally from composting toilet technologies. These systems can treat bodily waste, kitchen scraps, other organics (e.g. paper wastes), and allow excess liquid to pass through. The composting chamber in these systems is directly connected to the toilet seat, via a shoot, as in composting toilets (Figure 6.8).

The batch systems are used similarly to the compost system. They involve a container with aeration vents which is filled with bodily wastes (urine and excrement) and stored for up to 12 months to vermicompost, before reuse of the waste e.g. garden applications. When one container is filled, another one replaces it on a rotational basis. Batch systems require areas to store the filled containers to avoid low moisture conditions (< 60% moisture content), extreme temperatures (>35°C and <15°C) and to allow vermicomposting to complete.

The flushing vermicomposting systems are the next progression from dry vermicomposting systems. These systems have been devised for the whole domestic wastewater stream and can treat effluent from the laundry, kitchen and bathroom. They often require no secondary treatment system, and in some systems can treat the wastewater up to a high secondary standard e.g. <5mg SS and BOD. These types of vermicomposting systems include filter beds, which provide areas to gradually clean the effluent before UV disinfection, chlorination and/or filtration and reuse. These types of systems offer all the features of other modern urban onsite technologies (e.g. AWTS) but often require less maintenance and electricity. Figure 6.7 is a diagram of a Biolytix system, a new flushing vermicomposting system which has been developed in Australia.

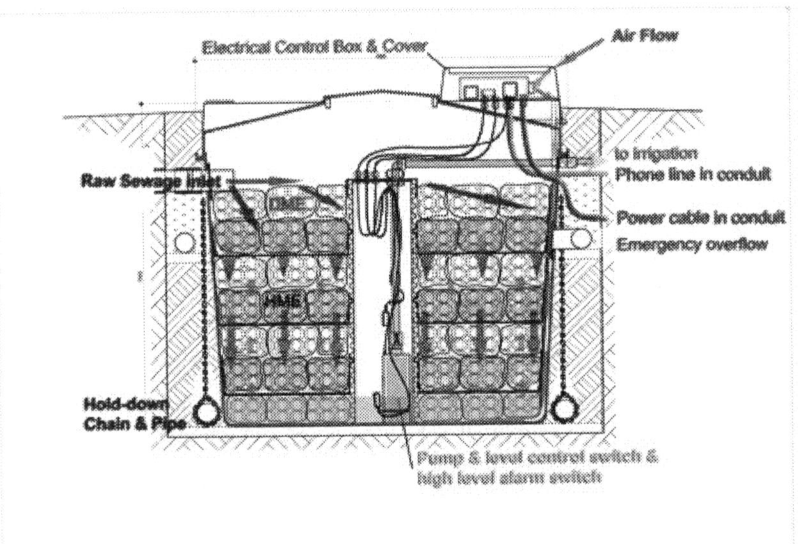

Figure 6.7 The biolytix filter BF6® is the latest development in vermicomposting treatment systems.

6.5.3 Composting toilets

Composting toilets use aerobic composting for the degradation of the faecal wastes to produce 'compost'. Air is often provided to a composting toilet though a pipe vent, although some models have additional 12 volt fans. Some composting toilets may be elevated to increase aeration through the compost chamber beneath the toilet. The aeration of the waste provides the oxygen needed by the aerobic bacteria that decompose the waste. There are three methods for the degradation of wastes in compost toilets.

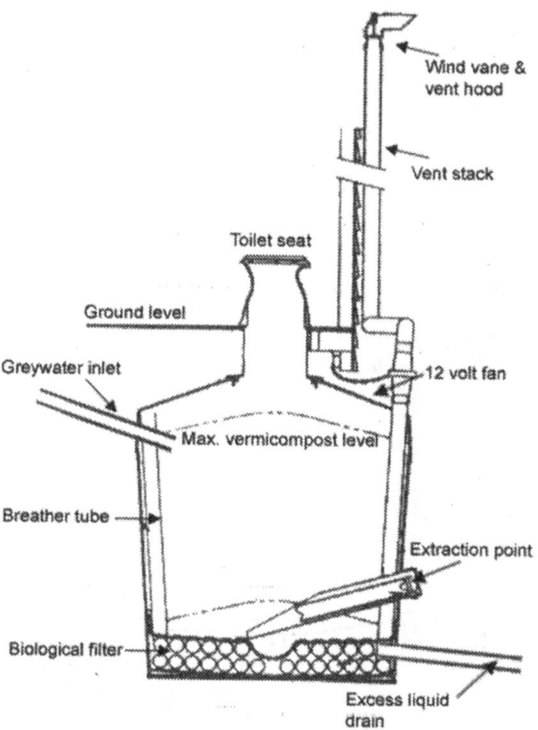

Figure 6.8 Schematic diagram of the below ground Dowmus aerobic composting toilet, showing extraction and leachate points.

The first method uses a batch system through wheelie bins that can be removed on a rotation basis, allowing the full bins to compost for a period up to 12 months. The second method is usually below ground where compost is removed through auguring. The third method is above ground and removal of compost occurs via a cleanout chamber e.g. shovelling. The final compost produced is often applied to the local garden. When excess liquid is applied to the compost chamber leachate is often disposed either through vegetation (e.g. garden), a disposal trench or sub surface reed bed. Some compost toilets can allow greywater reuse (Figure 6.8) as for wet vermicompost systems, although the technologies involved are not as advanced when compared to vermicompost systems

6.5.4 Pour flush toilets

Pour flush toilets have been developed on the basis of the traditional flush toilet, which relies upon a water seal that removes odour and insect infestations (Figure 6.9). The system works via a manual flush, where 2 to 3 litres of water is poured into the toilet after excretion.

The water, urine and faecal wastes from the pour flush toilet are collected in an anaerobic chamber, which may work similarly to a septic tank. The build up of faecal sludge within a pour flush chamber often necessitates regular emptying. The typical design of a pour flush toilet involves one chamber.

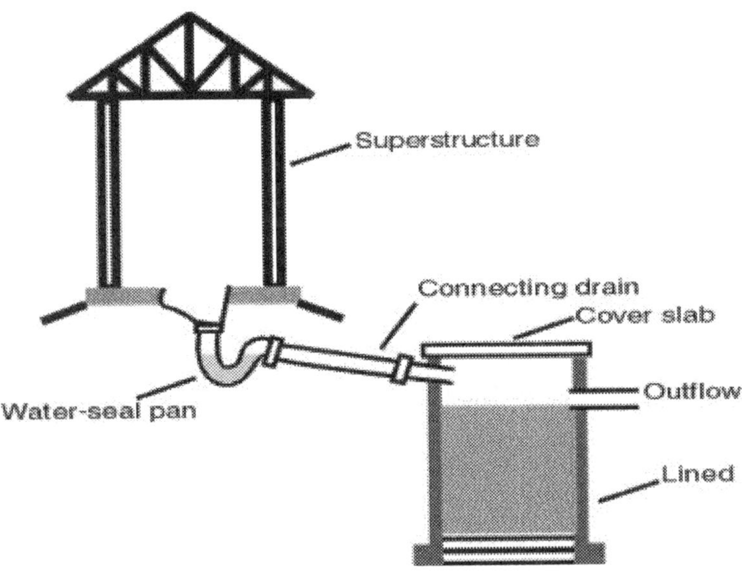

Figure 6.9 The typical setup for a pour flush toilet, including collection chamber and water seal (UNEP, 2002).

6.5.5 Septic tanks (including Imhoff tanks)

Septic tanks are holding and collection chambers for primary treatment of wastewater. They are designed to remove about 50% of the BOD and SS from the wastewater with a residency of 2 to 4 days. They are water sealed and can be made from bricks, mortar and rendered, or of concrete. They can receive wastewater from a pour or cistern flush toilet.

The shapes of septic tanks can vary between cylindrical or rectangular and the holding capacity is the main factor in their design. The traditional septic tank is now becoming baffled with an insert half way to ensure, solids have adequate residence time for removal and floating materials (e.g. oil and grease) do not flow through the system into the disposal area.

Imhoff tanks are a latter advent than septic tanks and have been developed to overcome many of the problems in septic tank design. The tanks can be rectangular or cylindrical but the chambers are pointed downwards to a point (Figure 6.10). They contain two chambers, one inside the other, which are joined by a small opening in the bottom of the first chamber. The two chambers allow depositing sludge in the first chamber to fall into the second (lower) chamber. Wastewater flows in through a pipe in the upper chamber and out the other side of this chamber through another pipe.

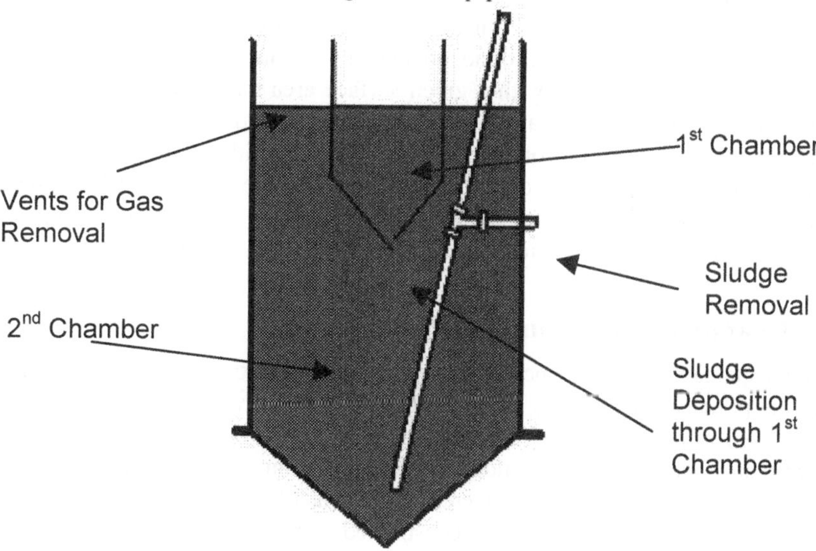

Figure 6.10 An Imhoff tank showing the sludge deposition pathway.

The two chambers one for sludge deposition and one for sedimentation separate the sludge from incoming wastewater and remove the major problem in septic tank design, where deposited sludge is mixed and allowed to resurface with incoming wastewater. Imhoff tanks have gas vents on either side of the lower chamber to allow gaseous emissions (e.g. methane and carbon dioxide) from anaerobic degradation of the sludge.

Imhoff tanks and septic tanks require a disposal method, such as a leach trench or evapotranspiration bed to dispose of the primary treated wastewater. Appropriate methodologies for disposal are described below.

6.5.6 Leach drains

A leach drain or seepage trench is used for the disposal of settled wastewater. The seepage trench allows uniform disposal of the wastewater over a given area. The size of a trench is often developed in respect to the wastewater load and the environmental conditions e.g. soil type, groundwater depth, and precipitation. When the groundwater table is close to the surface (e.g. <0.5m from the surface) or the soil has a low permeability (e.g. $\leq$ 3mm day) leach drains may be inadequate for wastewater treatment, due to either groundwater pollution, or surface ponding from inadequate infiltration. In soils with a moderate permeability, leach drains may be increased in size to ensure adequate disposal over a larger area, thus allowing for adequate soil infiltration.

Leach drains are designed with a given surface area that usually accounts for the bottom and sides of the drain. The drain is often filled with gravel or highly permeably material and a perforated pipe placed in the middle. The pipe is paced about 0.2m below the soil surface and is surrounded by the gravel, which can then act to convey the wastewater to the sides of the trench. The perforated pipe is typically around 0.1m in diameter.

6.5.7 Evapotranspiration beds

Evapotranspiration beds are principally designed for evaporative losses of wastewater where the primary soil is highly impermeable e.g. clay or rock (Figure 6.11). They may be used in permeable soils nevertheless to increase water losses and increase nutrient uptake via plants. In situations where space is limited, they may be used in conjunction with a seepage trench to increase wastewater loading over a given area.

Raised evapotranspiration beds are constructed with the use of highly permeable materials (e.g. sand) that elevate the disposal area above the impermeable primary soil. The perforated pipe used to diffuse the wastewater through the bed is often surrounded by gravel to increase dispersal of the wastewater throughout the whole bed, thus increasing water availability to the vegetation.

Vegetation is planted on top of the bed to ensure loss of water through evapotranspiration. The plants take up the nutrients, essential elements, and some organic compounds from the wastewater. The design of the bed should ensure it is large enough for wastewater loading coupled with precipitation. Conversely, the design should ensure enough water and nutrients for plant health.

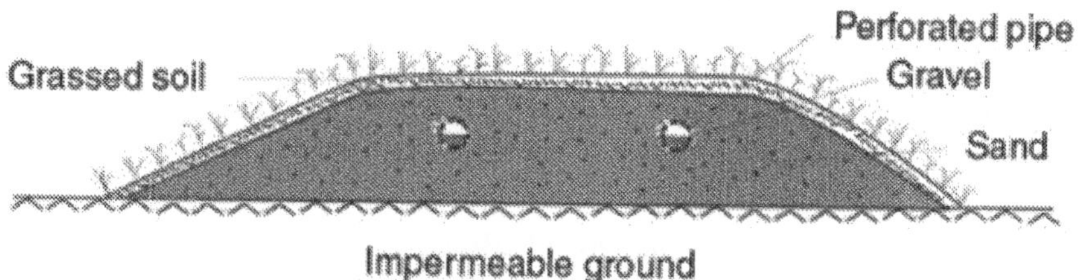

Figure 6.11 A raised evapotranspiration bed with grassed vegetation. (UNEP, 2002)

6.5.8 Digesters (small anaerobic systems)

Small anaerobic digesters utilise the same processes as larger scale anaerobic digestion. The systems are typically closed in-vessel devices that allow the microbial degradation of liquid organic waste in an anaerobic (oxygen free) environment. The major by-product (90%) of anaerobic digestion is biogas (60% methane and 40% carbon dioxide) that is a useful energy source (Mata-Alvarez *et al.*, 2000). Small amounts of other gases, such as ammonia and hydrogen sulfide, may be present with the biogas.

The anaerobic digestion of liquid organic wastes requires temperatures above 20°C to ensure the bacteria are active enough for subsequent waste degradation. The degradation of common contaminants and pathogens in anaerobic digestion has been found to occur, but requires high temperatures in the thermophilic range e.g. > 50°C (Hartmann and Ahring, 2003). These high temperatures are not always encountered in small systems, although most systems reach temperature high enough for efficient production of biogas.

A small anaerobic digester may employ one of two designs. The first type is the batch process, where a system is specifically designed to digest a single load of wastewater. Once the wastewater has been digested, another load is entered into the system. The second type of design is a continuous system. These systems are loaded so that digestion is continuous. Digested material is pushed through these systems from the loading of new material or by a mechanical process.

The common types of anaerobic digesters are cylindrical in-vessel designs. Septic and Imhoff tanks may be used in some cases. Lagoon systems are another design but may be more affected by eternal influences e.g. temperature. Anaerobic digesters are sealed with either a cover (e.g. lagoons) or are built as a closed vessel. Small anaerobic digesters have a compressor for collection of the biogas and may use a number of technologies, such as reciprocating engines, and steam, gas and micro turbines, to produce electricity once the biogas is free from moisture and contaminants e.g. ammonia. In some cases, the biogas may be piped to a regional facility equipped to produce electricity from the biogas. Some more advanced small anaerobic digesters may use a mixing and distribution system for even distribution of the liquid organic waste.

6.5.9 Reuse of wastewater and sludge

The reuse of onsite wastewater from small systems may take place for aquaculture, agriculture, natural flow allocation, water recharge in aquifers and water storage in surface waters. Many of the applications for water reuse in aquaculture and agriculture have been described above and involve the use of water and nutrients as well as essential elements for growth of terrestrial and aquatic macrophytes. These plants are subsequent food for animals in the agricultural and aquacultural systems. For natural flow allocation, the wastewater is usually treated to a high level similar to the natural chemistry of water body and then discharged to ensure that the aquatic environment has available water for aquatic health and ecological functioning. The types of systems used for this purpose may include a collection system such as a septic or Imhoff tank that conveys wastewater to a suitable wetland or lagoon, which then provides suitable treatment (e.g. removal of nutrients and pathogens) for discharge into the aquatic environment.

The recharge of aquifers involves the application of wastewater for recharging groundwater, where bores may remove water for potable, industrial, public or agricultural supplies. Many onsite systems dispose of wastewater through the soil and consequently dispose of wastewater into the groundwater. This is an unintentional reuse scenario and may not be pertinent to longevity of groundwater supplies. The reuse of wastewater for replenishment of surface water bodies will require a certain level of treatment depending on the water bodies classification, but may be used for potable, industrial and ecological classifications.

Onsite systems, including those described above, offer many other options for reuse of wastewater. The separation of greywater in many systems, such as the pit latrine and compost toilets, can allow domestic greywater reuse onto fruit trees or non vegetable garden areas. Due to the nature of greywater, it often requires less treatment than blackwater, as it is lower in pathogen and nutrient levels. Greywater is commonly reused through a subsurface irrigation system. The use of urine separation in onsite systems offers a viable option for reuse in agriculture and requires less treatment than blackwater. Urine offers many essential nutrients and elements in appropriate concentrations for plants. Urine separation can also reduce the nutrient loading into the wastewater system and therefore may reduce treatment costs.

The solid waste produced in small systems can be reused by a number of ways, some requiring no treatment. In the anaerobic systems, such as pour flush latrines, and septic and Imhoff tanks the build up of faecal sludge will be removed for proper performance of the system. In these cases, the sludge may be treated using the same methodology as for Sludge Management (section 6.4.6 above). In many aerobic systems, the solid products of treatment can be immediately reused, such as the vermicompost and compost from vermicomposting and composting toilets.

6.6 SELECTION OF SMALL AND ONSITE SYSTEMS

The selection of a wastewater system should be within the context of the people and environment it is aiming to protect. In this sense, a good understanding of the technologies capable of offering the appropriate level of treatment necessary is vital to the provision of an adequate sanitation system.

Small and onsite systems selection should consider the economic and environmental factors associated with a sanitation program, as well as social factors that can lead to the acceptance and maintenance of a system in the long term. Selection may additionally involve reuse considerations, such as for aquaculture and agriculture, where the nutrients and water may be recycled through a production system.

6.6.1 Decision support tools for selection of small and onsite systems

The criteria for selecting a suitable sanitation system are numerous. Generally, onsite or small systems are chosen due to their cost-effectiveness with larger systems preferred where funds and water is available (Mara, 1996). The choice between onsite and small systems, such as those described above is largely determined by the availability of water and resources within a given area. A simple decision making flowchart developed by Pickford (1995) is provided below (Figure 6.12), and outlines the decision making process with available resources. The factors considered include method of anal cleansing, water availability, cost affordability, and population density.

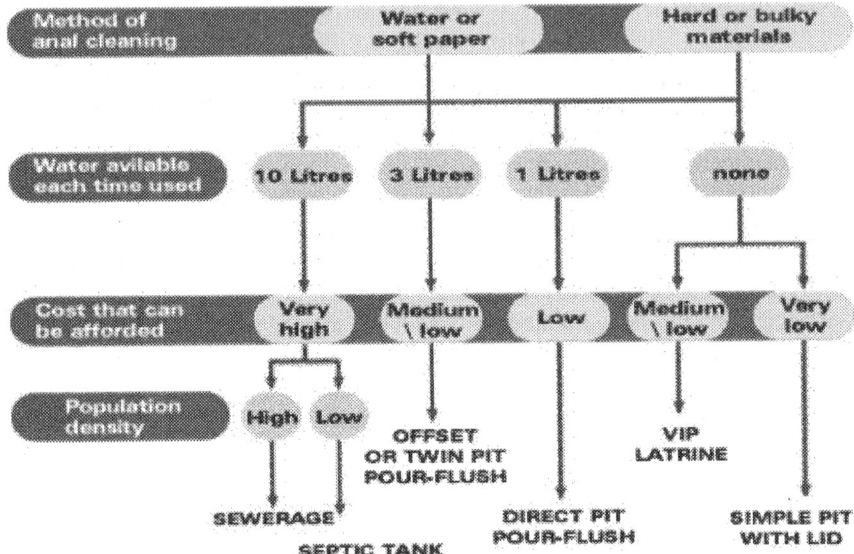

Figure 6.12 The simplified version of the decision making process for selection of a sanitation system (Pickford, 1995).

Mara (1996) has a more comprehensive flow chart that can be viewed in UNEP (2002). Additional factors included in the chart were the impact from sanitation technologies on the environment. An example is where a pit latrine was assessed as being unsuitable in areas with a high groundwater table, as this can lead to pollution in the immediate groundwater and subsequent contamination of wells and bores. These factors are obviously site specific in nature and may require additional modelling such as that described below in 6.6.2 Computer based decision support tools.

6.6.2 Computer based decision support tools

There are a number of computer based decision support tools to guide a sanitation program towards the most appropriate system. Examples are WAWTTAR[©] (Finney and Gearheart, 1998) and SANEX[©] (Loetscher, 1998). The decision support process for SANEX[©] and its application are detailed below to illustrate how sanitation software operates e.g. criteria importance and analysis.

SANEX©

The software SANEX© was developed for the MS Windows operating environment. The SANEX© program has been developed to aid decision makers and the community by identifying probable sanitation alternatives to centralised schemes and the community support that they are likely to receive. The knowledge base of this software contains more than 80 sanitation alternatives, which are combinations of the technologies outlined below. The program uses around 50 technical, socio-cultural and financial criteria for a multi-amalgamation assessment. The costing component employs approximately 50 functions.

Sanitation alternatives considered in SANEX©

Toilet facilities:

- Pour-flush toilet
- Cistern-flush toilet

On-site facilities:

- Simple pit latrine
- VIP latrine
- Pour-flush latrine
- Aquaprivy
- Septic tank
- Vault (vacuum cartage)
- Seepage pit
- Drain field

Public facilities:

- Public toilet block
- Overhung latrine

Resource recovery:

- Double-vault composting toilet
- Excreta-fed fish pond
- Septic tank for excreta reuse
- On-site biogas digester

Sewerage:

- Covered stormwater drains
- Conventional sewerage
- Simplified sewerage
- Settled sewerage

Off-site treatment:
- Communal septic tank
- Imhoff tank primary treatment
- Waste stabilisation ponds
- Activated sludge treatment

Two-Stage Evaluation

The program has two distinct evaluation stages. The first stage involves a screening process where non-feasible alternatives are eliminated principally based on technical criteria. During the second stage, the comparative process, remaining alternatives are compared using indicators such as implementability and sustainability. Sustainability is the probability that the sanitation will be beneficial over its required period of use. Implementability is expressed as the probability that sanitation facilities can be constructed within time and economic constraints.

Multi-Level Amalgamation

The main problem in this program, the amalgamation of numerous criteria outcomes, was found when developing an unbiased comparison of alternatives. The use of traditional algorithm methods would have caused a diminishing effect on each criterion. To preserve the effect on each criterion a new method for amalgamating criteria on multiple levels was developed. The combined advantages of multi-level amalgamation are reflected in more conceivable sanitation alternatives.

Costing of Sanitation Alternatives

The program assumes that the planners in developing countries have sufficient access to financial expertise so no criteria to assess the affordability of sanitation systems are formulated. Instead, a costing model that enables the proposed decision aid to estimate the capital costs as well as the recurrent costs of all alternatives is used. Further, a simple method based on the local residential building cost is utilised to convert these estimates to costs in local currency units, as they would occur in the project area.

6.7 CASE STUDY FROM AFRICA

This case study from Africa does not attempt to detail the use of onsite and small systems from the whole continent. It summarises the use of small and onsite systems in sections of the continent to gain an understanding of their importance for sanitation and it highlights the appropriate application of small and onsite systems in specific areas.

In Africa, although large scale sewer systems are highly sought, the cost of these systems coupled with low availability of water supplies in many centres, has reduced large scale sewer application.

Table 6.1 Coverage of full waterborne sewerage connections in some African countries (UNEP, 2002).

Country/urban centre	% population connected	Remarks
Botswana/Gaborone	50	Available access is 100%; length of sewer is 1.85 km /1000 persons
Ethiopia	7	
Ethiopia/Addis Ababa	12	
Ghana/Accra	3.3	
Kenya/four towns	100	DANIDA project: Busia, Homa Bay, Isiolo and Nyahururu
Malawi/all urban areas*	15	1987
Nigeria/Abuja Wupa	100; whole territory (75)	Abuja City in Wupa drainage district houses 20,000 people.
Nigeria/Lagos	5	Metropolitan Lagos.
Nigeria/Lagos/Festac Town	100	Population of 90,000
South Africa/all towns and cities	64.2	Ratio of all urban population (totalling 24.5 million in 1990.
South Africa/North Cape	29.9	For the whole province.
South Africa/Western Cape	85	Out of a total urban population of 3.156 million in the province.
Zimbabwe/Harare	~100	High income dwellers who have large plots of >2,000m^2 are allowed to have septic tanks, length of sewer is 4.1 km/1000 persons

*Note: Blantyre and Lilongwe each has a water board. Donor environment appears favourable, as many donor organizations have provided financial assistance through grants and loans. The donors include World Bank, UNICEF, UNDP, USAID, DANIDA, EEC, ADB, CIDA, British and French governments

The coverage of onsite and small systems varies with the region and its economic status. Most of the areas with high land values will have access to sewer or septic tanks. In some instances, cities with a lower economic status may implement exceptional sewerage planning, such as the treatment facilities in Gaborone. For a majority of the areas with a lower economic status they will generally rely upon small and onsite systems. Table 6.1 details the extent to which sewer systems are available to residents of selected regions across Africa.

The use of a sewer system has posed numerous problems across the African continent. The first problem relates to the malfunction of flush toilets, which can reduce the quality and increase the cost of sanitation provided. For example, in Kumasi, Ghana, more than 50% of households preferred a ventilated latrine to a water flushed toilet, because the ventilated latrine is more reliable and does not depend on water.

The second problem relates to water consumption and the ability of sewer piping to self-cleanse. For instance, it is known that the higher socio-economic groups within a population typically have more water access and water consuming white goods, such as clothes and dish washing machines. These white goods produce greywater that aids in the dilution of blackwater and thus helps to cleanse sewer pipes. However, with the addition of a high population of lower socio-economic groups to the sewer system, less dilution may occur through greywater additions and blocking through solid build up is highly probable.

The table (Table 6.2) below details the regional distribution of onsite systems in use across Africa. As can be noted by the figures within the table the reliance on pit latrines, either shared or private, is considerably high. In areas with greater access to water, the use of septic tanks is also high.

6.7.1 Onsite technologies employed in Africa

Septic Tanks

The common septic tank is used through out Africa as shown in Table 6.2. The system is relied upon where water availability or water use is high, thus allowing comparatively high wastewater loading from flush toilets and household white goods. In many of the locations that rely upon septic tanks, access to sewers may also be available. In these areas, the use of septic tanks removes extra loading and pressure on the sewer system.

Reid's Odourless Earth Closet (ROEC)

One of the earlier pit latrine designs was known as the ROEC improved pit latrine. The system was developed in South Africa and patented in 1944 (du McPherson, 1994). The dimensions of the pit are 1x2m and it is at least 3m deep. It is covered with a concrete slab fitted with a 75mm diameter vent pipe.

Double-vault VIP latrine

The Botswana's Ministry of Local Government and Lands, together with the Building Research Establishment in the UK, developed the double vault VIP latrine. About 10,000 units have been built in Botswana since 1978. The system has been used in squatter areas, site and service schemes and many urban areas. Details of the system can be found above in section 6.5.1 Ventilated improved pit (VIP) latrine.

Table 6.2 Access to onsite sanitation facilities in selected countries (UNEP, 2002).

Country/Urban Centre	% using septic tanks	% using VIP toilets	% using other pit latrines	% using public/share toilets	% using pail/bucket systems	%without access to toilet facility
Botswana/ Gaborone	34	-	52	14	-	-
Ethiopia/Addis Ababa	8	-	69[b]	42[a]		33
Ethiopia/all other towns (17 nos)	3	-	49[b]	37[a]	-	48
Ghana/Accra (1990)	31	-	56[b]	41[a]	-	13
Malawi/all urban areas (1987)	10	-	65[b]	-	-	10
Nigeria/Lagos	60	30	3	-	0	2
South Africa	1.8	1.1	21.4	(1.6)	7.9	2.1
South Africa/ North Cape	5.9	8.0	11.4	0.1	28.0	1.1
South Africa/ Western Cape	1.1	0.1	0	0.4	8.9	3.8
Zimbabwe	5	25	16	-	0	54
Zimbabwe/ Harare Province	37	16	36		0	11

a = a portion of total population using pit toilets; b: this figure includes VIP toilets; c = all types of latrines, including VIP and unimproved latrines
Sources: (World Bank, 1989), (World Bank, 1997).

Composting Toilets

The composting toilet is a recent technology in South Africa, having come into existence only since 1993. During this short period, its evaluation showed that Enviro Loo latrines performed adequately and were accepted by users as a satisfactory alternative to communal chemical toilets. In the assessment safety, privacy and accessibility were preferred aspects of the system and no adverse health impact was found among the users. Unlike the composting toilets described in section 6.5.3 (composting toilets), the technology in South Africa works in a prolonged process of dehydration rather than aerobic composting. Further treatment of the organic waste material maybe required based upon evaluation, once removed from the toilet chamber (Scott, 1998).

6.7.2 Onsite system application

The availability of water and sanitation in Mozambique presents serious problems. Only 12% of the dwellings in Mozambique have direct supplies of drinking water. The majority ($\geq$ 70%) of households in urban centres source their water through a communal network (public fountains), with the remainder relying on rivers, ponds, wells and tank trucks. Due to the low availability of water and low price of onsite treatment systems, a rigorous latrine construction campaign multiplied the number of latrines in Mozambique from 60,000 to 1.2 million. In 1984, 72% of urban dwellings were equipped, 93% in Maputo. A number of problems were encountered during this sanitation program including the instability of the soils, pollution of groundwater through high water tables, and the failure of the covers to keep out insects.

The squatting slab is the major component of all pit latrines, as is the drop hole in the squatting slab. Due to this, the government of Mozambique adopted a standard model of latrine cover that was shown to be more effective than the former wooden covers. It was 1.5m in diameter and made of reinforced concrete, circular and slightly conical. A single person could easily manage the slab. The shape was designed to fit the exact dimensions of a pit. The pit could be raised up if the water table was high or built on breeze-block if the soil was highly unstable.

Local authorities through the Directorate of Construction and Town Planning of Maputo set up cooperatives in order to produce the pit covers. Local authorities (the Facilitator Groups) that required greater sanitation provision chose members of the cooperative. This coordinating committee (Inter-Cooperative Management Committee) carried out many functions that were originally performed by the government. Annual production of latrine covers was 18,000 in 1986 at a cost of Mt 500 (US$12) each (Brandenberg, 1985). The present need is 2,500,000 in urban and peri-urban areas. The number of cooperatives and production centres continues to increase, and the project has been replicated in other towns with a total annual production of more than 10 million.

The project is largely dependent on international aid and is overly dependent on government policy which could compromise its long term viability. To ensure its future success the population must be more engaged in government direction towards sanitation to give the organizations involved in the project greater autonomy.

REFERENCES

Ali A. (2002) Operational Problems of Waste Water Treatment Plants in Developing World. WAPDEC: Water and Wastewater Perspectives of Developing Countries, New Delhi, India, pp. 933-937.

Bajsa O., Nair J., Mathew K. and Ho G. (2003) Vermiculture as a tool for domestic wastewater management. Water Science and Technology, 48(11/12), 125-132.

Bapat M. and Agarwal I. (2003) Our needs, our priorities; women and men from the slums in Mumbai and Pune talk about their needs for water and sanitation. Environment and Urbanisation, 15(2), 71-86.

Benitez E., Nogales R., Elvira C., Masciandaro G. and Ceccanti B. (1999) Enzyme and earthworm activities during vermicomposting of carbyl-treated sewage sludge. Journal of Environmental Quality, 28(4), 1099.

Brandenberg. (1985) The Latrine Project, Mozambique. International Development Research Centre (I.D.R.C), Ottawa.

Dallas S. and Ho G. (2004) Performance of subsurface reedbeds for the treatment of domestic greywater. 6th Specialist Conference on Small Water and Wastewater Systems, 1st International Conference on Onsite Wastewater Treatment and Recycling, Perth, Australia, 71.

Dawes L. and Goonetilleke A. (2003). "An investigation into the role of site and soil characteristics in onsite sewage treatment." Environmental Geology, 44, 467-477.

du McPherson H.J. (1994) Operation and Maintenance of Water System and Sanitation Systems: Case Studies. Operation and Maintenance Working Group, Who and Water and Sanitation Collaborative Council, Geneva.

Fane S.A., Asholt N.J. and White S.B. (2002) Decentralised urban water reuse: The implications of system scale for cost and pathogen risk. Water Science and Technology, 46(6-7), 281-288.

Finney B.A. and Gearheart R.A. (1998) A decision support model for prefeasibility analysis of water and wastewater treatment technologies appropriate for reuse. The US agency for International Development.

Foley J., Kasper T. and Cameron D. (2004) Biowater TM decentralised wastewater management for a small community in the Southern Moreton Bay Islands. 6th Specialist Conference on Small Water and Wastewater Systems; 1st International Conference on Onsite Wastewater Treatment and Recycling, Perth, Australia.

Hartmann H. and Ahring B.K. (2003) Phthalic acid esters found in municipal organic waste: enhance anaerobic degradation under hyper-thermophilic conditions. Water Science and Technology, 48(4), 175-183.

Ho G. (2003) Small water and wastewater systems: pathways to sustainable development? Water Science and Technology, 48(11/12), 7-14.

Hoehn J P. and Krieger D J. (2000) An economic analysis of water and wastewater investments in Cairo, Egypt. Evaluation Review, 24(6), 579-608.

Jonsson H. (2002) Urine separating sewage system - environmental effects and resource usage. Water Science and Technology, 46(6-7), 333-340.

Keraita B., Drechsel P. and Amoah P. (2003) Influence of urban wastewater on stream water quality and agriculture in and around Kumasi, Ghana. Environment and Urbanisation, 15(2), 171-178.

Lange J. and Otterpohl R. (1997) Oekologie Aktuell ABWASSER Handbuch zu einer zukunftsfaehigen Wasserwirtchaft., MALLBETON GmbH, Donaueschingen-Pfohren.

Loetscher T. (1998) Appropriate Sanitation in Developing Countries. (http://daisy.cheque.uq.edu.au/awm/manage/thomas12.html).

Maboeta M.S. and van Rensburg L. (2003) Vermicomposting of industrially produced woodchips and sewage sludge utilising Eisenia fetida. Ecotoxicology and Environmental Safety, 56, 265-270.

Mara D.D. (1996) Low Cost Sewerage, Wiley, Chichester.

Mara D.D. (2001) Appropriate wastewater collection, treatment and reuse in developing countries. Proceedings of the Institution of Civil Engineering-Municipal Engineer, 145(4), 299-303.

Mata-Alvarez J., Mace S. and Llabres P. (2000) Anaerobic digestion of organic solid wastes. An overview of research and achievements and perspectives. Bioresource Technology, 74(1), 3-16.

Maurer M., Schwegler P. and Larsen T.A. (2003) Nutrients in urine: energetic aspects of removal and recovery. Water Science and Technology, 48(1), 37-46.

Mulder A. (2003) The quest for sustainable nitrogen removal technologies. Water Science and Technology, 48(1), 67-75.

Parkinson J. and Tayler K. (2003) Decentralised wastewater management in peri-urban areas in low-income countries. Environment and Urbanisation, 15(1), 75-90.

Pickford J. (1995) Low-cost Sanitation: A survey of practical experience. Technology Publications, London.

Randall C.W. (2003) Changing needs for appropriate excreta disposal and small wastewater treatment methodologies or the future of small wastewater treatment systems. Water Science and Technology, 49(11-12), 1-6.

Scott S. (1998) An evaluation of Enviro Loo Composting Latrine in an informal settlement are in Greater Johannesburg. WRC report No KV 112/98.

Senzia M A., Mashuari D A. and Mayo A W. (2003) Suitability of constructed wetlands and waste stabilisation ponds in wastewater treatment: nitrogen transformation and removal. Physics and Chemistry of the Earth, 28(20-27), 1117-1124.

Stephenson D. (2001) Problems of Developing Countries. In: Frontiers in Urban Water Management: Deadlock or Hope, Maksimovic C. and Tjada-Guibert, J.A. (eds). IWA Publishing, London, England, pp. 264-312.

UNDP (2002) Human Development Report 2002: Deepening Democracy in a Fragmented World. United Nations Development Department.

UNEP (2002) International Source Book on Environmentally Sound Technologies for Wastewater and Stormwater Management., United Nations Environment Programme -

International Environmental Technology Centre, Osaka and International Water Association Publishing, London.

UNEP (2004) Guidelines on municipal waste management. Hague, Netherlands.

Vinneras B., Bjorklund A. and Jonsson H. (2003) Thermal composting of faecal matter as treatment and possible disinfection method - laboratory scale and pilot scale studies. Bioresource Technology, 88(1), 47-54.

Wilderer P.A. and Schreff D. (2000) Decentralised and centralised wastewater management: a challenge for technology developers. Water Science and Technology, 41(1), 1-8.

World Bank (1989) Demonstration Project - Regional Water and Sanitation Group, (People's Democratic Republic of Ethiopia). World Bank, Nairobi.

World Bank (1997) Participary Hygiene and Sanitation. WHO and Zimbabwe Institute of Water and Sanitation, UNDP-World Bank Wayer and Sanitation Program, Washington, D. C.

7

Waste stabilization ponds

Thomas Curtis and Duncan Mara

7.1 INTRODUCTION

It is not enough to design and build a treatment plant that works on paper: it must also work in practice. Many electromechanical treatment plants do not work well: they often breakdown because of untenable assumptions made at the design stage about the supply of spare parts, electricity and skilled labour and/or the availability of funds to pay chemical and electricity costs.

Moreover, when effluent standards are made by the local environmental regulatory agency, it will not generally be sufficient simply to consider the ecological status of the receiving water. Where water is scarce or aquaculture is commonly practised, wastewater reuse is likely to be practised by local farmers with or without the sanction of the relevant authorities. Consequently, the health status of the local communities will be at least as important as the ecological status of the receiving water. However, meeting health-based effluent standards by electromechanical treatment processes requires advanced tertiary treatment which needs skilled labour, reliable power supplies and funds

© 2006 IWA Publishing. *Municipal Wastewater Management in Developing Countries: Principles and Engineering* edited by Zaini Ujang and Mogens Henze. ISBN: 9781789065367

for spare parts, chemicals and electricity costs. These are demands that many hard pressed water utilities will struggle to meet.

This is why engineers should consider waste stabilization ponds (WSP) – because they are extremely simple to operate and maintain, have low construction and operating costs and yet achieve very high effluent standards, especially with respect to pathogens.

WSPs are a singular form of wastewater treatment; their managerial simplicity and unique properties make them the system of first choice in a wide variety of situations – and not just in tropical developing countries. For example, in temperate climates they are particularly suitable in rural areas not least because of their very low maintenance costs and simple format. Many thousands have been built in Germany, Austria, France, Germany, New Zealand and the USA.

Nevertheless, the demand for the combination of low running costs, simple operating requirements and high effluent quality is greatest in developing countries. Design engineers should never consider a more complex or more expensive design until they are satisfied that WSP are less suitable than an alternative solution. Moreover, in arriving at this conclusion designers must carefully consider whether their client has the financial and human resources to manage the alternative they propose.

However, the design of waste stabilization ponds is often poorly covered, if at all, in standard textbooks, especially those written for temperate climates. As a result many design engineers will not have seen good state-of-the-art WSP design procedures. This chapter should go some way to rectifying this situation. In addition some excellent WSP design manuals are available both online and in print (Environmental Protection Agency, 1983; Mara, 1997, 2003, 2004; Mara and Pearson, 1998).

The correct design of WSP systems requires a good understanding of their advantages, ecology and design procedures. Fortunately, there are tried and tested protocols for the design of WSP. Though these procedures are generally empirical, they are sufficiently conservative to ensure success. However, the designer needs to have an appreciation of the ecological principles that the empirical design procedures necessarily avoid. Consequently, the purpose of this chapter is to present both the tried and tested design procedures and the ecological principles (where known) which govern success.

7.2 WHAT ARE WASTE STABILIZATION PONDS?

WSPs are physically very simple. They are typically shallow (usually about 1.5 m deep) basins with earthen sides, and they are best arranged in a series in which wastewater flows from one pond to the next by gravity, improving in quality as it progresses through the system.

The apparent physical simplicity of WSP hides their biological complexity. Resembling shallow eutrophic lakes, they are characterised by high algal concentrations with high daytime dissolved oxygen concentrations and pH values. The combination of light, dissolved oxygen and humic substances creates a high concentration of activated oxygen species, such as singlet oxygen, that are normally associated with advanced oxidation processes. In WSP these activated oxygen species induce rapid die-off of bacterial pathogens in the pond water column (Curtis *et al.*, 1992).

The first ponds were constructed about 100 years ago in San Antonio, Texas, and formalised designs have been available since the late 1940s. An excellent overview of the historical development of WSP can be found in the classic monograph on WSP by Gloyna (1971).

7.3 ADVANTAGES AND DISADVANTAGES OF WSP

WSPs have many advantages that make them the system of first choice in a very wide variety of situations. They particularly lend themselves to situations requiring low costs, process reliability and simple maintenance. The advantages of WSP, which come from their unique combination of physical simplicity and biological complexity, include:

- low cost,
- simplicity of construction,
- excellent pathogen removal,
- ability to treat a variety of wastes,
- toleration of organic and hydraulic shock loads,
- low maintenance requirements,
- low sludge production,
- reliability of operation, and
- simple land reclamation

The most outstanding technical advantage of WSP is their ability to remove pathogens. Any excreted pathogen can be removed to essentially any required level, and this includes the removal of viruses, bacteria, protozoan cysts and oocysts, and helminth eggs.

These advantages must be set against the major perceived disadvantage of WSP: their relatively large land requirements. However, money spent on land is an investment (see below), whereas money spent on electricity to power electromechanical treatment systems is money gone forever.

A high level of nitrogen removal is achieved in WSP, although the mechanisms for this are still not well understood. Phosphorus removal is less satisfactory, but by no means negligible.

7.4 FINANCIAL AND ECONOMIC ASPECTS OF WSP

Whilst capital costs and running costs are clearly situation specific, some insight can be achieved from a French study (Berland and Cooper, 2001) (Table 7.1). This demonstrates the low capital and operational costs of WSP systems.

Table 7.1 Capital and annual operational costs (€/person) of small wastewater treatment plants for 1000 p.e. in France. (Berland and Cooper, 2001)

Treatment Process	Capital Costs	Operational Costs
Activated sludge	230	11.5
RBCs	220	7
Settler digester + reed bed	190	5.5
Biofilters	180	7
Aerated lagoons	130	6.5
Waste stabilisation ponds	120	5.5

A more rigorous economic analysis has been undertaken for a hypothetical large WSP system in the Yemen (Arthur, 1983) (Table 7.2). The overall cost of each treatment system alternative is given as a net present value (NPV), which is a figure that combines capital costs with future operational costs and discounts the latter at a given rate. The rationale for this is that a dollar spent on capital today has a greater effect on the NPV than a dollar spent on electricity in ten years time. This least-cost analysis found that WSP costs are dictated by the cost of land, the discount rate used and the possibility of future sale of the land (at the end of the project).

Table 7.2 The 25-year costs in millions of 1983 US dollars at a 12% discount rate of alternative wastewater treatment systems for 250,000 p.e. at a design temperature of 20°C and for an effluent concentration of ≤10,000 faecal coliforms/100 ml.

System	Capital Cost (US$ m)	Operating Cost (US$ m)	Income (US$ m)	Net Present Value (US$ m)
Waste stabilization pond	5.68	0.21	0.73	5.16
Aerated lagoons	6.98	1.28	0.73	7.53
Oxidation ditch	4.8	1.49	0.43	5.86
Biological filter	7.77	0.86	0.43	8.2

Table 7.3 The 25-year costs for the systems in Table 7.2 but with the future sale of the land included in the costs.

System	Net Present Value (US$ m)
Waste stabilization ponds	0.57
Aerated lagoons	2.55
Oxidation ditch	3.89
Biological filter	5.73

This study reveals an extremely important and universal aspect of WSP economics: their main capital item (land) is recoverable. Furthermore, since land is an appreciating capital item, it can presumably appear as such in company accounts. Future resale radically alters the net present values in favour of WSP (Table 7.3). Moreover, resale of land used for WSP is not just hypothetical. For example, it has proven extremely profitable in California (Oswald, 1976). Alternatively, and especially in industrialized countries, the land needed for WSP can be leased, rather than bought; thus a farmer can lease part of his land to the water company without the emotional consequences of selling the land.

Finally, when making comparisons between WSP and electromechanical treatment systems, it is important to recognise the superior ability of WSP to cope with infrequent maintenance and the inherent fail-safe nature of a non-mechanized system.

7.5 MAIN TYPES OF WSP

There are three main types of WSP: anaerobic facultative and maturation ponds. The ponds are usually arranged in a series (Figure 7.1), with each type of pond having a distinct function and design requirements.

Anaerobic ponds provide the first stage of treatment system and remove BOD and SS. Anaerobic conditions are maintained by virtue of a high organic load and thus they have few or no algae. Anaerobic ponds do not smell when designed correctly. However, anaerobic ponds are generally not popular in temperate climates, where septic or Imhoff tanks are commonly used to serve the same function. (Upflow anaerobic sludge blanket reactors (UASBs) could also be used, but their smaller land requirements must be set against their much higher construction costs.)

Facultative ponds remove both organic matter and excreted pathogens. Aerobic conditions exist in the upper layers and anaerobic conditions exist in the lower layers. Oxygenation is achieved by virtue of the large algal population. Algal photosynthesis leads to large diurnal variations in the oxygen concentrations and pH values. A facultative pond may be labelled primary or secondary depending on whether it receives raw wastewater or anaerobic pond effluent.

Maturation ponds are usually used to reduce pathogen levels to those set in the effluent quality requirements for the WSP system. Typically, aerobic conditions existing throughout the water column of maturation ponds. These ponds support large algal populations and consequently the dissolved oxygen concentration and pH values vary diurnally, reaching very high levels during daylight hours.

Waste stabilization ponds should always be used as part of a series. Ponds in series are hydraulically much more efficient. Additional series may be added in parallel in order to accommodate the increased populations.

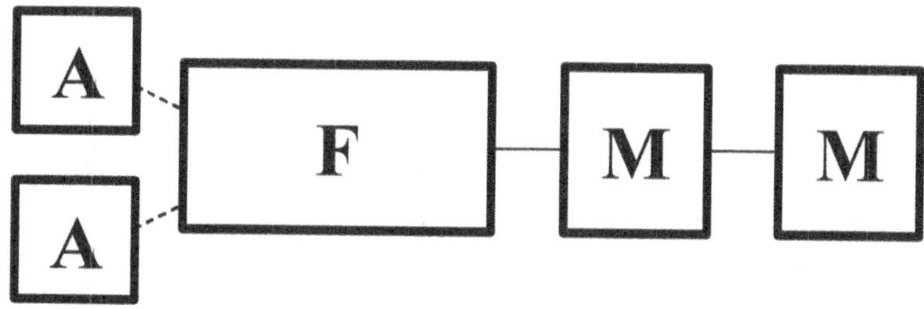

Figure 7.1 Typical layout of a waste stabilization pond system, A = anaerobic, F = facultative, M = maturation.

7.6 OTHER WSP FORMATS

There are a number of other design formats for waste stabilization ponds. Their use is restricted to certain specialised situations and in particular design engineers and their clients should be very wary of enthusiasts overselling any particular format, especially those which are patented or bear registered trademarks (as this will suggest that their advocates have a very strong pecuniary motive to propose the selection of their own proprietary solution). Of the four alternative WSP formats considered here, only fish ponds and effluent storage reservoirs are viable in the correct setting; macrophyte ponds and high-rate algal ponds have little to recommend them.

Fish ponds. WSP are an excellent format for aquaculture (Edwards, 1992). Large yields of fish, up to 10 tonnes per ha per year, are possible from well managed ponds. Important design considerations include the organic and total nitrogen loadings and the absence of pathogenic organisms. Design criteria for fish ponds are given by Mara *et al.* (1993) and in Mara (1997).

Effluent storage reservoirs. Where the demand for irrigation water is highly seasonal and the water is in short supply, the effluent from pond systems may be stored in large, deep storage reservoirs (Mara and Pearson, 1992, 1999; Juanicó and Dor, 1999). If managed correctly, such storage reservoirs bring about a significant increase in water quality prior to irrigation (Pearson *et al.*, 1996).

Macrophyte ponds. These are ponds in which either rooted aquatic plants (e.g., *Phragmites australis, Typha latifolia*) or floating plants (*Eichhornia crassipes, Lemna* spp.) are grown. They have few algae in their effluent but pathogen removal is typically much lower than in conventional WSP (presumably because of lower dissolved oxygen concentrations and substantially reduced light intensities). Macrophyte ponds also require much higher levels of maintenance and have the disadvantage of encouraging insects, especially mosquitoes, to breed.

High rate algal ponds. These were originally designed to produce high levels of algal protein (Oswald, 1988a, 1988b). However, this has never been commercially successful and high-rate algal ponds have been "re-branded" as Advanced Integrated Ponds Systems (Oswald, 1991) or, more simply (but equally misleadingly), Advanced Pond Systems (Craggs *et al.*, 2003). The growth of algae is encouraged through the use of stirrers to mix the water column, typically in some form of raceway. However, in most designs the high-rate pond has to be preceded by some form of facultative pond and followed by an algal separation system (which must be operated efficiently and maintained regularly) and one or more maturation ponds. It is apparent therefore that high-rate algal ponds lack the simplicity of conventional WSP and at their present stage of development they cannot be considered as a viable alternative to them.

7.7 ANAEROBIC PONDS

Anaerobic ponds are anaerobic reactors in which the waste is removed by a combination of sedimentation and anaerobic processes. They frequently appear black or dark grey, although they may have a slight green colour due to the growth of the sulphide-tolerant alga *Chlamydomonas*.

The principal function of an anaerobic pond is to reduce organic matter and thus reduce the size of the subsequent facultative pond. The inclusion of an anaerobic pond may reduce the size of the pond system by as much as 50%, with a corresponding reduction in land requirements and thus cost.

Pathogen removal is also observed in anaerobic ponds, typically of the order of 90%; this is thought to be a "one off" removal of particle-associated organisms. However, the superior survival of anaerobic organisms Oragui et al., (1987) and the inferior survival of *V. cholerae* O1 (Oragui et al., 1993; Arridge, 1995; Curtis, 1996). It is likely, but not proven, that metal removal occurs due to the precipitation of metal sulphides. as the metals precipitate out as metal sulphides.

Though the microbiology of anaerobic ponds is not well characterised, the little we know suggests that they are biologically similar to anaerobic digesters with the hydrolysis of complex organic compounds leading to the production of hydrogen and acetate which are then utilised by methanogens (Paing et al., 2000). Sulphate reduction also occurs. Anaerobic conditions are maintained by the organic load, and odour control is easy to achieve if the appropriate loads are maintained (see below).

7.7.1 Anaerobic pond design

Design is on the basis of the volumetric BOD loading (λ_v, g/m^{3}d).

$$\lambda_v = L_i Q / V_a$$

where $\quad L_i$ = influent BOD, mg/l
Q = flow, m^3/d
V_a = anaerobic pond volume, m^3

The value for λ_v depends on the temperature of the water. The designer must ensure that the load is high enough to ensure that anaerobic conditions are maintained without overloading the system. Universal design guidelines have yet to be developed, but the recommendations in Table 7.4 have been found to be of value. They are based on experience in Brazil, Europe and Kenya.

Table 7.4. Recommended BOD loadings on anaerobic ponds (T = temperature, °C).

Temperature (°C)	Volumetric BOD loading (g/m^{3}d)	BOD removal (%)
≤10	100	40
10–20	$20T - 100$	$2T + 20$
20–25	$10T + 100$	$2T + 20$
≥25	350	70

Once the organic loading has been determined the volume and then the retention time (θ_a) are easily calculated:

$$\theta_a = V_a/Q$$

Anaerobic ponds should have retention times of 1 day or more in order to discourage short-circuiting and the washout of slower growing methanogens. The pond may be of any depth, although the range is usually 2–5 m. The minimum temperature at which anaerobic ponds may be used is uncertain, but little methanogenesis occurs below 10°C and this is often regarded as the lower limit for anaerobic ponds. Sulphate levels of > 500 mg/l in the raw sewage should be avoided in order to discourage odour generation (Gloyna and Espino, 1969).

7.7.2 Anaerobic pond layout and maintenance

The removal of excess sludge is the principal maintenance task for those managing anaerobic ponds. As some operating authorities find this difficult; the designer must attempt to facilitate this task. Building two anaerobic ponds in parallel allows the operator to drain and desludge one of the ponds manually. It is best to plan to partially desludge the ponds in the same month of every year, rather than wait until the sludge reaches a particular depth in the pond. This is because experience suggests that annual tasks are more likely to be undertaken than those to be done less frequently or on an *ad hoc* basis.

7.8 FACULTATIVE PONDS

Facultative WSP are typically dark green in colour due to the high algal concentrations. The organic load is selected on the basis of surface loading to ensure a healthy standing population of algae. The algal concentration decreases with increasing load, decreasing sunlight and temperature and increasing depth, a phenomenon successfully modelled by Weatherell (2001). Chlorophyll *a* concentrations in a typical healthy pond vary from ~100 µg/l (in winter in a temperate climate) to > 2000 µg/l (in the tropics or in a temperate summer).

BOD removal is thought to be achieved by a mixture of anaerobic digestion and the aerobic breakdown by the heterotrophic bacteria also associated with other wastewater treatment plants. The oxygen for the aerobic processes is supplied by the algae which in turn benefit from the release of carbon dioxide and nutrients by the bacteria (Figure 7.2).

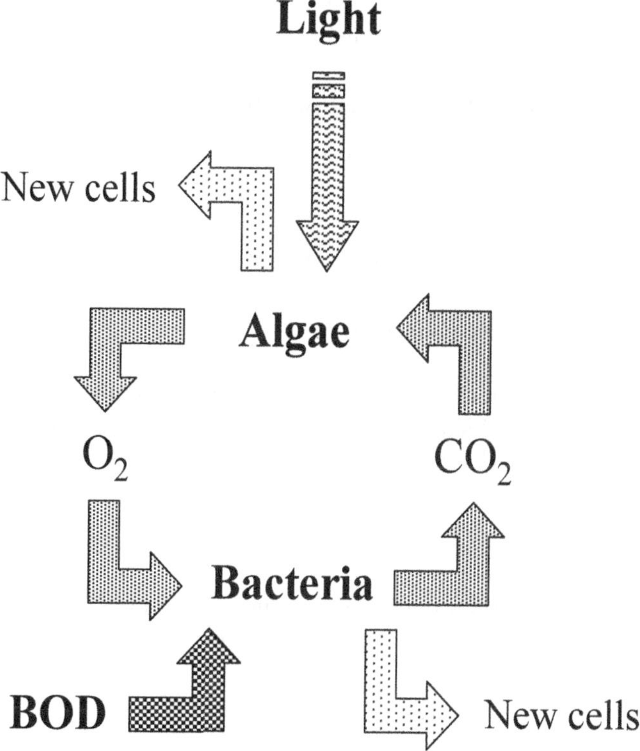

Figure 7.2 The mutualistic relationship between bacteria and algae in WSP.

However, some have suggested that the BOD removal is actually independent of the number of algae. It seems likely that at least some organic matter is removed anaerobically and that in temperate climates sophisticated patterns of sludge accumulation and feedback occur. The matter is undoubtedly worthy of further study.

Because the algae take up carbon dioxide they cause the pH of the water to rise by initiating a shift in the carbonate-bicarbonate buffering system:

$$CO_2 + H_2O \Leftrightarrow H_2CO_3 \Leftrightarrow H^+HCO_3^- \Leftrightarrow H^+ + CO_3^{2-}$$

The pH of the pond water column varies diurnally, often reaching a peak of 10 or more in the afternoon. The intense production of oxygen also varies diurnally; during daylight hours the water becomes supersaturated, with concentrations in excess of 20 mg/l being common. These high pH values and dissolved oxygen concentrations are characteristic of a 'healthy' pond and are prerequisites for effective bacterial pathogen removal.

7.8.1 The design of facultative ponds

Facultative ponds are best designed on the basis of a real BOD loading (λ_s, kg/ha d).

$$\lambda_s = 10 L_i Q / A_f$$

Where A_f = facultative pond area, m^2

The permissible areal load is related to the monthly mean ambient air temperature (T, °C), which is a proxy for the climate. A number of formulae have been proposed. The best known is that of McGarry and Pescod (1970):

$$\lambda_s = 60(1.099)^T$$

This equation describes the point of failure. Mara (1987) suggested a global design equation which incorporates a safety factor:

$$\lambda_s = 350(1.107 - 0.002T)^{T-25}$$

This equation should be used if no local data are available.

The value of T should usually reflect the coldest month in the year. However, for highly seasonal populations, it is generally more economical to have separate winter and summer designs.

The pond area may be calculated from this equation. A depth of around 1.5–1.8 m is usually chosen. This depth is selected because shallower ponds encourage emergent vegetation and deeper ponds may be more prone to failure, although both points are debatable. The retention time (θ_f, days) may be calculated from:

$$\theta_f = A_f D / Q$$

where D is the pond depth, m.

The value of Q may be modified to allow for seepage and evaporation.

The BOD removal of a primary facultative pond is in the range of 70–80% for unfiltered samples and > 90% for filtered samples. The combined performance of an anaerobic pond and a secondary facultative pond is similar or better.

7.9 WHY PONDS DO NOT SMELL

Properly designed ponds do not smell. Unfortunately, ill founded fears about odour nuisance sometimes limit the use of WSP. It is therefore important to understand how odour is caused and controlled in WSP. There are in essence two factors: pH and dissolved oxygen.

Most bad smells in WSP come from the reduction of sulphates in the raw wastewater to sulphides by sulphate-reducing bacteria. Sulphide ions exist in three forms: hydrogen sulphide, the bisulphide ion and the sulphide ion. The equilibrium of the three forms is dictated by pH:

$$H_2S \Leftrightarrow HS^- \Leftrightarrow S^{2-}$$

Only one form, dissolved H_2S, causes odour. This form constitutes about 50% of the sulphide at pH 7, but only 8% at pH 8. Consequently if the pH of the pond water is maintained at a value of 7.5 or above and the sulphate concentration in the influent is not abnormally high (> 500 mg SO_4/l), then no odour problems will be experienced. The pH of the pond is dictated by the organic load; provided the guidelines on loading given in Table 7.4 are adhered to, the pH should be at 7.5 or above.

The oxygen in the water column in facultative ponds provides further protection against odour by oxidising sulphide to sulphate or elemental sulphur. This may occur both chemically and biologically, with the latter due to anaerobic photosynthetic bacteria, and as a result elemental sulphur is occasionally visible on the surface of facultative ponds operated at high loads.

7.10 MATURATION PONDS

It is frequently desirable to further improve the quality of facultative pond effluent, either to reduce the concentration of pathogens in the effluent or to reduce the concentration of nitrogen. This may be achieved by the provision of maturation ponds.

Superficially, maturation ponds are very similar to facultative ponds: they are simple basins typically 1–1.5 m deep and rich in algae. However, there are subtle but important differences. Maturation ponds have fewer algae than facultative ponds in summer, but relatively more algae in the winter; moreover, the water column is aerobic throughout. The algae promote and improve the water quality by promoting the loss of ammonia through algal uptake and volatilization and by increasing pathogen removal by photo-oxidation. In addition, it is likely that bacterial pathogens may be lost through intense predation and adsorption to the algae.

7.10.1 Nutrient removal

Nutrient removal is still poorly understood and until recently the best available studies were from the USA. Recently more detailed studies have been undertaken in the UK and these suggest that good nitrogen and tolerable phosphorus removals can be attained.

An explicit design for phosphorus removal has not yet been developed; however, the pilot-scale maturation ponds in Bradford, England achieved effluent values of 0.5mg PO_4/l and at present the simple rule of thumb is that a 50% removal of P will be removed in a pond achieving 90% BOD removal. Approximately, one third of the P is organic and the remainder is inorganic. It would seem that, at least in principle, P removal could be designed into a WSP system.

Nitrogen removal can be designed into a WSP system (Figure 7.3). Nitrite and nitrate are virtually never found in WSP and it seems likely (though not certain) that nitrifying bacteria are essentially absent from the system. Consequently nitrogen removal is essentially synonymous with that of ammoniacal nitrogen. The mechanisms of ammonia loss are uncertain but are most likely to be a combination of volatilisation and algal uptake. Naturally algal uptake implies that a substantial proportion of the reduced nitrogen present is in the algae. Thus the removal of nitrogen in the algae is best achieved by treating the WSP effluent in rock filters, reed beds or deep ponds.

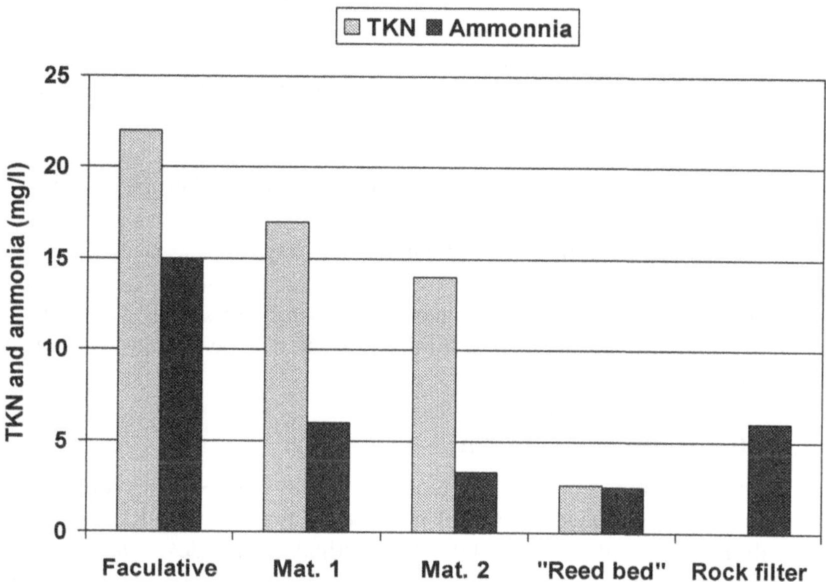

Figure 7.3 Ammonia and TKN removal in maturation ponds in the UK. (Johnson & Mara, 2002).

Ammonia removal has been described by Pano and Middlebrooks (1982) using the following formulae:

(a) for temperatures below 20°C:

$$C_e = C_i/\{1+[(A/Q)(0.0038 + 0.000134T)]\exp[(1.041+ 0.041T)(\text{pH} - 6.6)]\}$$

(b) for temperatures above 20°C:

$$C_e = C_i/\{1+[5.035\times10^{-3}(A/Q)][\exp(1.54 \times (\text{pH-6.6}))]\}$$

where C_e = ammoniacal nitrogen concentration in the effluent, mg/l
     C_i = ammoniacal nitrogen concentration in the influent, mg/l

The pH value may be estimated very crudely by using the following relationship.

$$\text{pH} = 7.3\exp(0.0005A)$$

where A is the influent alkalinity, mg $CaCO_3$/l.

The Pano and Middlebrooks procedure has been found to be somewhat conservative when compared to actual pond performance in the UK, possibly because it neglects the uptake of ammonia by the algae.

The following relationship has been proposed for the removal of total nitrogen (Reed, 1985):

$$C_e = C_i \exp\{-[0.0064(1.039)^{T-20}]\,[(\theta + 60.6(\text{pH} - 6.6)]\}$$

where θ is mean hydraulic retention time, days.

However, it is not at all clear that this relationship is valid for all WSP. Indeed, given that about one third of the N will be present as algae it seems likely to be flawed.

7.10.2 Pathogen removal

Pathogen removal is typically required to allow the safe use of wastewater for crop irrigation. Two classes of pathogens must be considered in this context: helminths (specifically the human intestinal nematodes) and enteric bacterial pathogens. The former comprise *Ascaris lumbricoides* (the human roundworm), *Necator americanus* and *Acylostoma duodenale* (hookworms) and *Trichuris trichiura* (whipworm). *Ascaris* eggs are usually monitored and used as an indicator of all helminths. The bacterial pathogens (e.g., salmonellae and campylobacters) are monitored by monitoring faecal coliforms or *Escherichia coli*. Both monitoring assumptions have been shown to be valid in WSP. Viruses have also been shown to be removed at a slightly slower rate than bacterial indicators (Oragui *et al.*, 1987).

Table 7.5 WHO guidelines for the microbiological quality of treated wastewaters used for crop irrigation.

Parameter	Restricted irrigation*	Unrestricted irrigation*
Human intestinal nematode eggs per litre	≤1 but ≤0.1 if children under 15 are exposed**	≤1 but ≤0.1 if children under 15 are exposed**
Faecal coliforms or *E. coli* per 100 ml	$\leq 10^5$	≤1000

* Restricted irrigation is the irrigation of crops that will not be eaten raw, and unrestricted irrigation includes the irrigation of salad crops and vegetables eaten uncooked.

**For restricted irrigation exposure is via working or playing in wastewater-irrigated fields; for unrestricted irrigation exposure is limited to the consumption of crops brought home by parents working in wastewater-irrigated fields which are eaten uncooked.

The World Health Organization recommends that only treated wastewaters be used in agriculture and that they should be treated to the microbiological quality given in Table 7.5 (WHO, 1989, 2005).

7.10.2.1 Helminth egg removal

Helminth eggs are removed by sedimentation. Removal may be estimated on the basis of mean hydraulic retention time using the equations of Ayres *et al.* (1992) which are based on egg removal data from WSP in Brazil, India and Kenya. The following equation represents the lower 95-percentile confidence limit of the relationship between percentage removal (R) and mean hydraulic retention time (θ, days):

$$R = 100[1 - 0.41\exp(- 0.49\theta + 0.0085\theta^2)]$$

This relationship should be used for design. The number of eggs in the raw wastewater is somewhat variable, although a range of 100–1000 eggs/litre is common. In the absence of field data a value of 1000 eggs/litre should be used. However, up to 15,000 eggs/litre have been reported in newly sewered poor communities.

7.10.2.2 Bacterial and viral pathogen removal

The mechanisms of removal of are the same in facultative and maturation ponds. The death of bacterial pathogens is dependent on the existence of certain biological conditions, notably high pH and dissolved oxygen concentrations, which the designer must seek to promote.

The mechanisms of removal can be divided into the very fast light-dependent mechanisms and the slower light-independent or dark processes (Figure 7.4).

There are two principal light-dependant mechanisms: photo-oxidation and high pH. Photo-oxidation is the product of a synergistic interaction of light, humic substances, oxygen and high pH. Light is absorbed by humic substances, which are inherently present in the water; the humic substances then react with the oxygen to form singlet oxygen and oxygen radicals, both of which are very toxic faecal bacteria. Elevated pH is synergistic with this process: high pH values allow the less powerful, but deeply penetrating, red light to kill bacteria. This process is also capable of killing enteric viruses. When pH values rise above approximately 9.0, the hydroxyl ions alone can kill the bacteria. For the light-dependant mechanisms to function the pond must have high pH values and high dissolved oxygen concentrations. Consequently, the fast light-dependent mechanisms will only function if there is a healthy and active algal population; pathogen removal is slow in overloaded facultative and maturation ponds. Most of the light-dependant removal occurs during daylight hours (Figure 7.4).

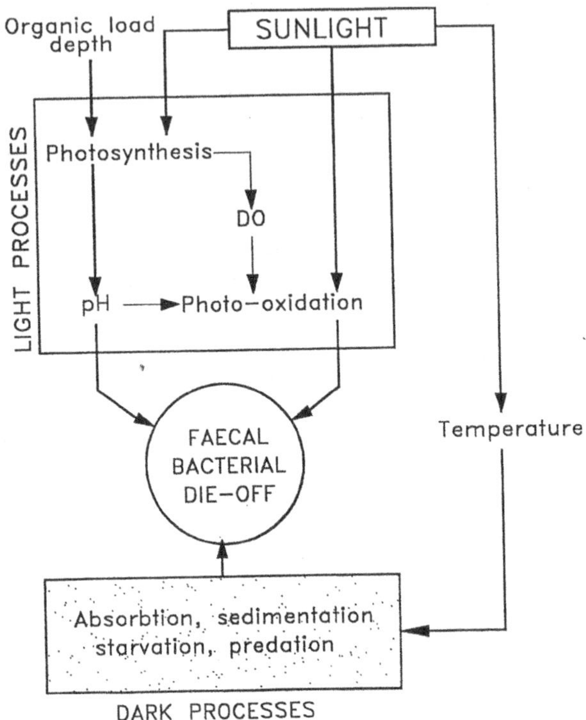

Figure 7.4 Faecal pathogen removal mechanisms in waste stabilization ponds.

The dark processes are less well defined and essentially a matter of speculation. However, they may be of considerable importance in certain circumstances and are presumably what sustains the indicator removal observed in winter (Figure 7.4). The slow rate at which such removal occurs may be offset by their occurrence at all depths and for 24 hours a day. Possible dark mechanisms include sedimentation, starvation, predation and adsorption to solids. Sedimentation almost certainly plays an important role in anaerobic ponds and primary facultative ponds.

Though most of the studies on pathogen removal mechanisms in WSP have used faecal coliforms, it appears that this organism is a remarkably good indicator, dying more slowly than most other bacterial pathogens (Figure 7.4). Viruses are removed, but a little more slowly; however, there is no reason why WSP cannot be explicitly designed for the removal of viruses. Pathogenic forms of *V. cholerae* (serotypes O1 and O139) are also removed from WSP more quickly than faecal coliforms (Oragui *et al.*, 1993; Curtis in preperation), although the non-pathogenic forms survive and even grow in WSP and other wastewater treatment plants (Curtis, 1996; Curtis in preparation).

7.10.2.3 Maturation ponds: design for bacterial pathogen removal

To design the removal of a bacterial pathogen or indicator from a WSP, one needs to know or assume two things: the hydraulic characteristics of the ponds and the death rate of the organism.

The hydraulic characteristics of the pond will typically fall between the extremes of plug flow and complete mixing. In principle, the removal can be described using the Wehner-Wilhelm equation. However, in practice predicting the value of the pond dispersion number appears to be problematic and requires considerable skill and possibly specialist hydrodynamic modelling. Consequently, the standard design procedure is that of Marais (1974) which assumes a completely mixed reactor. This approach (given below) has the advantage of simplicity; however, it is conservative. Therefore, where land or performance is at a premium, determining and improving the hydraulic characteristics of the system is likely to be very beneficial (Shilton and Harrison, 2003).

The equation below uses faecal coliforms as an indicator, assumes first-order kinetics for bacterial removal and that the pond is a completely mixed reactor. The first-order removal constant is related to the mean ambient air temperature (used as a proxy for climate). Important factors known to affect the removal rate, notably the organic load and depth, are not considered in this equation. The designer should ensure that the maturation ponds (and facultative ponds) are not overloaded and be aware that deeper systems may have lower removal rates. The basic design equation is:

$$N_e = \frac{N_i}{(1 + k_t \theta)}$$

Where: N_e = faecal coliform count per 100 ml in the effluent
N_i = faecal coliform count per 100 ml in the influent
k_t = faecal coliform removal rate constant, day^{-1}
θ = retention time, days

For a series of ponds in which the maturation ponds are all the same size this equation becomes:

$$N_e = \frac{N_i}{\left[(1 + k_t\theta_a)(1 + k_t\theta_f)(1 + k_t\theta_m)^n\right]}$$

where a, f and m represent the anaerobic, facultative and maturation ponds, and n represents the number of maturation ponds.

The equation is particularly sensitive to temperature as the value of k_t is strongly dependant on the climate and is related to the mean monthly ambient air temperature by the equation:

$$k_T = 2.16(1.19)^{T-20}$$

There is a problem of using the same k_T value throughout the series. It is possible for k_T to either increase or decrease. This equation is based on a relatively limited dataset and the first-order removal coefficient will be affected by high organic loads (which decrease the algal concentration and thus pH and dissolved oxygen) as well as temperature. It is therefore hardly surprising that removal rates are reported to be somewhat variable. Alternative methods for estimating k_t have been proposed – see, for example, von Sperling (1999, 2002, 2003).

The unknowns in the design procedure are θ_m and n. The value of n refers to the fact that more than one maturation pond is often necessary and the mathematics (and field observation) suggests that a series of small ponds work better than one large pond. The correct values of n and θ_m, can be derived by trial and error, remembering that the recommended minimum retention time in a maturation pond is 3 days (otherwise there is a high risk of short-circuiting).

For raw wastewater, N_i is in the region of 10^7–10^8/100 ml. Thus the faecal coliform concentration in the effluent of the last maturation pond can be estimated and checked, if required, for compliance with the WHO guideline for either restricted irrigation ($\leq 10^5$ faecal coliforms/100 ml) or unrestricted irrigation (≤ 1000/100 ml).

Although these procedures represent the best available conservative design procedures, they could be relatively easily improved. Retention time is the key physical parameter in pathogen removal. It has recently become possible relatively easily to estimate the diffusion coefficient in lagoons and so the assumption of complete mixing may be revisited and the more complex but less conservative design procedures using first order decay in conjunction with dispersed flow. In addition, it should be possible to estimate the removal coefficient with greater accuracy. In addition there is no reason in principle why this approach should not be extended to faecal streptococci or viruses.

7.11 Operation and maintenance of WSP

WSP are a low-maintenance system, but not a no-maintenance system. The maintenance tasks are simple and can be undertaken by relatively unskilled staff if adequate supervision is available. However, failure to undertake maintenance may lead to fly and odour problems and damage to the embankments. Nevertheless WSP are far more tolerant of failures in maintenance than are conventional electromechanical treatment plants. The main tasks are.

- the removal of screenings and grit;
- ensuring that vegetation does not grow or overhang the sides of the pond (this simple task will prevent the growth of mosquito larvae);
- scum removal from facultative and maturation ponds (scum should be left on anaerobic ponds);
- checking inlets and outlets for blockages;
- repair of damage to embankments.

Routine monitoring or observation of the WSP may also be instituted at a level appropriate to the system. This may range from simple visual observations to routine analysis of the chlorophyll a levels and effluent quality. The staff required will reflect the size of the pond system.

The pond base should be impermeable (the in-situ coefficient of soil permeability should be 10^{-7} m/s or less; if the soil is more permeable than this, the pond should be lined with puddled clay or a plastic membrane). The embankments should be carefully constructed from the excavated soil (on the basis of balancing cut and fill) by compacting 200mm layers to at least 90 percent of the soil's maximum dry density (as determined by the modified Procter test). Embankment slopes should be 1:3 and there should be at least 0.5m freeboard. Concrete slabs, in-situ concrete or stone rip-rap are often used at the water surface to prevent embankment erosion due to wind-induced wave action and to prevent vegetation developing down into the pond.

WSP usually employ high length-to-breadth ratios, commonly 3 (or more) to 1, to minimize short-circuiting. The longest dimension should be located along the direction of the prevailing wind to encourage natural mixing, with the flow from inlet to outlet against the prevailing wind.

7.11.1 WSP start-up

Pond start-up is relatively straightforward if the following guidelines are followed. Ideally start-up should be timetabled for the start of the warm/hot season. All ponds should be free from vegetation and the facultative pond should be commissioned first. If possible, the facultative and maturation ponds should be filled with fresh water (from a river, lake or spring) and the load on the plant gradually increased to permit the gradual development of the algal population. If freshwater is unavailable the facultative pond may first be filled with raw wastewater and allowed to stand for 3–4 weeks; however, some odour release is inevitable during this period.

Anaerobic ponds should be filled with raw wastewater and seeded wherever possible with digester or septic tank sludge and the load gradually increased over 2–4 weeks. The optimum pH for methanogenic bacteria and odour control is >7. Soda ash may be added during the start-up to help control the pH. Anaerobic ponds should not be commissioned until sufficient flow is available to maintain a BOD load of at least $100g/m^3$ day.

7.11.2 Pond desludging

Facultative ponds rarely require desludging, especially in the tropics, perhaps only every 10–20 years. By contrast, anaerobic ponds require regular desludging at much more frequent intervals, usually when the pond is ⅓ full (or annually pro rata). The frequency of desludging frequency may be calculated from the equation:

$$Y = V_a/3Ps$$

where:	Y	=	time, years
	P	=	population served
	s	=	sludge accumulation rate, $m^3/$ person year

The value of s depends on the climate. A nominal value of 0.04 $m^3/$ person year is used in the tropics; in temperate climates it may be as high as 0.1 $m^3/$ person year. Sludge removal may be undertaken by draining the pond if there is an additional anaerobic pond provided in parallel for this purpose or by using a raft-mounted sludge pump. Care should be taken to ensure that an excessive build up of grit does not add to desludging problems; therefore all WSP serving populations > 1,000 should have a simple grit removal facility ahead of the anaerobic pond.

REFERENCES

Arridge, H.J. (1995) The influence of geometry on the performance of waste stabilisation ponds with special reference to pathogen removal, PhD Thesis, University of Liverpool.

Arthur, J. P. (1983) Notes on the Design and Operation of Waste Stabilization Ponds in Warm Climates of Developing Countries (Technical Paper no 7). The World Bank, Washington, DC;

available at: http://www-wds.worldbank.org/servlet/WDS_IBank_Servlet?pcont =details&eid=000178830_98101904165457.

Ayres, R. M., Alabaster, G. P., Mara, D. D. and Lee, D. L. (1992) A design equation for human intestinal nematode egg removal in waste stabilization ponds. Water Research **26** (6), 863–865.

Berland, J-M. and Cooper, P. (2001) Extensive Wastewater Treatment Processes Adapted to Small and Medium Sized Communities. Office of Official Publications of the European Community, Luxembourg; available at http://europa.eu.int/comm/environment/water/water-urbanwaste/waterguide_en.pdf

Craggs, R. J., Davies-Colley, R. J., Tanner, C. C. and Sukias, J. P. (2003) Advanced pond system: performance with high rate ponds of different depths and areas. Water Science and Technology **48** (2), 259–268.

Curtis, T.P. (1996) The fate of Vibrio cholerae in wastewater treatment systems. In: Cholera and the Ecology of Vibrio cholerae, Drasar, B.S. and Forrest, B.D. (eds). Chapman and Hall, London, UK.

Curtis, T. P., Mara, D. D. and Silva, S. A. (1992) Influence of pH, oxygen and humic substances on ability of sunlight to damage faecal coliforms in waste stabilization pond water. Applied and Environmental Microbiology **58** (4), 1335–1343.

Edwards, P. (1992) Reuse of Human Wastes in Aquaculture (Water and Sanitation Report No. 2). The World Bank, Washington, DC; available at http://www-wds.worldbank.org/servlet/WDS_Ibank_Servlet?pcont=details&eid=000009265_3961003 0757.

Environmental Protection Agency (1983) Design Manual: Municipal Wastewater Stabilization Ponds (Report No. EPA-625/1-83-015). Environmental Protection Agency, Cincinnati, OH.

Gloyna, E. F. (1971) Waste Stabilization Ponds (Monograph No. 60). World Health Organization, Geneva.

Gloyna, E. F. and Espino, E. (1969) Sulfide production in waste stabilization ponds. Journal of the Sanitary Engineering Division, American Society of Civil Engineers **95** (SA3), 607–628.

Johnson, M. and Mara, D. D. (2002) Research on waste stabilization ponds in the United Kingdom – II. Initial results from pilot-scale maturation ponds, reedbed channel and rock filters. In: Pond Technology for the New Millennium, New Zealand Water and Wastes Association, Auckland.

Juanicó, M. and Dor, I. (1999) Hypertrophic Reservoirs for Wastewater Storage and Reuse: Ecology, Performance, and Engineering Design. Springer Verlag, Heidelberg.

Mara, D. D. (1987) Waste stabilization ponds: problems and controversies. Water Quality International (1), 20–22.

Mara, D. D. (1997) Design Manual for Waste Stabilization Ponds in India. Lagoon Technology International, Leeds; available at http://www.leeds.ac.uk/civil/ceri/water/tphe/publicat/pdm/indiaall.pdf.

Mara, D. D. (2003) Design Manual for Waste Stabilization Ponds in the United Kingdom. School of Civil Engineering, University of Leeds, Leeds; available at http://www.leeds.ac.uk/civil/ceri/water/ukponds/pdmuk/pdmuk.html.

Mara, D. D. (2004) Domestic Wastewater Treatment in Developing Countries. Earthscan Publications, London.

Mara, D. D. and Pearson, H. W. (1992) Sequential batch-fed effluent storage reservoirs: a new concept of wastewater treatment prior to unrestricted crop irrigation. Water Science and Technology **26** (7–8), 1459–1464.

Mara, D. D. and Pearson, H. W. (1998) Design Manual for Waste Stabilization Ponds in Mediterranean Countries. Lagoon Technology International, Leeds; available at http://www.leeds.ac.uk/civil/ceri/water/tphe/publicat/pdm/medall.pdf.

Mara, D. D. and Pearson, H. W. (1999) A hybrid waste stabilization pond and wastewater storage and treatment reservoir system for wastewater reuse for both restricted and unrestricted irrigation. Water Research 33 (2), 591–594.

Mara, D. D., Edwards, P., Clark, D. and Mills, S. M. (1993) A rational approach to the design of wastewater-fed fishponds. Water Research 27 (12), 1797–1799.

Marais, G. v. R. (1974) Faecal bacterial kinetics in waste stabilization ponds. Journal of the Environmental Engineering Division, American Society of Civil Engineers, 100 (EE1), 119–139.

McGarry, M. G. and Pescod, M. B. (1970) Stabilization pond design criteria for tropical Asia. In: Proceedings of the Second International Symposium for Waste Treatment Lagoons, McKinney, R. E.(ed). University of Kansas, Lawrence, KS, pp. 114-132.

Oragui, J. I., Curtis, T. P., Silva, S. A. and Mara, D. D. (1987). The removal of excreted bacteria and viruses in deep waste stabilization ponds in northeast Brazil. Water Science and Technology 19 (Rio), 569–573.

Oragui, J. I., Arridge, H., Mara, D. D., Pearson, H. W. and Silva, S. A. (1993) Vibrio cholerae O1 (El Tor) removal in waste stabilization ponds in northeast Brazil. Water Research 27 (4), 727–728.

Oswald, W. J. (1976) Experiences with new pond designs in California. In: Ponds as a Wastewater Treatment Alternative, Gloyna, E. F., Malina, J. F. and Davis, E. M.(eds). University of Texas Press, Austin, TX, pp.257-272.

Oswald, W. J. (1988a) The role of microalgae in liquid waste treatment and reclamation. In: Algae and Human Affairs , Lembi, C. E. and Waaland, J. R.(eds). Cambridge University Press, Cambridge, pp. 255-281.

Oswald, W. J. (1988b) Micro-algae and wastewater treatment. In: Micro-algal Biotechnology, Borowitzka, M. A. and Borowitzka, L. J.(eds). Cambridge University Press, Cambridge, pp.305-328.

Oswald, W. J. (1991) Introduction to advanced integrated wastewater ponding systems. Water Science and Technology 24 (5), 1–7.

Paing J, Picot B, Sambuco JP, Rambaud A (2000) Sludge accumulation and methanogenic activity in an anaerobic lagoon. Water Science and Technology, 42, 247-255.

Pano, A. and Middlebrooks, E. J. (1982) Ammonia nitrogen removal in facultative waste stabilization ponds. Journal of the Water Pollution Control Federation 54 (4), 344–351.

Pearson, H. W., Mara D. D., Oragui, J. I., Cawley, L. R. and Silva, S. A. (1996) Pathogen removal dynamics in experimental deep effluent storage reservoirs. Water Science and Technology 33 (7), 251–260.

Reed, S. C. (1985) Nitrogen removal in wastewater stabilization ponds. Journal of the Water Pollution Control Federation 57 (1), 39–45.

Shilton, A. N. and Harrison, J. (2003) Guidelines for the Hydraulic Design of Waste Stabilization Ponds. Massey University, Palmerston North; available at http://www.leeds.ac.uk/civil/ceri/water/tphe/publicat/pdm/pdm.html

von Sperling, M. (1999) Performance evaluation and mathematical modeling of coliform die-off in tropical and subtropical waste stabilization ponds. Water Research 33 (6), 1435-1448.

von Sperling, M. (2002) Relationship between first-order decay coefficients in ponds for plug flow, CSTR and dispersed flow regimes. Water Science and Technology 45 (1), 17–24.

von Sperling, M. (2003) Influence of the dispersion number on the estimation of coliform removal in ponds. Water Science and Technology **48** (2), 181–188.

Weatherell, C.A. (2001) Predicticting Algal Concentrations in Waste Stabilisation Ponds, PhD Thesis, University of Newcastle, UK.

WHO (1989) Health Guidelines for the Use of Wastewater in Agriculture and Aquaculture (Technical Report Series No. 778). World Health Organization, Geneva; available at http://whqlibdoc.who.int/trs/WHO_TRS_778.pdf

WHO (2005) Health Guidelines for the Use of Wastewater in Agriculture. World Health Organization, Geneva; to be available via http://www.who.int/ water_sanitation_health/index.html

8

Design and operation of constructed wetlands for wastewater treatment and reuses

Chongrak Polprasert

8.1 INTRODUCTION

Wetlands are areas that are inundated or saturated by surface or groundwater at a frequency and duration sufficient to maintain saturated conditions (U.S. EPA 1988). They are comparatively shallow (typically less than 0.6 m) bodies of slow-moving water in which dense stands of water-tolerant plants such as cattails, bulrushes, or reeds are grown. In man-made systems, these bodies are artificially created and are typically long, narrow trenches or channels.

Constructing a wetland where one did not exist before avoids many of the environmental concerns and user conflicts associated with natural wetlands and allows design of the wetland for optimum wastewater treatment. Unlike natural wetlands, which are confined by availability and proximity of the wastewater source, constructed wetlands can be built almost anywhere, including lands with limited uses. Typically, a constructed wetland should perform better than a natural wetland of equal area since the bottom is usually graded and the hydraulic regime in the system is controlled. Process reliability is also improved because the vegetation and the other system components can be managed as required (Reed *et al.*, 1995). If properly designed and operated, effluents of constructed wetlands can have characteristics comparable to those of secondary or tertiary effluents, making them suitable for discharging into receiving water body or reuse in agriculture and aquaculture practices (Proprasert, 1996). The harvested vegetation can be used to make compost or some furniture products, hence providing an economic return to the constructed wetland operators.

8.2 TYPES AND FUNCTIONS OF CONSTRUCTED WETLANDS

Constructed wetlands are classified into two types: free water surface (FWS) systems with shallow water depths and subsurface flow (SF) or vegetated submerged bed (VSB) systems with water flowing laterally through the sand or gravel.

8.2.1 Free water surface (FWS) systems

These systems typically consist of basins or channels, with a natural or constructed subsurface barrier of clay or impervious geotechnical material to prevent seepage, soil or another suitable medium to support the emergent vegetation, and water at a relatively shallow depth flowing over the soil surface (Figure 8.1). The shallow water depth, low flow velocity and presence of the plant stalks and litter regulate water flow and, especially in long, narrow channels, ensure plug-flow conditions to minimize short-circuiting (U.S. EPA, 1988).

8.2.2 Subsurface flow (SF) systems

An SF system typically consists of a trench or a bed underlain by impermeable material to prevent seepage and containing a medium that supports the growth of emergent vegetation (Figure 8.2). The media used have included rock or crushed stone (10-15 cm diameter), gravel, and different soils, either alone or in various combinations (Reed *et al.*, 1995). The wastewater flows laterally through the medium and is purified during the contact with the surfaces of the medium and the root zone of the vegetation. This subsurface zone is continuously saturated and therefore is generally anaerobic. However, the plants can convey an excess of oxygen to the root system, so there are aerobic microsites adjacent to the roots and rhizomes.

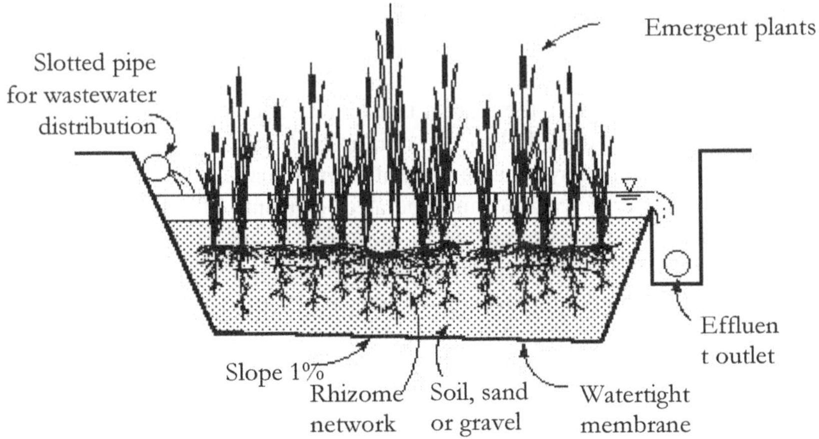

Figure 8.1. Free water surface (FWS) system.

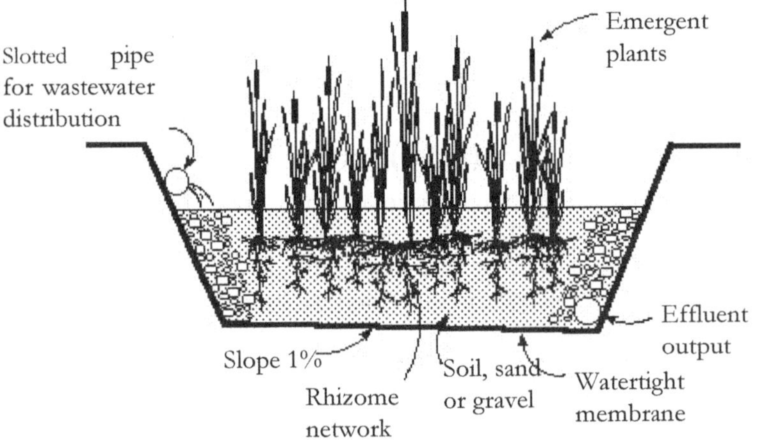

Figure 8.2. Subsurface flow (SF) system.

8.2.3 Advantages and disadvantages

Constructed wetlands offer several potential advantages as a wastewater treatment process. These potential advantages include site location flexibility, less rigorous pre-application treatment, no alteration of natural wetlands, simple operation and maintenance, process stability under varying environmental conditions, lower construction and operating costs, and in the case of FWS systems, the possibility to create a wildlife habitat. The quality of constructed wetland effluent is usually comparable to the secondary or tertiary effluent, suitable for applying to agricultural lands or aquacultural ponds for nutrient reuses and recycling. The potential problems with FWS constructed wetlands include mosquitoes. Start-up problems in establishing the desired aquatic plant species can be a problem with FWS and SF constructed wetlands alike (Bastian *et al.*, 1989; U.S. EPA, 1988).

8.3 TYPES AND FUNCTIONS OF VEGETATION

The major benefit of plants in constructed wetlands is the transferring of oxygen to the root zone. The plant roots in the system penetrate the soil or support medium, and transport oxygen deeper than it would naturally travel by diffusion alone.

Perhaps most important in the FWS constructed wetlands are the submerged portions of the leaves, stalks, and litter, which serve as the substrate for attached microbial growth. It is the responses of this attached biota that is believed responsible for much of the treatment that occurs.

The emergent plants most frequently found in wastewater wetlands include cattails, reeds, rushes, bulrushes and sedges. It is estimated that these plants transfer 5-45 g O_2/day-m^2 depending on plant density and oxygen stress levels in the soil (Reed *et al.*, 1995). Table 8.1 lists some of the major environmental requirements of each as well as their maximum depths of root penetration

8.4 WASTEWATER TREATMENT MECHANISMS

Wetland systems can significantly reduce biochemical oxygen demand (BOD_5), suspended solids (SS), and nitrogen, as well as metals, trace organics, and pathogens. The basic treatment mechanisms include sedimentation, chemical precipitation and adsorption, and microbial interactions with BOD_5, SS and nitrogen, as well as some uptake by the vegetation (U.S. EPA, 1988).

Table 8.1. Emergent aquatic plants for wastewater treatment (modified from U.S. EPA, 1988)

Common name	Scientific name	Temperature, °C		Max. salinity tolerance, ppt[a]	Root penetration, cm	Effective pH range
		Desirable	Seed germination			
Cattail	*Typha* spp.	10-30	12-24	30	30	4-10
Common	*Phragmites*	12-23	10-30	45	60	2-8
reed	*communis*	16-26	-	20	-	5-7.5
Rush	*Juncus* spp.	16-27	-	20	76	4-9
Bulrush	*Scirpus* spp.	14-32	-	-	-	5-7.5
Sedge	*Carex* spp.					

[a] ppt = parts per thousand

8.4.1 BOD removal

The removal of settleable organics is very rapid in all wetland systems and is due to the quiescent conditions in the FWS systems and to deposition and filtration in the SF systems.

In FWS constructed wetlands, removal of the soluble BOD is mainly due to the attached microbial growth. The major source of oxygen for these reactions is re-aeration at the water surface, since algae are typically not present. Any excess oxygen transmitted by the plant to the root zone is likely to be consumed in the soil profile and not to contribute significantly to oxygen levels in the water. Wind-induced water turbulence and mixing will also be reduced or eliminated if a dense stand of vegetation is present.

The major oxygen source for the subsurface components (soil, gravel, rock and other media, in trenches or beds) is the gases transmitted by the vegetation to the root zone. In most cases the system is designed to maintain flow below the surface of the bed, so there can be very little direct atmospheric re-aeration. The selection of plant species can therefore be an important factor.

Table 8.2 reports the removal efficiencies of BOD_5 and SS from constructed wetlands in Canada, U.S.A. and Australia

8.4.2 Suspended solids removal

Suspended solids removal is very effective in both types of constructed wetlands (see Table 8.2). Most of the removal occurs within the first few meters beyond the inlet, owing to the quiescent conditions and the shallow depth of liquid in the system. Controlled dispersion of the influent flow with proper diffuser pipe design can help to ensure low velocities for solids removal and even loading of the constructed wetland so that anoxic conditions are prevented at the upstream end of the channels.

8.4.3 Nitrogen removal

Nitrogen is mainly removed by nitrification/ denitrification. Other removal mechanisms include plant uptake and volatilization as ammonia. In constructed wetlands, nitrogen removal ranges from 25-85 percent (U.S. EPA, 1988).

Table 8.2. Summary of BOD_5 and SS removal from constructed wetlands (U.S. EPA, 1988)

Project	Flow, m^3/d	Wetland type	BOD_5, mg/L		SS, mg/L		% reduction		Hydraulic surface loading rate, $m^3/ha\text{-}d$
			Inf.	Eff.	Inf.	Eff.	BOD_5	SS	
Listowel, Ontario	17	FWS[a]	56	10	111	8	82	93	-
Santee, CA	-	SF[b]	118	30	57	5.5	75	90	-
Sydney, Australia	240	SF	33	4.6	57	4.5	86	92	-
Arcata, CA	11,350	FWS	36	13	43	31	64	28	907
Emmitsburg, MD	132	SF	62	18	30	8.3	71	73	1,543
Gustine, CA	3,785	FWS	150	24	140	19	84	86	412

[a] Free Water Surface system
[b] Subsurface Flow system

8.4.4 Phosphorus removal

Phosphorus removal in constructed wetlands is not very effective because of the limited contact opportunities between the wastewater and the soil. The principal mechanisms for phosphorus removal are plant uptake or retention in the soil (U.S. EPA, 1988).

If phosphorus removal is required, clay with iron and aluminum content should be considered. However, soils with a high phosphorus removal capacity are finer textured, and sand may be added to improve hydraulic conductivity. Also, iron or aluminum added to the substrate or fed into the wastewater can improve phosphorus removal (Steiner and Freeman, 1989).

8.4.5 Heavy metals removal

The predominant removal mechanisms in constructed wetlands are precipitation and adsorption. Precipitation is enhanced by wetland metabolism which increases the pH of inflowing acidic waters to near neutrality. Removal of Cu, Zn and Cd at the rates of 99, 97 and 99 percent respectively, for a residence time of 5.5 days in the Santee, California, constructed wetlands have been reported (U.S. EPA, 1988). However, metals removal will likely be finite due to exhaustion of the soil cation exchange capacity (CEC).

8.4.6 Trace organics removal

Municipal and industrial wastewaters contain variable concentrations of synthetic organic compounds. Adsorption of trace organics by the organic matter and clay particles present in the treatment system is thought to be the primary physicochemical mechanism for removal of refractory compounds in constructed wetlands. Other mechanisms can be biological degradation of easily degraded organic compounds, sedimentation and volatilization (U.S. EPA, 1988).

8.4.7 Pathogen removal

The pathogens of concern in constructed wetlands are parasites, bacteria and viruses. Pathogenic bacteria and viruses are removed by such mechanisms as: predation, sedimentation, absorption, and die-off from unfavorable environmental conditions, including UV in sunlight and temperatures unfavorable for cell reproduction (U.S. EPA, 1988). Table 8.3 reports performance data on pathogen removal for both FWS and SF constructed wetlands in the U.S. and Canada.

Table 8.3 Pathogen removal in constructed wetland systems. (modified from Reed et al., 1995)

Location (vegetation)	Winter season			Summer season		
	Influent	Effluent[a]	% reduction	Influent	Effluent[a]	% reduction
Santee, CA (bullrush)[b]						
Total coli, no./100 mL	5×10^7	1×10^5	99.80	6.5×10^7	3×10^5	99.54
Bacteriophage, PFU/mL	1900	15	99.21	2300	26	98.87
Iselin, PA (cattails & grasses)[c] Fecal coli, no./100 mL	1.7×10^6	6200	99.64	1.0×10^6	723	99.93
Arcata, CA (bullrush)[d] Fecal coli, no./100 mL	4300	900	79.07	1800	80	95.56
Listowel, Ont. (cattails)[d] Fecal coli, no./100 mL	556,000	1400	99.75	198,000	400	99.80

[a] Undisinfected

[b] Gravel bed, subsurface flow

[c] Sand bed, subsurface flow

[d] Free water surface

8.5 DESIGN EQUATIONS

All constructed wetland systems can be considered to be attached growth biological reactors, and their performance can be described with first order plug-flow kinetics. Design equations given below for both FWS and SF systems should give a reasonable and conservative estimate of design requirements. However, a pilot test is strongly recommended for large-scale projects.

8.5.1 FWS constructed wetlands

BOD_5 removal in plug-flow reactors such as a constructed wetland can be described by first-order model as follows (Reed *et al.*, 1995):

$$\frac{C_e}{C_0} = exp(-K_T t) \tag{8.1}$$

where

C_e = effluent BOD_5, mg/L
C_0 = influent BOD_5, mg/L
K_T = temperature-dependent first-order reaction rate constant,
$= k_{20}(1.06)^{T-20}$ day^{-1}
t = hydraulic residence time, day *
T = water temperature, ^0C
k_{20} = 0.678 day^{-1}

* In calculating the t value, the constructed wetland bed porosity and water loss by evapotranspiration must be taken into account.

For an unrestricted flow system, hydraulic residence time can be expressed as:

$$t = \frac{L \cdot W \cdot d}{Q} \tag{8.2}$$

where

L = length of system, m
W = width of system, m
d = depth of system including the bed depth and water depth, m
Q = average flow rate = $(Q_{inf} + Q_{eff})/2$

In an FWS constructed wetland, because a portion of the available volume will be occupied by the bed media and the vegetation, the actual detention time will be a function of the porosity (η), which can be defined as follows:

$$\eta = \frac{V_v}{V} \tag{8.3}$$

where V_V and V are volume of voids and total volume, respectively.

The value of η for FWS constructed wetlands is generally taken as 0.75. So the actual detention time is:

$$t = \frac{L \cdot W \cdot d \cdot \eta}{Q} \tag{8.4}$$

Reed *et al.* (1995) developed another general model for FWS constructed wetland design by combining the relationships in Eqs. 8.2 and 8.3 with Eq. 8.1:

$$\frac{C_e}{C_0} = A \exp\left[-\frac{0.7 K_T (A_V)^{1.75} LWd\eta}{Q}\right]$$

(8.5)

where

C_e	= effluent BOD_5, mg/L
C_0	= influent BOD_5, mg/L
A	= fraction of BOD_5 not removed as settleable solids near headworks of the system (as a decimal fraction)
K_T	= temperature-dependent rate constant, day^{-1}
A_V	= specific surface area for microbial activity, m^2/m^3
L	= length of system (parallel to flow path), m
W	= width of system, m
d	= design depth of system, m
η	= porosity of system (as a decimal fraction)
Q	= average flow rate = $(Q_{inf} + Q_{eff})/2$

The rate constant K_T (in day^{-1}) at water temperature T (in °C) and is defined by:

$$K_T = K_{20}(1.1)^{(T-20)}$$

(8.6)

where, K_{20} is the rate constant at 20°C.

Other coefficients in Eq. 8.5 have been estimated (Reed *et al.*, 1988):

A	= 0.52
K_{20}	= 0.0057 day^{-1}
A_V	= 15.7 m^2/m^3
η	= 0.75

When the bed slope or hydraulic gradient (S) is greater than 1 percent, it is necessary to adjust the design model accordingly:

$$\frac{C_e}{C_0} = A \exp\left[-\frac{0.7 K_T (A_V)^{1.75} LWd\eta}{4.63 S^{1/3} Q}\right]$$

(8.7)

Table 8.4 lists typical values used to test the equation against actual values at Listowel, Ontario, Canada. Table 8.5 summarizes design criteria for FWS constructed wetlands. It should be noted in this table that an aspect ratio (L/W) of at least 3:1 is needed for FWS constructed wetland systems to approach plug-flow conditions and optimum treatment.

Table 8.4 Predicted versus actual C_e/C_0 values for constructed wetlands at Listowel, Ontario. (Reed *et al.*, 1995)

Distance along channel, m	*Fall*		*Winter*		*Spring*		*Summer*	
	Predicted	*Actual*	*Predicted*	*Actual*	*Predicted*	*Actual*	*Predicted*	*Actual*
0	0.52	0.52	0.52	0.52	0.52	0.52	0.52	0.52
67	0.38	0.40	0.40	0.40	0.47	0.30	0.38	0.36
134	0.27	0.23	0.31	0.20	0.42	0.28	0.27	0.41
200	0.20	0.19	0.24	0.19	0.38	0.22	0.20	0.30
267	0.14	0.18	0.18	0.17	0.34	0.23	0.14	0.27
334 (final effluent)	0.10	0.15	0.14	0.17	0.30	0.26	0.10	0.17

Table 8.5 Summary of design criteria for FWS constructed wetlands. (adapted from Reed et al., 1995).

Organic loading, kg BOD_5/ha-day	< 112
Actual detention time as determined by Eq. 8.5 or 8.7, days	3-15
Specific surface (A_V in Eq. 8.5 or 8.7) for attached microbial growth, m^2/m^3	15.7
Porosity (η value in Eq. 8.5 or 8.7) of constructed wetland flow path	0.75
Aspect ratio (L/W)	$\geq 3:1$
Water depth, cm	
Warm months	< 10
Cool months	< 45

Substituting these factors in Eq. 8.5 or 8.7 and rearranging terms to solve for detention time or for the required surface area produced the following equations for FWS constructed wetlands.

Hydraulic residence time is given by:

$$t = \frac{\left(\ln C_0 - \ln C_e\right) - 0.6539}{65 K_T} \tag{8.8}$$

If the bed slope or hydraulic gradient, is greater than 1 percent, then

$$t = \frac{(\ln C_0 - \ln C_e) - 0.6539}{301 K_T S^{1/3}} \qquad (8.9)$$

where, S is the bed slope in decimal fraction (e.g., for a 2% bed slope, S is equal to 0.02).

The design surface area of the constructed wetland is given by

$$A = \frac{Q(\ln C_0 - \ln C_e - 0.6539)}{65 K_T d} \qquad (8.10)$$

If the bed slope or hydraulic gradient is greater than 1 percent

$$A = \frac{Q(\ln C_0 - \ln C_e - 0.6539)}{301 K_T S^{1/3}} \qquad (8.11)$$

Equations 8 to 11 are only valid for FWS constructed wetlands meeting the conditions defined in the criteria summarized in Table 8.5.

Example 1 Design an FWS constructed wetland to treat facultative lagoon effluent in a warm climate with a mean annual temperature of 25°C. The design flow is 760 m³/day, the facultative lagoon effluent has a BOD$_5$ concentration of 130 mg/L, and the required effluent BOD$_5$ of the FWS constructed wetland is 10 mg/L.

Solution
1. Assume the slope of the constructed wetland bed will be 1 percent to allow drainage when required. Use Eq. 8.8 to estimate required detention time. At 25°C,

$$K_T = K_{20}(1.1)^{(T-20)} = 0.0057(1.1)^{(25-20)} = 0.0092 \text{ day}^{-1}$$

$$t = \frac{(\ln C_0 - \ln C_e) - 0.6539}{65 K_T}$$

$$t = \frac{(\ln 130 - \ln 10) - 0.6539}{(65)(0.0092)} = 3.2 \text{ days}$$

2. For the warm climate site, use a 10-cm water depth on a year-round basis. If cattail plants are to be grown, the bed depth should be 30 cm, the total bed depth or d is 40 cm. Use Eq. 8.10 to estimate the surface area required.

$$A = \frac{Q(\ln C_0 - \ln C_e - 0.6539)}{65\,K_T\,d}$$

$$= \frac{(760\text{ m}^3/\text{day})(\ln 130 - \ln 10 - 0.6539)}{(65)(0.0092)(0.40\text{ m})(10{,}000\text{ m}^2/\text{ha})}$$

$$= 0.60\text{ ha}$$

3. Use an aspect ratio (L/W) of 10:1 and determine the dimensions for the constructed wetland channels, assuming a square plot is available.

$$A = LW = (10W)W = 600\text{ m}^2$$

Thus W = 7.75 m, L = 77.5 m. Divide the square plot into 10 parallel channels, each 77.5 m long.

4. The design procedure assumes, and experience at operational systems confirms, that SS and nitrogen concentrations in the FWS constructed wetland effluent will also be at satisfactory levels suitable for discharge or reuse. Some of the remaining nitrogen will be in the ammonia form. If stringent ammonia limits prevail, recycle of constructed wetland effluent should be incorporated in the system design. The amount of recycle required will depend on the evapotranspiration (ET) losses for the area and should be sufficient to maintain the 3.2-day detention time.

8.5.2 SF constructed wetlands

The major oxygen source for the subsurface components (soil, gravel, rock, and other media in trenches or beds) is the oxygen transmitted by the vegetation to the root zone. In most cases, the SF system is designed to maintain flow below the surface of the bed, so there can be very little direct atmospheric reaeration. The selection of plant species is therefore an important factor. The root penetration of the various plants (see Table 8.1) makes it possible to remove BOD_5 in the expanded aerobic zone.

Root penetration is important for the oxygen transfer and treatment of organics and nitrogen at the full bed depth. The maximum treatment potential and possibly the maximum design hydraulic loading may not be realized until the roots and rhizomes have penetrated to their full potential depth. Reed *et al.* (1995) recommended that the depth of root penetration, as listed in Table 8.1, be used as the design depth for SF systems.

BOD$_5$ removal in SF systems can be described with first-order kinetics, as described in Eq. 8.1 for free water surface systems. Equation 8.1 can be rearranged and used to estimate the required surface area for an SF system. Both forms of the equation are shown below for convenience.

$$\frac{C_e}{C_0} = exp\left(-K_T t\right)$$

(8.1)

$$A_s = \frac{Q\left(\ln C_0 - \ln C_e\right)}{K_T d \eta}$$

(8.12)

where,

C_e = effluent BOD$_5$, mg/L
C_0 = influent BOD$_5$, mg/L
K_T = temperature-dependent first-order reaction rate constant, $= k_{20}\left(1.06\right)^{T-20}$ day^{-1}
t = hydraulic residence time, day
Q = average flow rate through the system, m^3/d
d = depth of submergence, m
η = porosity of the bed, as a fraction
A_s = surface area of the system, m^2

The saturated cross-sectional area for flow through an SF system can be calculated according to Darcy's law:

$$A_c = \frac{Q}{k_s S}$$

(8.13)

where

A_c = $d \cdot W$, cross-sectional area of constructed wetland bed, perpendicular to the flow direction, m^2
d = bed depth, m
W = bed width, m
k_s = hydraulic conductivity of the medium, m^3/m^2-d
S = slope of the bed, or hydraulic gradient (as a decimal fraction)

Bed cross-sectional area and bed width are independent of temperature (climate) and organic loading since they are controlled by the hydraulic characteristics of the media.

The value of K_T can be calculated from suggested k_{20} values for SF constructed wetland systems (Table 8.6). Table 8.6 lists estimated porosities (η), hydraulic conductivity (k_s) and k_{20}.

To avoid disruption of the medium-rhizome structure and to ensure sufficient contact time for treatment, the unit flow velocity (Q/A_c, which is equal to k_sS according to Eq. 8.13) through a cross-section of the medium should not exceed 8.6 m/d (Reed *et al.,* 1988).

Table 8.6 Media characteristics for subsurface flow systems. (adapted from Reeds *et al.,* 1995)

Media type	Effective size [a], mm	Porosity (η)	Hydraulic conductivity (k_s), m^3/m^2-d	k_{20} [b], day^{-1}
Coarse sand	2	0.28-0.32	100-1,000	1.35
Gravelly sand	8	0.30-0.35	500-5,000	0.86
Fine gravel	16	0.35-0.38	1,000-10,000	0.60
Medium gravel	32	0.36-0.40	10,000-50,000	<0.60

Note: [a] is the maximum 10 percentile grain size

[b] k_{20} values are site specific and should be experimentally validated

Example 2 Design an SF constructed wetland to treat a facultative lagoon effluent in a warm climate with a mean annual temperature of 25°C. The design flow is 760 m^3/day, the facultative lagoon effluent has a BOD_5 concentration of 130 mg/L, and required effluent BOD_5 of the SF constructed wetlands is 10 mg/L. (*Note:* These are the same conditions used in Example 1, so comparisons can be made.) The predominant constructed wetland plant type in surrounding marshes is cattail.

Solution
1. Choose cattail for this SF system since it is successfully growing in local constructed wetlands. From Table 8.1, cattail rhizomes penetrate about 30 cm into the medium. So the bed media depth (d) should be 30 cm.
2. Choose a slope of 1 percent ($S = 0.01$) for ease of construction.
3. Choose a gravelly sand as the medium. From Table 8.6, $k_s = 500$ m^3/m^2-d. Check the suitability of a 1 percent slope.

$$k_sS = (500 \text{ m}^3/\text{m}^2\text{-d})(0.01) = 5.0 \text{ m/d} < 8.60 \text{ m/d}$$
$$(OK)$$

4. From Table 6, $k_{20} = 0.86$. Solve for K_T using Eq. 8.12.

$$K_T = k_{20}\,(1.06)^{(T-20)} = 0.86\,(1.06)^{(25-20)} = 1.34 \text{ day}^{-1}$$

5. Determine the cross-sectional area of the bed using Eq. 8.13.

$$A_c = Q/(k_s S) = 760/5.0 = 152 \text{ m}^2$$

6. Determine the bed width.

$$W = A_c/d = 152/0.3 = 507 \text{ m}$$

7. Determine the surface area required using Eq. 8.12.

$$A = \frac{Q(\ln C_0 - \ln C_e)}{K_T d\eta}$$

$$= \frac{(760 \text{ m}^3/\text{day})(\ln 130 - \ln 10)}{(1.34)(0.3)(0.35)}$$

$$= 13{,}854 \text{ m}^2, \quad \text{requiring more area than FWS constructed wetlands due to low porosity}$$

8. Determine the bed length (L) and the detention time (t) in the system.

$$L = A_s/W = 13{,}854/507 = 27.3 \text{ m}$$

$$t = LWd\eta/Q = (27.3)(507)(0.3)(0.35)/(760) = 1.91 \text{ days}$$

8.6 OTHER CONSIDERATIONS

8.6.1 Hydraulic budget

For a constructed wetland, the water balance can be expressed as follows:

$$Q_i - Q_0 + P - ET = \frac{dV}{dt} \qquad (8.14)$$

where

$$
\begin{aligned}
Q_i &= \text{influent wastewater flow, volume/time} \\
Q_0 &= \text{effluent wastewater flow, volume/time} \\
P &= \text{precipitation, volume/time}
\end{aligned}
$$

ET   = evapotranspiration, volume/time
V = volume of water
t = time

Groundwater inflow and infiltration are excluded from Eq. 8.14 because of the impermeable barrier. If the system operates at a relatively constant water depth ($\mathrm{d}V/\mathrm{d}t = 0$), the effluent flow rate can be estimated using Eq. 8.14. Historical climatic records can be used to estimate precipitation and evapotranspiration.

8.6.2 Site selection

A constructed wetland can be constructed almost anywhere. Because grading and excavating represent a major cost factor, topography is an important consideration in the selection of an appropriate site. In selecting a site for an FWS constructed wetland, the most desirable soil permeability is 10^{-6} to 10^{-7} m/s. Sandy clays and silty clay loams can be suitable when compacted. In heavy soils, additions of peat moss or top soil will improve soil permeability and accelerate initial plant growth. In areas where highly permeable soils (e.g. sandy soil) exist, the constructed wetland beds should be lined with heavy-duty plastic sheets to avoid groundwater contamination (see section 8.6.5).

8.6.3 Flow patterns

A constructed wetland cell is designed to use one or more of three types of flow patterns: plug flow, step feed, or recirculation. Plug flow (Figure 8.3a) is once-through flow down the cell length. Plug flow is now used for most municipal and acid drainage systems and requires minimal piping, energy use, operation, and maintenance.

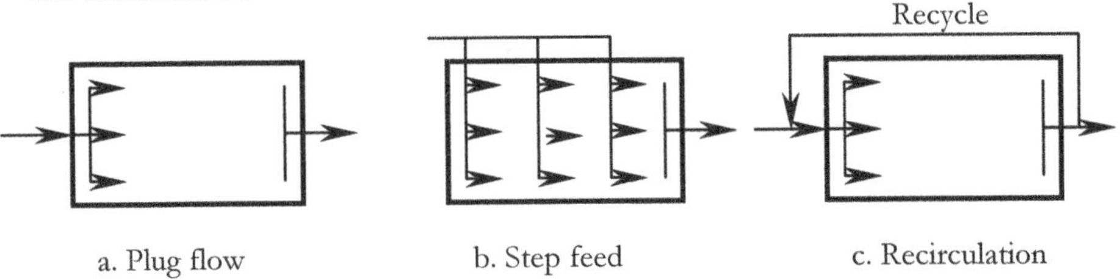

Figure 8.3 Flow patterns for constructed wetlands.

Step feed (Figure 8.3b) may benefit pollutant removal by using more of the constructed wetland area for solids removal and by providing carbon for nitrogen removal in the lower bed area. Step feed is typically combined with recirculation. The cost-benefit of step feed needs investigation.

Recirculation (Figure 8.3c) should be considered, and its potential needs further investigation. Recycling treated effluent will dilute influent BOD_5 and suspended solids, decreasing odor potential and increasing dissolved oxygen concentration and retention time, which will enhance nitrification and subsequent nitrogen removal (Steiner and Freeman, 1989).

8.6.4 Slope

A slope of 0.5% or less, as limited by construction tolerances, is recommended for an FWS system. Some slope is needed to drain the cell for maintenance and possible mosquito control. SF bed slopes should be 2% or less based on the initial hydraulic conductivity of the substrate. A level substrate surface is recommended for an SF system to facilitate vegetation planting and to control weed by flooding the bed. Bed slopes greater than 4% would cause most of the flow to be on the surface due to poor hydraulic conductivity, with channeling, treatment, and vegetation management problems (Steiner and Freeman, 1989).

8.6.5 Liners

If groundwater contamination or water conservation is a concern, an impermeable liner below the substrate is required. Possible materials are compacted in situ soil (permeability less than 10^{-6} cm/sec); bentonite; asphalt; synthetic butyl rubber; or plastic membranes. The liner must be strong, thick, and smooth to prevent root penetration and attachment (Steiner and Freeman, 1989).

8.7 OPERATION AND MAINTENANCE

Two operational considerations associated with constructed wetlands for wastewater treatment are mosquito control and plant harvesting. In addition, system disturbances can occur from time to time.

8.7.1 Mosquito control

Mosquito problems may occur when wetland treatment systems are overloaded organically and under anaerobic conditions. Strategies used to control mosquito populations include effective pretreatment to reduce total organic loading; step feeding of the influent wastewater stream with effective influent distribution and effluent recycle; vegetation management; natural controls, principally by mosquitofish (*Gambusia affinis*), in conjunction with the above techniques; and application of man-made control agents. In general, natural controls are preferred because of a concern that man-made control agents might develop resistant strains of mosquito (Wieder *et al.,* 1989).

8.7.2 Plant harvesting

The usefulness of plant harvesting in constructed wetland treatment systems depends on several factors, including climate, plant species, and the specific wastewater objectives. Harvesting plants to remove wastewater contaminants taken up by the plants is inefficient. However, plant harvesting can affect treatment performance of constructed wetlands by altering the effect that plants have on the aquatic environment. Further, because harvesting reduces congestion at the water surface, control of mosquito larvae using fish is enhanced. Where a segmented wetland system is used, drying each segment separately allows harvesting with conventional equipment. Depending on location, burning the dried plant mass in place may be most economical (*Wieder et al.,* 1989). The frequency of plant harvesting should be in accordance with the plant's area doubling time. In the tropics, under favourable condition, the area dubling times of most emergent aquatic plants (Table 8.1) are 1-2 months (Polprasert, 1996).

8.7.3 System perturbations and operation modifications

Deviations from "average" operating conditions will occur in the system lifetime with greater frequency than predicted or preferred. Perturbations generally are of two types (Girts and Knight, 1989): 1) predictable perturbations, which can be predicted and occur periodically; and 2) unpredictable perturbations, which are unforeseen in the design phase or which occur so infrequently that incorporation into the design would entail unnecessary expense. Tables 8.7 and 8.8 summarize, respectively, predictable and unpredictable events, along with operational features most affected, associated symptoms, and appropriate operation modifications.

Table 8.7 Predictable system disturbances. (Grits and Knight, 1989)

Disturbance	Features	Symptoms	Operation Modifications
Startup	Difficulties in vegetation establishment and microbial flora colonization	Widely fluctuating treatment efficiencies	Control loading rates, i.e. water flow rate, chemical concentration - water inflow rates - freshwater source Control water depths - critical for vegetation establishment and development of conditions suitable for target microbial populations Dilution/recirculation Chemical additions (e.g. lime to improve soil pH, fertilizer)
Seasonal	Extreme precipitation	High loading rates	Control loading rates, i.e. water flow rates, chemical concentration - water inflow rates - dilution - recirculation
		Decreased storage capacity	Increase storage capacity - stormwater diversion - detention pond - increase water depth
		Insufficient residence times; Channeling	Control outflow rates Install baffles
	Extreme low temperatures	Insufficient flow Freezing; sheet flow over ice surface	Freshwater source, recirculation, parallel cells Recirculation, aeration, control water depth Preheated water
	Vegetation growth/decay	Flushing of chemicals, nutrients, and microbes from decaying vegetation, sediment	Secondary treatment pond Vegetation harvest or burning Recirculation to increase nutrient retention
	Population composition changes (microbial, algal)	Gradual change in treatment efficiency	Secondary treatment pond Recirculation

Table 8.8. Unpredictable system disturbances (Girts and Knight, 1989)

Disturbance	Symptoms	Operation modifications									
		Water inflow	Water outflow	Water depth	Dilution	Recircu-lation	Pretreatment pond	Chemical addition	Vegetation harvest	Replant	Predator control
Record storm event	High hydraulic loading rates	X				X	X				
	Decreased storage capacity	X					X				
	Insufficient residence time	X	X	X	X	X	X	X			
	Channeling	X					X				
	High sediment loads			X			X				
	High chemical loads		X		X	X	X	X			
	High chemical loads	X			X	X	X				
Change in chemical constituents and concentrations	Increased toxicity (vegetation, wildlife)				X			X	X	X	
Vegetation damage	Release of chemicals from Sediments/vegetation		X	X		X			X	X	
	Change in chemical form			X	X						
	Increased debris, flow hindrance								X	X	
	Nutrient release from vegetation	X		X					X	X	
	Change in conditions for replanting		X	X		X			X	X[a]	

Disturbance	Symptoms	Operation modifications										
		Water inflow	Water outflow	Water depth	Dilution	Recircu-lation	Pretreatment pond	Chemical addition	Vegetation harvest	Replant	Predator control
Pests (rodents, mosquitoes, etc.)	Complaints from neighbors	x		x			x				
	Reduced flow and water level control			x				x			x
Malfunctions/ construction failures	Reduced flow and water level control	x	x	x	x	x	x	x			
Design flaw	Inability to respond to need for changes in operations[b]										
	Limited treatment capacity[b] Limited lifespan[b]										

[a] New species.
[b] All operation modif may need to be considered.

8.8 CASE STUDIES

8.8.1 Case Study A: Emmitsburg, Maryland, USA, SF Constructed Wetland

The town of Emmitsburg constructed an SF system with a design flow rate of 130 m^3/d in 1984 to treat a portion of its effluent flow. The system is a single basin, 76.3 m long, 9.2 m wide, and 0.9 m deep, filled with 0.6 m of crushed rock. Clay was used in the bottom of the basin to prevent groundwater contamination. Perforated pipes placed near the bottom of the basin are used for influent distribution and effluent collection. The water level during normal operation is approximately 5 cm below the surface of the gravel. The system was initially seeded with 200 broadleaf cattail plants in August 1984 and another 200 in July 1995. By March 1986, about 35 percent of the basin surface area was covered by cattails. The planting density used should have been at least an order of magnitude higher. Until the plants cover the entire basin, the performance will not be representative of an SF system.

The influent to the Emmitsburg system is trickling filter effluent. Influent flows vary between 95 and 132 m^3/d, which corresponds to a surface hydraulic loading rate of 1,420-1,870 m^3/ha-d. Effluent samples are collected and analyzed weekly for BOD$_5$, SS, TDS, DO and pH.

The influent BOD$_5$ concentrations to the system range between 10-180 mg/lwhile SS concentrations normally range between 10 and 60 mg/l Results from two years of operation are presented in Table 8.9, which show very good performance even with the limited plant coverage. Odors in the effluent have been an occasional problem but the frequency of noticeable odors is decreasing as cattail coverage increases. The engineering and construction costs of the Emmitsburg system were less than US$35,000.

8.8.2 Case Study B: The Eastern Seaboard Industrial Estate (ESIE), Rayong Province, Eastern Thailand, Vertical-flow Constructed Wetlands

The ESIE comprises of over 100 industrial factories and 14,000 staff. The industries are required to pre-treat their wastewaters to remove heavy metals and other toxic compounds according to the Thailand effluent standards prior to discharging into combined sewers and mixing with other domestic wastewaters. The current wastewater flow rate is about 7,000 m^3/day.

Table 8.9 Performance of the Emmitsburg SF system (U.S. EPA, 1988).

Season	Average flow, m^3/d	BOD$_5$, mg/l		SS, mg/l		Effluent DO, mg/l	Area covered with cattails, %
		Inf.	Eff.	Inf.	Eff.		
Fall 1984	117	29	12	25	7	1.0	< 5
Winter 1985	111	68	29	37	9	0.3	< 10
Spring 1985	130	117	38	37	13	0.0	< 20
Summer 1985	100	87	11	28	10	1.3	< 25
Fall 1985	97	28	7	29	7	2.1	< 30
Winter 1986	106	40	11	25	4	-	< 35

Table 8.10. Treatment performance of the CW unit (AIT, 2002).

Parameter	Average Concentration (mg/l)			Overall Removal
	Influent	Effluent Unit 1	Effluent Unit 2[b]	
BOD	88	19	4	95.8%
COD	229	52	19	91.5%
SS	98	16	10	89.5%
TKN	24.1	14.5	4.6	81.1%
NH$_3$-N	14.2	10.8	3.3	76.7%
NO$_3$-N	0.06	0.53	4.75	-[c]
TP	7.0	4.7	1.5	78.5%

[a] percolate of CW unit 1, [b] percolate of CW unit 2 in series
[c] Increased likely due to nitrification reaction

This combined wastewater is being treated by two constructed wetlands (CW) in series, each with a dimension of 35×18×0.8 m (length×width×depth). The CW beds are lined with high-density polyethylene sheet and filled with 1-cm gravel. Wastewater is applied intermittently over the CW beds in a vertical-flow mode, and the percolates are collected through underdrainage pipes. Cattails, bulrushes and canna are the primary vegetation grown in these CW beds.

During the period of September 2001-May 2002, these CW units were operated at the following conditions: hydraulic loading rates 60-160 l/m²-day; organic loading rate 57-140 kg BOD$_5$/ha-day and hydraulic retention times 1.4-4.0 days. The treatment performance of these CW units in series was found satisfactory (Table 8.10) with the effluent BOD$_5$ and SS concentrations being 4 and 10 mg/l, respectively. Because of the nitrification reactions occurring the CW beds, there was an increase in NO$_3$-N concentration from 0.06 mg/l in the effluent to 4.75 mg/l in the effluent. This treated water is being sold to some factories located in the ESIE for uses in factory air-cooling and other processes.

Due to prolific growth of the vegetation under tropical conditions, plant harvesting was done once in 4 months, with the annual yields of 130-150 tons/ha-yr (wet weight). These harvested plants can be used in making furniture and other decorations, another income generation for the ESIE.

8.8.3 Case Study C: Vertical-flow Constructed Wetlands for Septage Dewatering and Stabilization, Asian Institute of Technology (AIT), Bangkok, Thailand

Septic tank sludge (or septage) usually contains high solid, organic and enteric microorganism contents, with poor setting and dewatering characteristics. A pilot study was carried out at the AIT campus during 1998-2001 to demonstrate the feasibility of applying vertical-flow CW to dewater and stabilize septage collected from Bangkok city, Thailand.

Three pilot-scale CW beds, each with a surface area of 25 m^2, having 65 cm sand-gravel substrata, supported by a ventilated drainage system (Figure 8.4) were planted with narrow-leave cattails (*Typha augustifolia*). To operate in a vertical-flow mode, the Bangkok septage (Table 8.11) was uniformly distributed on the CW beds 1-2 times weekly at the solids loading rate (SLR) of 1.5-9.6 kg TS/m^2-week, while the percolate was collected from the drainage system for further treatment.

The experimental data obtained so far indicated the SLR of 4.8 kg TS/m^2-week to be the most suitable, resulting in the highest TS, COD and TKN removal of 80, 96 and 92%, respectively (Koottatep *et al.*, 2001). The TS contents of the dewatered septage remaining on the CW beds were increased from 1-2% to 30-60% with the one-week operation cycle. Because of the vertical-flow mode of operation and with the effectiveness of the ventilation pipes, there was a high degree of nitrification occurring in the CW beds – the NO_3-N contents were found to increase from about 4 mg/l (in the Bangkok septage) to 180-250 mg/l in the CW percolate.

Due to rapid flow-through of the percolates, there was little liquid retained in the CW beds, causing the cattail plants to wilt, especially during the dry season. To reduce the wilting effects, the operating strategies in the second year were modified by ponding the percolate in the CW beds for periods of 2 and 6 days prior to discharge.

Table 8.11 Characteristics of Bangkok septage samples[a]

Parameter	Range	Average	Standard Deviation
pH	6.7 – 8.0	7.5	0.6
TS (mg/L)	2,200 – 67,200	19,000	12,500
TVS (mg/L)	900 – 52,500	13,500	9,400
SS (mg/L)	1,000 – 44,000	15,000	10,100
BOD (mg/L)	600 – 5,500	2,800	1,400
TCOD (mg/L)	1,200 – 76,000	17,000	15,000
TKN (mg/L)	300 – 5,000	1,000	800
NH_4 (mg/L)	120 – 1,200	350	170
NO_3 (mg/L)	1 – 11	4.5	3.5
Helminth eggs, no./g of sample	0 – 14	4	1

[a] Based on 120 raw septage samples during August 1997-February 1999

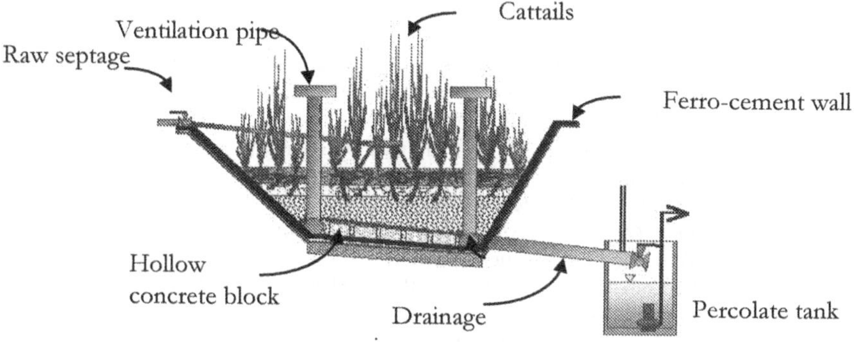

Figure 8.4 Schematic diagram of pilot-scale CW units.

This operating strategy was found beneficial not only for mitigating plant wilting, but also for increasing N removal through enhanced denitrification activities in the CW beds. During these 3 year-operations, the dewatered septage was not removed from the CW beds and no adverse effects on the septage dewatering efficiency were observed.

prior to discharge. This operating strategy was found beneficial not only for mitigating plant wilting, but also for increasing N removal through enhanced denitrification activities in the CW beds. During these 3 year-operations, the dewatered septage was not removed from the CW beds and no adverse effects on the septage dewatering efficiency were observed.

As can be expected, the dewatered septage, after 2-3 years of decomposition in the CW beds, was a well-stabilized sludge without the presence of viable helminth eggs, suitable to be used as a fertilizer or soil conditioner.

ACKNOWLEDGEMENT

The technical support provided by Mr Kusumakar Sharma in the preparation of this manuscript is gratefully acknowledged.

REFERENCES

Asian Institute of Technology (2002) *Water purification and reuse through constructed wetlands at the Eastern Seeboard Industrial Estate.* Thailand. Final report submitted to National Center for Genetic Engineering and Biotechnology and Hemaraj Land Development Public Co., Ltd., Thailand.

Bastian, R.K., Shanaghan, P.E., and Thompson, B.P. (1989) Use of constructed wetlands for municipal wastewater treatment and disposal—Regulatory issues and EPA policies. In: *Constructed Wetlands for Wastewater Treatment: Municipal, Industrial, and Agricultural,* Hammer, D.A. (ed). Lewis Publishers, Chelsea, Michigan.

Girt, M.A. and Knight, R.L. (1989) Operations optimization. In: *Constructed Wetlands for Wastewater Treatment: Municipal, Industrial, and Agricultural,* Hammer, D.A.(ed). Lewis Publishers, Chelsea, Michigan.

Koottatep, T., Polprasert, C., Oanh, N.T.K., Heinss, U., Montangero, A. and Strauss, M. (2001) Septage dewatering in vertical-flow constructed wetland located in the tropics. *Water Science and Technology* **44**, 181-188.

Polprasert, C. (1996) *Organic Waste Recycling, Technology and Management.* 2nd edition, John Wiley and Sons, Chichester, UK.

Reed, R. C., Middlebrooks, E. J., and Crites, R. W. (1995) *Natural Systems for Waste Management and Treatment.* 2nd edition, McGraw-Hill, New York.

Steiner, G.R. and Freeman, R.J.Jr. (1989) Configuration and substrate design considerations for constructed wetlands wastewater treatment. In: *Constructed Wetlands for Wastewater Treatment: Municipal, Industrial, and Agricultural,* Hammer, D.A. (ed). Lewis Publishers, Chelsea, Michigan.

U.S. EPA (1988) *Design Manual — Constructed Wetlands and Aquatic Plant Systems for Municipal Wastewater Treatment.* EPA/625/1-88/022, United States Environmental Protection Agency, Cincinnati, Ohio.

Wieder, R.K., Tchobanoglous, G., and Tuttle, R.W. (1989) Preliminary considerations regarding constructed wetlands for wastewater treatment. In: *Constructed Wetlands for Wastewater Treatment: Municipal, Industrial, and Agricultural,* Hammer, D.A. (ed). Lewis Publishers, Chelsea, Michigan.

9

Innovation and technology for sustainability

Robert Hughes, Goen Ho and Kuruvilla Mathew

9.1 INTRODUCTION

This chapter is designed to inform the reader on available wastewater innovations and technologies that provide sustainable outcomes in both the developed and developing world. The type of drivers used to develop these innovative approaches towards sustainability are explained, by emphasising how technologies have changed to fulfil certain focuses of the wastewater industry.

The use of policy and public perception to guide the wastewater industry from the 19th to the 21st century is explained to show how sustainability has been brought about by science and public demand. The areas that currently need more development and research are further explained and will be key factors for the wastewater industry in the future.

Appropriate wastewater management promises many options for the longevity of water resources and the environment, and is one of the most crucial aspects of environmental management for human health and well being. The wastewater sector unlike the water sector is not compromised by limited cheap sources of water and has a resource readily available for many beneficial uses after effective collection and treatment. The type of benefits wastewater offers includes its use as a source of water (and often nutrients) for agricultural, aquacultural, industrial and domestic uses. Wastewater can additionally provide energy to reduce costs in treatment plant operation. In many parts of the world, the reuse of wastewater may reduce the demands on water sources and ensure water is available for the majority of people, not only the higher income classes.

The collection and treatment of wastewater are the most crucial aspects of wastewater management, as these factors can be tailored to the local legislative requirements, monetary funds, availability of resources and technologies, and other site-specific factors. The types of systems chosen may further be influenced by public perception. Where possible a system should be designed to safely provide resources (e.g. irrigation nutrients and water) that are limited within an area.

Many countries across the world have utilised their wastewater to ensure that adequate resources are available, which may include high quantities of wastewater reuse for agriculture and industrial processes (e.g. Israel reuses up to 90% of its wastewater) or the use of wastewater to increase limited potable water supplies as occurs in Singapore. Most countries across the world are specifically focusing on wastewater reuse, as the finite nature of water becomes apparent with population growth and increasing consumption, to reduce water demand.

Currently, there are aspects of wastewater management that may reduce its ability to provide an equitable resource. These include public perception on specific types of wastewater reuse and the availability of technologies to provide appropriate treatment for the intended wastewater use. The technological developments and innovations to overcome these limitations are becoming increasingly important for current and future wastewater systems, as in many cases there may not be the available water resources or monetary funds to provide potable water for non-contact purposes. In the developing world, this problem is evident by the fact there are over 1.1 billion people without clean water and up to 2.4 billion people without adequate sanitation. Many water sources in developing nations are contaminated from inappropriate sanitation thus reducing their value as a resource. Appropriate wastewater management in these areas may improve water resources (mainly those contaminated by disease transmitting organisms from wastewater), although it cannot completely remove contaminants that are derived from other processes e.g. contamination of water supplies in Bangladesh from naturally occurring arsenic.

In developing nations there is a particular focus on cheap small scale systems that can provide beneficial resources, rather than large scale systems such as dual reticulation, which require considerable infrastructure costs and can provide significant health threats (e.g. cross contamination of water supplies) where they are not maintained appropriately e.g. appropriate plumber certification programs. The types of technologies may include small anaerobic digesters, onsite wastewater systems, onsite reuse technologies, urine separating toilets and greywater systems. Advances in collection systems may include the use of trenchless technology and simplified sewerage. The use of sludge management technologies is also an important factor required, for safe use of treatment by-products.

9.1.1 Sustainability as a context

The sustainability context for the wastewater sector has clearly changed overtime. The original wastewater systems were developed by early cultures, such as the Roman Civilisation, to convey wastewater to nearby watercourses. However, the beginning of the current sustainability framework for the wastewater sector was not properly established until the 19th century, when knowledge on the harmful effects of polluted water supplies from wastewater was found. The most sustainable option in the 19th century was to develop deep sewerage infrastructure and dispose of the wastewater away from living areas into the natural environment. Due to the comparative low populations during this time, the natural environment could assimilate most of the wastewater. Through time however, it was found that as more demand was placed on water supplies and as standards of living improved, wastewater production increased and traditional methods of disposal caused significant changes in the natural environment. During this period, sustainability, largely driven by public perception and cost, began to encompass a larger framework, which included aspects of wastewater management such as treatment before disposal to protect the environment and subsequent human health. Once appropriate treatment technologies became available more stringent standards for discharge where implemented. In rapidly rising population centres or areas of few water sources, the reuse of wastewater became a part of sustainability and was utilised for non-potable purposes.

Recently, the changes in public perception from scarcity of water resources have increased the focus on wastewater reuse, and the reuse of wastewater to supplement water supplies has been introduced in countries where it has been publicly accepted. The standards for discharge have also become more stringent and many developed nations have placed an emphasis on natural flow allocations for discharge into natural freshwater environments. In addition, the reuse of wastewater in some urban centres has taken into account a water cycle approach, where wastewater is used to ensure the hydrological cycle is maintain, by supplementing water sources (e.g. irrigation water) with appropriately treated wastewater. This approach takes into account natural purification processes, to ensure they are optimised to reduce the costs in wastewater treatment and allow maximum use of specific wastewater components e.g. water, nutrients and trace elements for horticulture.

Currently, sustainability within the wastewater sector is no longer limited to direct impacts from the wastewater but is integrated with other aspects of sustainability such as energy and material demands. The use of decision support tools such as life cycle assessment has greatly assisted the sector by allowing it to assess its sustainability as a part of a whole system. For example, as a part of this process the air, water and soil emissions from a wastewater plant may be assessed including aspects not directly related to plant location. In this case, the air emissions from a wastewater treatment plant may include emissions from energy production for plant operation, wastewater treatment, and disposal or reuse of sludge. The emissions to water may include the discharge of hazardous chemicals and nutrients and the emissions to soil may include those from the sludge or wastewater if reused. Emissions of persistent pollutants to sediments may also be included. The use of decision tools such as life cycle assessment will be used to show which method of wastewater treatment offers the best sustainable outcomes. The tools will often be further used to show which method of wastewater management has the greatest public support or public health risks.

Consequently, due to whole system support tools, there is a greater emphasis on wastewater composition, especially the reuse of non-renewable components and hazardous chemicals. Hazardous chemicals are slowly being phased out of the wastewater industry by priority listing and reporting. Those chemicals shown to have a significant environment consequence and without adequate treatment pathways may be removed from production. The non-renewable components, such as phosphorus (P), are important aspects of wastewater management that are currently subject to intense research. Obviously, reuse of non-renewable components is pertinent to their long-term sustainability, especially where they offer environmental and human health implications when disposed of improperly e.g. eutrophication from ortho-phosphate.

9.1.2 Drivers for technology innovation

The drivers for technology innovation have followed the development of sustainability within the wastewater sector. Traditionally, the main driver for innovation within the wastewater sector was pubic health. This traditional driver caused the development of sewerage infrastructure to remove wastewater away from human contact. Environmental health became a part of this driver once the world population and subsequent wastewater discharge increased, leading to an increased chance of contact between the public and the discharged wastewater and/or subsequent degraded and contaminated environments. Treatment systems were developed once it was discovered that the natural environment could not assimilate high loadings of wastewater and early treatment works attempted to reduce the load of harmful contaminants e.g. pathogens, suspended solids (SS) and biochemical oxygen demand (BOD). Once treatment technologies were starting to advance, there were further additions to environmental health criteria that included nutrients, heavy metals, organic compounds and specific indicators of pathogens. These criteria were treated to an appropriate standard depending on the classification and/or assimilation capacity of the receiving environment. Currently, due to the improvement in treatment technologies, there is no limit on the treatment provided as technology can treat wastewater to primary, secondary and tertiary levels. Much of the emphasis currently is placed on technological cost (e.g. capital and maintenance costs) and the environmental benefits the technology offers e.g. reuse of wastewater. Other influences on current technology innovation, such as footprint size, capacity to treat highly toxic or hazardous waste, and required buffer distances, are important in some site-specific sanitation projects.

The second factor that has been a driver for innovation is the cost of a treatment technology. Early deep sewerage was borne at high costs, but was developed due to public health risks from local sewage discharge. Hence, many developed nations paid exorbitant costs for their sewerage infrastructure. At present, governments are pursuing other means to provide cheaper alternatives, as many infrastructure networks are starting to reach their lifespan and may be beyond regional government capacity for replacement. Feinbaum (2001) and Dix (2001) both noted that in the US the use of systems that can be developed without deep sewerage are being pursued, due to the high cost of infrastructure development and maintenance. Cheaper methods of sewerage maintenance are also becoming pertinent, especially where sewerage infrastructure is required in larger cities due to the limited space and design of the city.

In terms of sustainability, the cost involved in sanitation may not only be represented by capital cost, but also maintenance and ongoing operation costs. Energy consumption is a major aspect of these costs and can include wastewater pumping in deep sewerage infrastructure and plant operational aspects such as aeration of the wastewater. The cost of personnel required to operate the system is included in operational costs. Additional costs to a system, such as upgrade costs, often occur when the standard level of treatment within an area is increased, through legislative requirements. In some developing nations ongoing, maintenance and upgrade costs are often unsustainable in the long term and need to be factored into the design of a wastewater system (Ali, 2002). In most developed nations these costs are subsidised by a central body (e.g.

government) to ensure sanitation is suitable for the population (Hoehn and Krieger, 2000).

The third driver for innovation is the environmental benefits, in terms of greenhouse gas emissions, ecological impacts (as opposed to environmental impacts leading to human health risks), and reuse potential. Many wastewater standards are beginning to implement values for ecological conservation, rather than specific public health risks. The level where ecological health is affected is often much lower than would be expected to protect public health (Ying *et al.*, 2004), which has brought about greater levels of treatment than would have been previously expected for public health protection. Thus, countries or regions with an ecological focus tend to implement more advanced treatment processes. The impacts on the wastewater process on other environmental aspects, such as the emissions of greenhouse gases, is starting to become assessed and will be part of the design of wastewater plants in the future. The other factor important to this driver is the reuse of wastewater, as nutrients and water offer numerous benefits. Discharge in many examples, does not offer a long term option and can be seen as wasting resources, where available technology can reuse the wastewater more effectively. Reuse offers a more appropriate method for pollution control and can be used as a treatment process, thus reducing the cost of wastewater treatment (Lange and Otterpohl, 1997).

In terms of wastewater system choice, it is apparent that many of the drivers are interlinked. The choice of the system may further be reliant on the technology availability within an area, the capacity of the developers to design an appropriate system, and the local government policy (Feinbaum, 2001; Ho and Anda, 2004). In many parts of the world unsustainable systems have been maintained due to inappropriate technology choice. In some developing nations, it may not be possible to utilise the most environmentally beneficial system, due to the lack of available funds and operational expertise. Therefore, an approach that provides the best available technology based on available resources can be applied and gradually upgraded as expertise and money become available (UNEP, 2002; UNEP, 2004).

9.2 CURRENT ADVANCES AND INNOVATIONS

The use of innovative technological solutions for many of the World's wastewater and water problems has been well documented. The original 19th century solutions and those that preceded them have all been innovative at their time of establishment, and were largely based on the knowledge of sustainable practice at the time of their development. Due to our increased knowledge on sustainability and greater enforcement of wastewater systems however, the drivers that are important for technological development are more complex currently than when the early wastewater systems were developed.

The use of innovative technologies is being pursued across the world to provide some of the framework for sustainability that may not be exhibited by older technologies. An example of this is the change in focus from centralised technologies to decentralised technologies. In terms of sustainability, decentralised systems offer the major advantage of reuse potential, because they do not concentrate wastewater in one location and can be placed to divert a particular stream of wastewater to an area where reuse potential is high. The concentration of wastewater in one location using centralised systems reduces the capacity to reuse the wastewater once treated, due to inadequate land space e.g. open public space or agricultural land (UNEP, 2002).

The cost of smaller scale technologies is currently another factor influencing their development within the current marketplace. This has occurred for a number of reasons. One reason is due to the high cost of infrastructure maintenance (which is largely subsidised by governments) and another is due to their greater ability to be implemented with limited resources and become upgraded when more resources are available. In particular, there is currently a focus on these types of technologies for in house reuse or source separation of a particular waste stream. These types of technologies might include advanced onsite wastewater treatment systems, urine and greywater separation and technologies and appropriate reuse systems. Many of these technologies have been devised for both developing and develop nations.

The development of new frameworks and approaches toward sanitation has occurred over the last few decades. These approaches are an important part of new technological developments and can work as tools to ensure the technology within an area is appropriate for its intended use. The best approaches towards sustainable development involve the use of capacity building and education of the community on the importance of sanitation. The development of simplified technologies that provide adequate sanitation and the features required by the local population are an important step in this process. The ownership of the sanitation problem and technological systems to overcome this problem is another step, which ensures their maintenance and management is implemented over a larger time frame than the initial development.

9.2.1 Innovations in onsite systems

The innovations in onsite systems include high tech systems, improvements to existing traditional technologies and source separation techniques. Many of the high tech treatment systems such as aerobic treatment units (ATUs) are detailed under section 10.2.3, Innovations in treatment systems. Many of the innovations have been developed for two principal factors, which are cost-effectiveness and reuse of specific waste stream streams.

The improvement to traditional onsite waterless technologies includes the use of the ventilated improved pit latrine (VIP) and compost toilets. Both the VIP and compost toilet technologies are detailed in Chapter 6. The VIP has the major advantage of permanency compared to traditional pit latrines, which may not be used continually in the one place. Compost toilets have been a later advent than pit latrines and in terms of sustainability offer many advantages. The use of compost toilets to provide a readily viable resource without faecal sludge production has been shown to have two advantages. Firstly, the sludge is safer for handling after composting has taken place and secondly the system has removed the possibility of ground-water pollution because it reuses the nutrients rather than disposes of them. Further advances to these compost toilets include the wet compost designs produced during the late 1990s that work for whole domestic wastewater flows. The current model produced from these early wet-compost systems is detailed below e.g. the Biolytix Filter. Many of the latter wet compost toilets rely upon worms and other soil microflora and are more appropriately termed 'vermifiltration technologies'.

Source separation of wastewater fractions from households has been shown to be an important innovation in decentralised and onsite system design (Jonsson, 2002; Maurer et al., 2003). Examples include the separation of urine (yellow water) and greywater. Both greywater and urine are relatively safe compared to blackwater, which contains most of the harmful pathogens. Greywater is often separated without much treatment. Traditionally greywater may pass through a small mesh grill or sedimentation tank to remove particulate matter, before reuse in trenches. New approaches use reed beds or small filters (e.g. sponge filters) so the greywater can be reused through subsurface irrigation. Reed bed designs may be developed using alternative resources, such as shredded plastic bottles, that ensure greater root development in plants (Dallas and Ho, 2004).

Greywater is separated with plumbing diversion to the treatment system from the laundry and bathroom. Commonly kitchen greywater may only be used where it can be treated through a biological process, as it contains a higher BOD and nutrient concentration when compared to the other household greywater (UNEP, 2002).

Urine is commonly separated using a urine-separating toilet. These toilets contain holes within the front portion to allow urine flows to be diverted separate from faecal wastes (Figure 9.1). Urine is high in many nutrients and minerals, including at least 50% of the nitrogen (N), P and potassium in the domestic wastewater stream (Jonsson, 2002; Maurer et al., 2003). Reuse of urine, due to its low levels of solids can take place through an appropriate treatment system, although depending on its constitution treatment may not be required. Irrigation of gardens with urine is an excellent example of resource recovery and can close the nutrient loop, thus reducing the need for fertilisers. Maurer *et al.* (2003) found when assessing the most sustainable options for urine reuse that reuse onsite is the most sustainable option and the resource recovery efficiency decreases as it is reused further away from the source e.g. household.

The reuse of urine onsite is obviously the most beneficial option, but city design is often not inclined towards this approach. In such cases, urine separation can be used for the production of fertilisers that can be applied to agricultural lands. The urine may be stored in appropriate holding tanks before it is removed for fertiliser production. This may reduce over 50% of the nutrients flowing to municipal treatment plants and ensures that the influent has a better ratio of ingredients for typical biological removal of nutrients (Matsui *et al.*, 2001).

Increasing the sustainability of current approaches may have to be used until cities become more sustainable in the long-term. Ho and Anda (2004) pointed out that cities can become sustainable in the long-term using among other things, planning that is designed for reuse of wastewater fractions as close to the source as possible for food production.

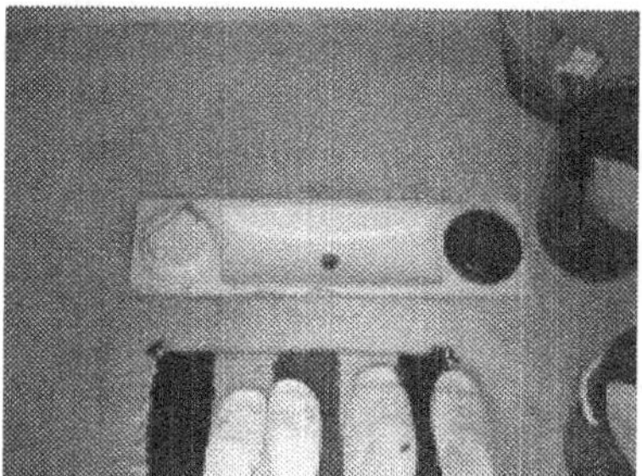

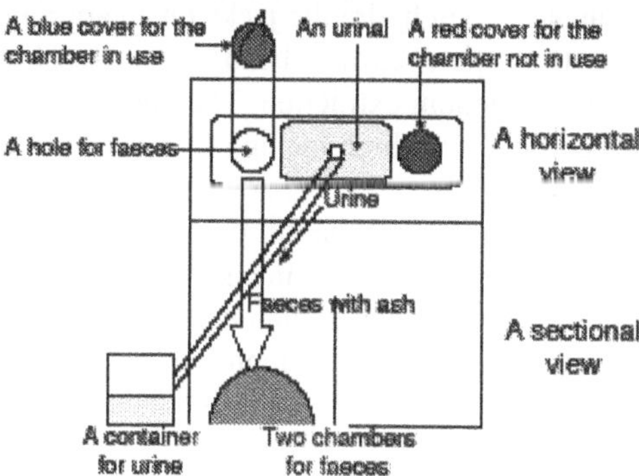

Figure 1. Urine separation toilet (From Harada et al. 2006)

9.2.2. Innovations in sewerage systems

Innovative sewerage systems have been designed to provide adequate sanitation services in specific situations that require new approaches to overcome economic, environmental and social problems. Examples include the use of trenchless technology and simplified sewerage (UNEP, 2002). Many of these innovative approaches have been developed for use in other sectors also, such as the use of trenchless technology in telecommunication and energy utility development.

Trenchless technology is a method to reduce many of the problems with traditional sewerage development, maintenance and upgrade. Previously, open trenches were the preferred method of developing underground infrastructure. Open trenches are labour intensive and provide many constraints to the surrounding environment, such as disruption to localised roads, business and living space. Such problems are exacerbated in high-density urban centres. Trenchless technology as the name implies, does not involve open trenches but is an underground method for the development of pipework infrastructure. This type of technology has been commonly used across the developed world over the last decade. The main constraint on this type of technology in developing countries is the lower price of labour, to dig trenches, thus reducing its feasibility. This technology however is becoming more common and countries with growing economies such as China have been using it in high-density urban areas for a number of years, because it causes no social disruption during development.

The main type of trenchless technology used for sewerage development is microtunneling. Microtunnelling is a method of developing sewerage pipework using a remotely controlled excavation system. It typically involves pipelines of 1000mm to 100mm and the length of tunnel from the launch pit (underground pit) to the reception pit (underground pit) is usually less than 200m. It involves the movement of an excavator head from the launch pit to the reception pit and pipework is placed behind the excavator using hydraulic jacks on an appropriately aligned jacking frame. Spoil media from the excavator head is removed from the laid pipe by auger, slurry conversion or vacuum extraction. The pipe sections used for microtunneling are designed specifically for high jacking forces. The accuracy of these systems is commonly less than 20mm.

The other innovative types of sewerage are based on small scale trench technology. These include advances in simplified and settled sewerage design. Both types can be used where space is available for trenches, although settled sewerage is more applicable to peri-urban and rural areas, whereas simplified sewerage is more suited to urban areas. The use of simplified sewerage has allowed many innovations to traditional sewerage infrastructure. This types of sewerage increases sewerage infrastructure building capacity through the use of small diameter piping and low depth trenches, and increases community participation because it can be developed using hand tools (Mara, 2001). The design allows the piping to be laid by the inhabitants of an area, and requires a low level of expertise (UNEP, 2002). For more information on these sewerage systems refer to Chapter 6.

9.2.3 Innovations in treatment systems

The innovations to treatment systems includes advances in small systems that incorporate conventional centralised processes, onsite systems that rely on increased treatment through soil processes and filter bed media, and innovation that provides new processes undiscovered in conventional treatment. The innovation in traditional onsite systems such as septic tanks has also taken place with the use of baffled designs and the use of pre-trench filters that provide greater solid removal than traditional designs.

The first type of system for small scale treatment has incorporated similar processes to those found in conventional activated sludge treatment. The technology may involve chambers for; sedimentation and separation of solids; aeration and subsequent nitrification; clarification and particulate settlement; chlorination; and pump out. The common types of these technologies are the aerated treatment units (ATUs) used throughout many peri-urban and urban areas of the world. Some of these systems treat wastewater to secondary treated wastewater standards, although the wastewater is typically reused for irrigation of private or public open spaces.

The soil treatment field has been the traditional approach to onsite wastewater disposal. The method employed has been to dispose of effluent after primary treatment through a septic tank. Innovations of disposal fields include the use of amended soil filters, which utilise material with a high nutrient adsorption (e.g. P retention index >20) and infiltration capacity. These systems allow greater percolation of the primary treated wastewater and greater adsorption of wastewater contaminants. The amended soil filters remove pathogens to high standards e.g. tertiary. The systems are employed where the primary soil is inappropriate for treatment or the space required unavailable, so the system can be used to allow more advanced soil treatment. The amendments typically consist of red mud (e.g. bauxite refining residue), bentonite and zeolite. These systems may have gypsum added to reduce problems of sodicity.

The use of peat bed filters in many disposal designs is becoming common in the US. These filters utilise approximately 0.5 to 1m of peat before the soil, and may involve a gravity fed or a modular pressurised system. Modular systems may re-circulate the wastewater to an anaerobic containment vessel to enhance nutrient removal. Peat bed filters rely on treatment processes similar to amended soil filters for adsorption of P, but may also involve more complex biochemical reactions, such as nitrification via specific fungi. Denitrification in these beds may occur in the lower layers, where anaerobic conditions and sources of carbon (e.g. peat) are present.

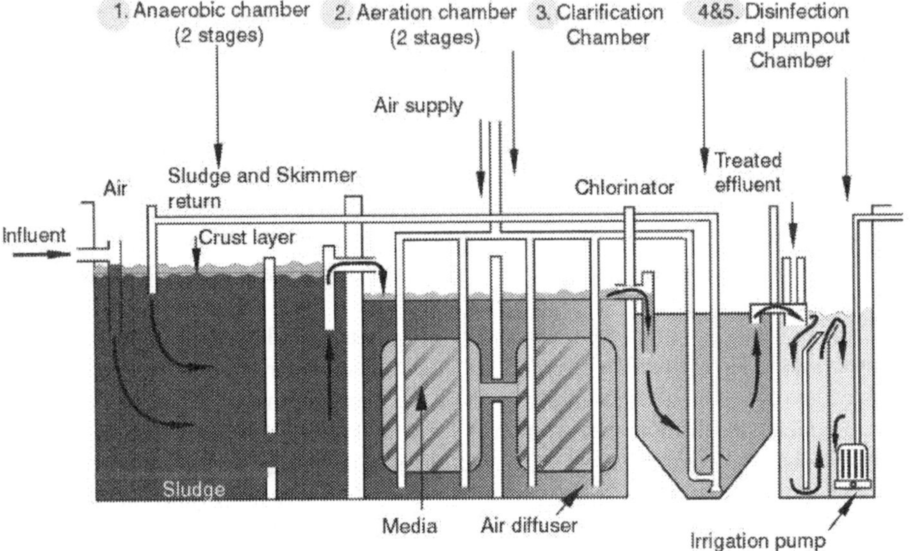

Figure 9.2 The treatment stages of a common ATU system showing chambers for each process. (Biomax, from UNEP 2002).

The final type of treatment process used in small-scale systems is a new innovation. The process is termed vermifiltration, but involves a community of organisms, including mites, worms, larvae, beetles and other unidentified organisms. The design of these systems involves a number of layers that capture faecal solids, which are broken down by the organisms to form a layer of humus. The humus is aerobic in nature and acts as a filter that removes nutrients and organic solids from the wastewater. The primary system involving this technology is the Biolytix Filter that produces wastewater of a high secondary standard. These types of systems are robust in design and can work on minimal energy inputs, as they largely rely on the organisms to break down the faecal matter into humus (Foley *et al.*, 2004).

9.2.4 Innovations in reuse systems

Innovation in reuse systems has been made necessary by the development of new treatment technologies and the importance of wastewater reuse for sustainability. The innovative reuse systems can offer a method to convey wastewater that is not suitable for direct human contact but is acceptable for reuse in environments, such as soils, that act as a further treatment process. Along with innovative reuse systems, standards have also changed to allow the use of innovation with wastewater reuse (Dix, 2001).

The innovative reuse technologies include advances in surface and sub surface irrigation drip lines. Drip lines technology is important due to the water efficiency it exhibits when compared to surface irrigation. Drip line technologies come in numerous sizes and water flow rates and allow even distribution of treated wastewater compared to subsurface disposal trenches. The smaller technologies termed 'microdriplines' (e.g. 5mm diameter at $1Lmin^{-1}$ drip emitters) can be used with onsite technologies e.g. ATUs. The microdrip systems, due to the low water output increase plant uptake and can be suitable for soils with low permeability e.g. clay soils (Dix, 2001). Larger scale drip lines may be designed with the use of geotextile coverings that negate the need for root intrusion herbicides and ensure uniform distribution. These systems have greater flow rates than the microdriplines and can be used to irrigate larger areas.

Along with advances in irrigation technologies, innovation has come in the form of greater timing control and water efficiency through control modules and soil moisture sensors. The design of reuse systems has advanced with greater use of irrigation filters, to remove organics and iron from drip line systems. The advances in design have often been developed due to system failures in traditional systems. An example is the use of alternating reuse fields (commonly 2) to allow the soil to recover after it has treated a specific quantity of wastewater.

9.3 RESEARCH NEEDS

The research needs for innovation and sustainability are diverse in nature. The range of research needs may include developing information for site-specific situations or the development of a treatment process for a particular waste stream. This may include gathering information directly in regions where it is to be applied, or regions with similar bioregional characteristics e.g. climate. In most cases, there is a need to develop new alternatives to technologies that are currently available to ensure they are cost effective to the local community.

The development of integrated designs involving source separation techniques will become part of the future where technology and design are used to ensure greater resource recovery. In these cases, there is a need to develop new management procedures and environmental health guidelines, because traditional methods may be inappropriate.

The development of new approaches to community development, participation and ownership of sanitation projects has been researched extensively. However, this area needs to continue with the development of new technologies, which offer many new approaches to sustainable development e.g. reuse of resources. The knowledge of the wastewater sector on new and emerging decentralised and source separation technologies is a pertinent part of this process.

9.3.1 Technology

In developing nations, the technologies chosen to provide sanitation can be simplistic in nature. Simplistic technologies may be widely implemented due to their cost-effectiveness without adequate research, which may cause many problems especially when they are transferred to different regions of the world. An example of this is the changes in compost toilet functioning between Australia and South Africa. In most regions of Australia, compost toilets readily break down faecal matter into pathogen free compost, whereas in South Africa compost toilets may dehydrate the faecal waste, which may require additional treatment before reuse. Hence, appropriate research of a technology should be carried out before it is widely implemented.

The increased use of high tech decentralised systems in the future will bring about a focus of research in this area, as these systems will be used in future urban developments across the world. There is less information on these systems when compared to conventional centralised technologies. The research on decentralised systems in the developing world may need to be focused on areas that are known to cause potential problems. These potential problems may include:

- changes in long term performance,
- changes in performance under different climates (e.g. many technologies are developed in colder climates countries and may fail in subtropical and tropical regions)
- and operational problems due to different influent characteristics and the availability of power.

Possible examples include problems with ATUs due to their reliance on electricity and the changes in humus production and subsequent wastewater treatment from key vermifiltration species in different regions.

The development of new robust decentralised systems may be required where high tech decentralised systems are inadequate for an area. These types of systems may be developed based on ATUs or different processes, such as occur in vermifiltration. The systems may be devised for a particular reuse situation, such as for an aquaculture system, where the level of treated wastewater matches the desired loading of the aquaculture system. The development of a small scale cost effective tertiary treatment system is a future step for onsite reuse of wastewater, although appropriate design and integration of technologies (e.g. in house source separation of waste streams) may be the present aim for many wastewater engineers, as they are currently more cost effective approaches.

The research on in house technologies may be focused in a number of areas. The first area is the development of guidelines and standards for the reuse of greywater and urine in developing nations, which are based on research into their potential human health and environmental risks. The second area is the development of appropriate technologies to reach these standard levels of treatment. The technologies may focus on particular in house processes such as the recycling of water from the bathroom to the laundry or toilet. The technologies may also be adapted to particular conventional technologies within a region, such as the development of urine and greywater separating technologies, with traditional soil based disposal systems, thus allowing the reuse of the nutrients and water and reducing the pressure on the conventional disposal system.

9.3.2 Technology management

The management of technology, especially for small systems, is inappropriate in many regions of the world. In developed nations, onsite systems such as septic tanks have been found to fail after a short time period and they can lead to significant environmental problems. For example, in the US, one of the leading sources of groundwater pollution are septic tanks, which have been shown to pollute water supplies with hazardous chemicals, pathogens and nutrients. Many of these small systems fail because the responsibility of management is placed on the owner of the system (Feinbaum, 2001). Ho and Anda (2004) noted from experience in implementing best management practices, bad management in decentralised systems is due to a number of additional factors including:

- A lack of government policy and regulatory framework
- Inappropriate accreditation framework, that is inhibitory to decentralised systems
- A lack of demonstration projects, that exhibit best management practices
- Low knowledge of regulators, developers, consultants and service providers on appropriate methods to manage decentralised systems

Many of the factors are linked due to the importance placed on centralised wastewater systems in the developed world. However, for sustainability, decentralised systems need to be managed better and their importance is gradually increasing. This is especially the case in the developing world where most of the technologies chosen for future sanitation are likely to be decentralised in nature e.g. treating wastewater from up to 5000 people (Randall, 2003; Wilderer and Schreff, 2000).

Currently some countries that are starting to rely on decentralised technologies (e.g. US) are providing some impetus for better management practices (Dix, 2001; Feinbaum, 2001). The management frameworks may include models based on technological design and environmental sensitivity (Feinbaum, 2001). The required treatment system and maintenance schedule may be based on the receiving environment. Feinbaum (2001) explained that this may include; contracts of maintenance (legally binding) and greater treatment in sensitive environments where pollution of valuable water supplies is imminent; low tech systems and quarterly maintenance in areas of low environmental importance; and intermediate technologies and quarterly maintenance where the receiving environment is moderately sensitive.

The certification of maintenance personnel is another factor in the management of decentralised systems, which is inadequate across the world. Accreditation allows specific levels of maintenance to be devised and will ensure that all systems within an area are provided with appropriate maintenance. A recognised institution commonly undertakes the certification of maintenance personnel. The maintenance procedures for decentralised systems are often devised by the developers of a system and are checked by the regulatory authority before accreditation of that system. Most maintenance procedures are system specific but the accreditation of maintenance personnel may follow an accredited course that is generic in content. Hence, many personnel may require regular training on new treatment technologies and processes.

Developing nations need to develop their own management frameworks which may be based on those developed across the world but should also involve finding management approaches that work based on available funding, expertise and technologies. Education of the wastewater industry and the public on small systems is an important part of this framework, as it will enhance standards of management.

9.3.3 Environmental health

The research needs for environmental health include information on different receiving environments for the development of standards. The standards may include values for specific reuse applications and wastes streams as well as values for localised problems such as diseases and contaminants. Additionally the standards maybe developed using information on cost effective environmental monitoring methods as well as indicators of specific diseases, which require less analysis and can be carried out by the local population.

Many standards, such as WHO (1989) reuse standard, rely on accurate identification of contaminated waters. The use of laborious techniques can often lead to inadequate monitoring, especially where the cost is high and education and equipment is unavailable. In these cases, potential human health risks occur, because without adequate monitoring it is likely that site-specific problems may cause changes in treated effluent quality. Hence, there is a requirement for cost effective monitoring techniques for environmental health.

Simple reaction-based presence tests may provide monitoring techniques that can be carried out by localised populations. An example is the hydrogen sulfide (H_2S) water testing technique that identifies the presence of harmful bacteria by their reaction with hydrogen sulfide, which turns the hydrogen sulfide water brown to black after incubation in yogurt culture containers. The darker the water colour, the higher the number of bacteria colonies present. Presence-base reaction tests are available for other wastewater contaminants, including nutrients. These types of tests may be standardised in an area and once standardised disseminated throughout the local community.

The final and most important factor in environmental health monitoring is education on the consequences of using poorly treated effluent. Capacity building along with community participation in standard preparation may ensure that environmental monitoring is viewed as an important factor in sustainable livelihoods and can lead to benefits such as increased food production and increased health.

REFERENCES

Ali, A. (2002) Operational Problems of Waste Water Treatment Plants in Developing World. *WAPDEC: Water and Wastewater Perspectives of Developing Countries,* New Delhi, India, pp. 933-937.

Dallas, S., and Ho, G. (2004) Performance of subsurface reedbeds for the treatment of domestic greywater. *6th Specialist Conference on Small Water and Wastewater Systems, 1st International Conference on Onsite Wastewater Treatment and Recycling,* Perth, Australia, p. 71.

Dix, S. P. (2001) Onsite wastewater treatment: a technological and management revolution: Part 2. *Water Engineering and Management* **148**(10), 22.

Fane, S. A., Asholt, N. J., and White, S. B. (2002) Decentralised urban water reuse: The implications of system scale for cost and pathogen risk. *Water Science and Technology* **46**(6-7), 281-288.

Feinbaum, R. (2001) New outlook for decentralised wastewater treatment. *BioCycle* **42**(5), 36-40.

Foley, J., Kasper, T., and Cameron, D. (2004) Biowater [TM] decentralised wastewater management for a small community in the Southern Moreton Bay Islands." *6th Specialist Conference on Small Water and Wastewater Systems; 1st International Conference on Onsite Wastewater Treatment and Recycling,* Perth, Australia, p. 57.

Harada, H., Matsui, S., Phi, D.T., Shimizu, Y., Matsuda, T. and Utsumi, H. (2006). Keys for successful introduction of ecosan toilets: experiences from an ecosan project in Vietnam. Paper presented at 7th Specialized Conference on Small Water and Wastewater Systems, 7-10 March, Mexico City.

Ho, G., and Anda, M. (2004) Centralised versus decentralised wastewater systems in an urban context: the sustainability dimension. *Leading Edge Conference on Sustainability in Water-Limited Environments*, Sydney, Australia.

Hoehn, J. P., and Krieger, D. J. (2000) An economic analysis of water and wastewater investments in Cairo, Egypt. *Evaluation Review* **24**(6), 579-608.

Jonsson, H. (2002). Urine separating sewage system - environmental effects and resource usage. *Water Science and Technology* **46**(6-7), 333-340.

Lange, J., and Otterpohl, R. (1997) *Oekologie Aktuell ABWASSER Handbuch zu einer zukunftsfaehigen Wasserwirtchaft.* MALLBETON GmbH, Donaueschingen-Pfohren.

Mara, D. D. (2001) Appropriate wastewater collection, treatment and reuse in developing countries. *Proceedings of the Institution of Civil Engineering-Municipal Engineer* **145**(4), 299-303.

Matsui, S., Henze, M., Ho, G., and Otterpohl, R. (2001) Emerging Paradigms in Water Supply and Sanitation. In: *Frontiers in Urban Water Management: Deadlock or Hope,* Maksimovic,C. and Tjada-Guibert, J.A. (eds). IWA Publishing, London, England, pp. 229-263.

Maurer, M., Schwegler, P., and Larsen, T. A. (2003) Nutrients in urine: energetic aspects of removal and recovery. *Water Science and Technology* **48**(1), 37-46.

Patterson, R. A. (1997) Household chemical impact on effluent re-use. *waterTECH conference convention centre*, Brisbane.

Randall, C. W. (2003) Changing needs for appropriate excreta disposal and small wastewater treatment methodologies or the future of small wastewater treatment systems *Water Science and Technology* **49**(11-12), 1-6.

UNEP. (2002) *International Source Book on Environmentally Sound Technologies for Wastewater and Stormwater Management.* United Nations Environment Programme - International Environmental Technology Centre, Osaka and International Water Association Publishing, London.

UNEP. (2004) *Guidelines on municipal wastes management.* Hague, Netherlands.

Wilderer, P. A., and Schreff, D. (2000) Decentralised and centralised wastewater management: a challenge for technology developers. *Water Science and Technology* **41**(1), 1-8.

Ying, G.-G., Kookana, R., and Waite, T. D. (2004) *Endocrine disrupting chemicals (EDCs) and pharmaceuticals and personal care products (PPCPs) in reclaimed water in Australia.* CSIRO Land and Water, Adelaide, Australia.

10
Sludge treatment and management

Blanca Jiménez and Lin Wang

10.1 INTRODUCTION

All over the world wastewater treatment ends with two products: treated water and a slurry that is often considered a byproduct. This slurry is a semisolid waste named "sludge" and contains all of the compounds removed from wastewater as well as those added during treatment. Sludges from developing countries differ a lot from the sludges generated in developed ones, due to divergent industrialization and public health levels. In developing countries, the metal and toxic content is much lower, while the microbiological content is much higher. Another big difference has to do with quantity. In low-income countries, sewerage coverage and wastewater treatment are so low that only a small amount of sludge is produced. Also, sediments that come from pluvial erosion owing to unplanned urbanization in cities are found in sludges and are conveyed to wastewater treatment plants.

In spite of their quality, in many cases sludges are pouring into sewers or are simply being discharged, without any treatment, into soils or water bodies. In some cases, they are sent to lagoons, landfills or non-controlled discharged sites. This is due not only to a lack of economic resources, but also to the lack of an appropriate legislation. In developing new legislation and sludge management programs, worldwide reuse and minimization are the key issues that should be considered. As due to economic and social factors there are increasingly fewer sites for new landfills to confine solid waste. Although incineration is another disposal option, developing countries cannot afford it.

Another factor to take into account when considering sludge disposal is land degradation and the need to increase food production in the developing world. On the one hand the poorest countries cannot afford fertilizers, while on the other, land degradation costs between 5 to 10% of the agricultural production and 5 to 6 million hectares of arable land are being lost through soil degradation (WHO-UNICEF, 2000). Thus, beneficial disposal of sludges in soils can be an interesting option. In this chapter, treatment stabilization techniques and guidelines for sludge minimization, disposal in soils and landfill are discussed. Also, general criteria for developing the appropriate legislation are presented as well as some examples of sludge management in different countries.

10.2 CHARACTERIZATION

10.2.1 Types of sludges

Sludges can be classified according to their origin: (a) drinking water sludges, (b) sludges or sediments extracted from cleaned sewers, and (c) sludges from wastewater treatment processes. These classifications can be broken down into the following sub-classifications: industrial, sewage and municipal sludges according to the type of treated wastewater. In general, sludges in developing countries are more of the municipal kind, as sewerage systems combine domestic, industrial and rainwater. Drinking water sludges are often discharged into sewers. And, if a treatment method is required, lime stabilization and dewatering are excellent options prior to dumping. But, the best option if economical is to recover alum to recycle it.

10.2.2 Sludge production

Due to insufficient sanitation system coverage and a lack of sewage treatment (30% in Asia, 15% in Latin-American and Caribbean and near 0% in Africa, WHO-UNICEF, 2000) sludge production is very low, and in very few cases is production data reported. As long as the level of sanitation will improve, sludge production can be expected to increase, and hopefully, its quantification, characterization and treatment will also.

Sludge production depends on the type of treatment applied (Table 10.1). Primary and physicochemical treatments produce greater quantities than biological ones, because the later mineralize organic matter. Generation also depends on other items, such as wastewater composition, rainwater composition, solids conveyed by pluvial erosion and operating conditions.

Table 10.1 Sludge generation by different wastewater treatment processes and solids content. (Metcalf and Eddy, 1991 and Jimenez and Chavez, 1997)

Process	Sludge generation, kg ST/10^3 m^3		% Total solids (TS)
	Range	Typical	
Primary sedimentation	108 – 168	150	4.0 – 10.0
Advanced Primary treatment	185 – 315	*	0.4 – 10.8
Activated sludges	72-96	84	0.5 – 1.5
Trickling filter	60 – 96	72	1.0 – 3.0

* Depends on the TS content in water and the coagulant dose

10.2.3 Quality

Sludge composition determines the type of treatment required and defines disposal options. Sludge characteristics are classified as physical, chemical and biological.

10.2.3.1 Physical characteristics

The most important physical characteristic is the total solids content (TS), it indirectly measures dryness or water content. Solid content is the ratio between the solids' dry weight and the total weight of the sludge, as follows:

$$C = \frac{\text{g dry solids}}{\text{g sludges}}$$

(10.1)

This number multiplied by 100 is known as the "solids percentage" and is the most common way of expressing solids concentration. Generally, it varies from 0.25% to 12%, depending on the kind of sludge (Table 10.2), and the lower the solid content the higher the volume to be handled. The volatile fraction (TVS, determined at 550°C) represents the organic matter in sludges that can be biodegradable, and has the potential to cause nuisance odors. Considering the TS content, sludges can be classified as liquid, paste-like, solid or dry (Table 10.3).

Another physical characteristic is the particle size distribution, which has to do with the tendency to loose water. It is difficult to remove water from particle sludges with sizes between 1 to 10 microns. Also, how water is bound to sludges is important to the method used in removing it. *Free* water can be eliminated by simple thickening, while capillary water requires mechanical methods, such as filtration or pressing. *Constitution* water is only removed by thermal methods.

Table 10.2 Typical solid concentration in sludge from different sewage treatment processes.(Metcalf and Eddy, 1991).

Operation or processes	TS concentration (% weight/weight)	
	Range	Typical
Primary sedimentation	4.0 a 10.00	5.00
Secondary sedimentation		
Primary sedimentation with activated sludge	0.5 a 1.5	0.8
Trickling filter sludge	1.0 a 3.0	1.5
Biological rotating contactor sludge	1.0 a 3.0	1.5
Gravity thickener		
Primary sludge	5.0 a 10.0	8.0
Mix of primary and activated sludge	2.0 a 8.0	4.0
Mix of primary and trickling filter sludge	4.0 a 9.0	5.0
Belt gravity thickener		
Chemical Sludge with chemical addition	3.0 a 6.0	5.0
Anaerobic digestion		
Primary Sludge	5.0 a 10.0	7.0
Primary sludge with activated Sludge	2.5 a 7.0	3.5

Table 10.3 Characteristic of sludges with different water content, OTV (1997).

Class	Liquid	Paste-like	Solid	Dry
% ST,	1 to 10	10 to 30	30 to 90	> 90
Physical characteristics	Non dewatered sludge	Dewatered sludge. Lost its form due to its own weight with a tendency to spread on surface	Piles with 1 m height and a 45° angle can be formed	Powder or granular
Transport	Centrifuge or volumetric pumping	Volumetric or screw pump	In vehicles and belt conveyers. Volumetric or screw pump.	Belt conveyers, pneumatic systems

10.2.3.2 Chemical

The chemical composition of sludges is not well known due to variation and a lack of research. It depends on the sludge's origin more that on the wastewater treatment process. With the exception of odor, organic and metal content cannot be economically modified. For reutilization the main chemical properties of sludges are organic matter, nitrogen, phosphorus content and, for land application, those that generate odor. Metals and toxic compound content limit both sludge valorization and disposal.

Compounds in sludges are presented as precipitates (sulfides, oxides or bicarbonates) or are adsorbed chelated with organic matter. Five classes of chemical compounds are distinguished: (a) metals and cyanides (b) volatile organic compounds, (c) semi-volatile organic compounds (d) pesticides and BPCs, and (e) others. In general, for developing countries toxic compounds are present in lower quantities than in developed countries unless sewers collect mainly specific industrial wastewater (such as wastewater from tanneries). Concerning heavy metals, the little literature available (from Brazil, Chile, China, Mexico and South Africa) is consistent in pointing out that concentrations are well below international standards (Andreoli *et al.*, 1999; Barrios *et al.*, 2001; Martín del Campo *et al.*, 2002; Cardoso and Ramirez, 2002; Texeira *et al.*, 2002; Leppe *et al.*, 2002; Mena, 2002; Smith and Vasiloudis, 1989, in Snyman *et al.*, 2000 ; Wang, 1997).

Calorific content depends on the presence of combustible fractions, such as greases and scums. For raw sludges it varies from 11.16 MJ/kg to 23.24 MJ/kg and in digested ones from 5.81 MJ/kg to 12.78 MJ/kg (Lue-Hing *et al.*, 1992). Municipal sludges are not hazardous wastes, but sometimes, when they are subjected to the hazardous test, positive results are obtained because sulfates, normally present in important quantities in sludges, are released as H_2S in great concentrations in the conditions established for the reactivity test. But this does not imply that sludges are really hazardous wastes.

10.2.3.3 Biological characteristics

Pathogen content in sludges is the main limitation. Pathogens are often spread through hand contact and thus can cause serious health problems. Pathogens belong to four groups of organisms: bacteria, viruses, protozoan and helminths. Table 10.4 shows pathogens that have been isolated from wastewater and that may consequently be present in sludges. The type and quantity of pathogens in sludges depends on the epidemiological conditions of the community that produces the wastewater. Table 10.5 shows the pathogen concentration in sludges of several countries.

In terms of both quantity and quality, biological data useful for developing countries is still very poor. Nevertheless, there is a big difference with concentrations found in the sludges of developed countries. Magnitude orders vary from 10^5 - 10^6 UFP/gTS to 10^7 to 10^{10} MPN/g TS for fecal coliforms, 10^3 *to* 10^7 MPN/g TS for *Salmonella Typhi*, 10^2 *to* 10^4 for Giardia lamblia cysts /g TS) and from 177 to < 1 viable helminth ova/ g TS. This means that there is an urgent need to gather more data and to develop efficient, economical and appropriate sludge treatment processes to deal with this problem. In doing this, certain characteristics of different types of organisms must be taken in to account.

Viruses. – It is impossible to measure all kinds of viruses that may be present in sludges; as with bacteria, an indicator group is often monitored. Bacteriophages are the most commonly selected group, and is supposed to be an indicator of enteroviruses that really constitute the group of health concern. The reason is that they are easier to detect. Anyway, very little data is available for developing countries in order to define which the convenient groups are.

Bacteria. - As in water, fecal coliforms are the indicator group. It has been extensively demonstrated that fecal coliforms are a good indicator of *Salmonella spp.*, a bacteria very often present in sludge in high concentrations. Due to the widespread use of the analytical technique for determining them, a lot of data is available; nevertheless, fecal coliforms are not useful indicators of other groups of organisms, such as viruses, protozoan and helminth ova. All these groups must be analyzed in separate tests.

Protozoan.- Some protozoan are very dangerous pathogens because they have the ability to produce very resistant structures (cysts) when they are exposed to adverse environmental conditions, such as temperature, pH, dryness, and low food concentrations. Protozoan recover their original state when conditions turn favorable. *Giardia, Cryptosporidium, Entamoeba* and *Paramecium* are some species of this group that cause dangerous gastrointestinal diseases. There is little available data about their content in the wastewater and sludge of developing countries, due to difficulty in detecting them. However, this is also true for developed countries, where they are not a main health concern. Cysts more often present in sludge are *Giardia sp.,* and *Entamoeba sp.* (2.6×10^2 - 3.8×10^3 No/g). There is not an indicator of this group, but it is generally accepted that they are less resistant than helminth ova to adverse environmental conditions.

Table 10.4 Pathogens found in wastewater and sludge (EPA, 1994 and Jimenez, 2003).

Organism	Species	Disease
Bacteria	*Salmonella sp.*	Salmonellosis and typhoid
	Shigella sp.	Dysenteric disease
	Yersinia sp.	Gastroenteritis
	Vibrio cholerae	Cholera
	Campylobacter jejuni	Gastroenteritis
	Clostridium sp.	Gangrene, tetanus, botulism.
	Mycobacterium sp.	Tuberculosis
	Escherichia coli	Gastroenteritis
Viruses	Hepatitis A	Infective hepatitis
	Rotaviruses	Gastroenteritis with acute diarrhea
	Reoviruses	Respiratory infections, gastroenteritis
	Adenoviruses	Eye infections and respiratory problems
	Polioviruses	Poliomyelitis, meningitis, paralyses.
Protozoa	*Cryptosporidium*	Gastroenteritis
	Entamoeba histolytica	Acute enteritis
	Giardia lamblia	Giardiasis (diarrhea and weight loss)
	Balantidium coli	Diarrhea and dysenteria
	Toxoplasma gondii	Toxoplasmosis
Helminths	*Ascaris lumbricoides*	Digestive and nutritional disorders
	Anchylostoma duodenale	Anchylostomiasis
	Necator americanus	Anchylostomiasis
	Enterobius vermicularis	Enterobiasis
	Ascaris suum	Coughs, pain chest and fever
	Trichuris trichiura	Abdominal pain, diarrhea, and anemia.
	Toxocara canis	Fever, abdominal pains and neurological symptoms
	Taenia saginata	Digestive and nervous disorders, insomnia, anorexia
	Taenia solium	Taeniasis
	Hymenolepis nana	Taeniasis
	Fasciola hepatica	Fasciolasis

Table 10.5 Organism content in the sludge of different countries (All concentrations are in log/gram of total solids, but helminth ova is in an ova/gram of TS)

Country	Fecal Coliforms	*Salmonella*	*Pseudomona aeruginosa*	Bacteriophages	Protozoan cysts	Helminth Ova	References
Germany						<1.1	Schuh *et al.*, 1985
Australia		2-3					Sidhu *et al.*, 2001
Brazil	5			<1 - 3	1-3	75	Thomaz-Soccol *et al.*, 2000.
Chile		3.5		2.7			Castillo *et al.*, 2002
Egypt					Mean: 1.4, Max 2.6	Mean: 67; Max 735	Hall, 2000
Mexico	10	7-8	5 - 7	3 - 6	2 - 4 [G]	73 - 177 [V]	Barrios *et al.*, 2001; Jiménez *et al.*, 2002.
France					3	4.4 - 7.7	Stien, 1989; Gaspard *et al.*, 1997.
Ghana						76	Hall, 2000
Japan	5	1		3 - 4			Hays, 1977; Watanabe *et al.*, 1997
Great Britain	4 - 6	2 - 4	3 - 5			< 6	Crewe, 1984; Michel and Rooksby, 2000.
United States	7	2	3	4-6 [E]	2 [G]	2 - 13	Reimers *et al.*, 2001

(E): Enteric viruses (G): *Giardia* (V): Viable ova

Helminth Ova.- Helminth ova are a special challenge in the field of sludge treatment processes, due to their very high resistance to extreme conditions. This is particularly true for the genus *Ascaris* (Carrington and Harman, 1984), which is the most frequent one found in sludges (Table 10.6) (Gaspard *et al.*, 1997 and Hays, 1997). Helminth ova (Figure 10.1) can infect animals and humans if consumed. There are a lot more genera than *Ascaris*; for instance, *Trichuris, Hymenolepis, Toxocara, Taenia and Trychozomoides.* Compared to bacteria, infectious doses are very low. They vary from 1 to 10 ova, according to the genus.

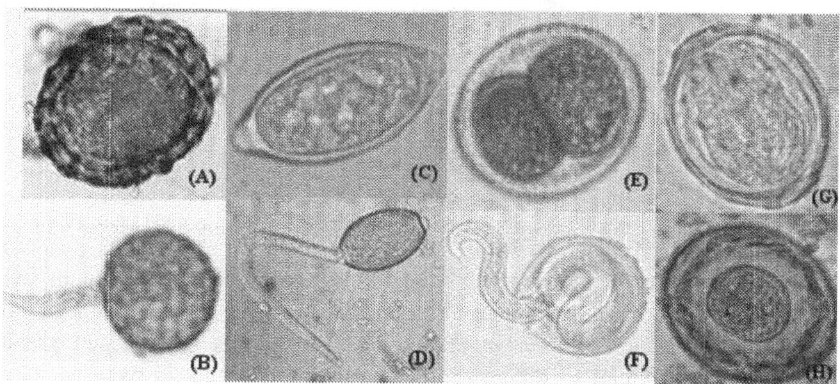

Figure 10.1 Ova or *Ascaris sp.* (A and B), *Trichuris sp.*, (C and D), *Toxocara sp.*, (E and F), *Trichosomoides sp.* (G) and *Hymenolepis diminuta* without larvae (H). Courtesy of the: Treatment and Reuse Group, Engineering Institute, UNAM

Table 10.6 Helminth Genus distribution in physiochemical sludges (Jimenez *et al.*, 2000 and 2002).

Genus	Percentage of the total
Ascaris	90.6
Trichuris	3.8
Hymenolepis	3.5
Toxocara	1.7
Taenia	0.4

Around 94% of the more than 4 billion cases of diarrhea in the world are caused by helminths (Murray and López, 1996). Figure 10.2 shows the morbidity rates in countries grouped by regions. *Ascaris* has been directly related with the reuse of wastewater and sludges in agriculture (WHO, 1997 and EPA, 1991).

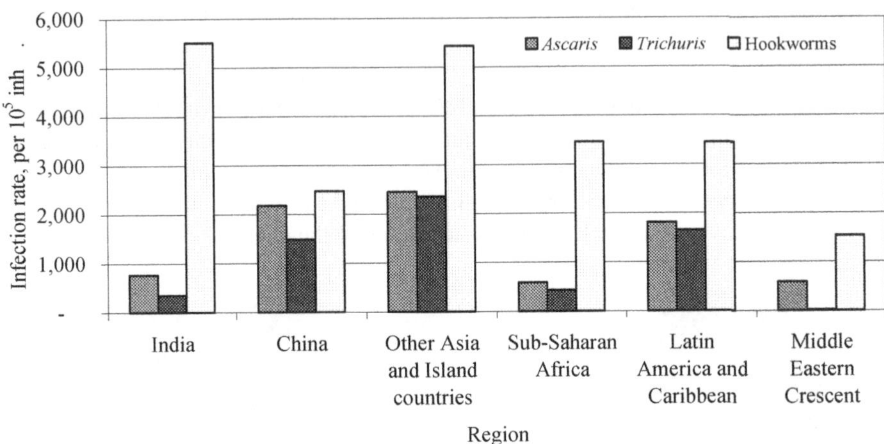

Figure 10.2 Helminth morbidity rates for different regions for 1990 (Murray and Lopez, 1996).

Helminths ova can be inactivated through processes that use high temperature, create high dry conditions or add disinfectant compounds, such as lime, ammonia or acids.

Pathogen inactivation in treatment processes.- Each species of pathogen tolerates environmental conditions in different ways, hence the difficulty in reducing all the populations present in sludge.

One of the main concerns of people that handle biosolids is the regrowth of microorganisms when they are being applied to fields. This only happens with bacteria, since protozoan and helminths (that is not a microorganism) require a host (EPA, 1999). In order to control this problem it is necessary to select the suitable conditions that not only permit the desired efficiency but also prevent bacterial regrowth. Anyway, it is important to remember that in the long term, pathogenic organisms perish when they are exposed to environmental conditions (temperature, sunlight and humidity; Table 10.7).

Table 10.7 Pathogen survival in environmental conditions (EPA, 1992).

ORGANISM	SOIL		VEGETATION	
	Absolute maximum	Common maximum	Absolute maximum	Common maximum
Bacteria	1 year	2 months	6 months	1 month
Viruses	1 year	3 months	2 months	1 month
Protozoan cysts	10 days	2 days	5 days	2 days
Helminth ova	7 years	2 years	5 months	1 month

Note: Periods can increase in favorable climatic conditions.

10.3 TREATMENT

For sludge treatment two terms must be defined: stabilization and digestion. Stabilization looks to reduce or inactivate pathogens in sludges, control disagreeable odors and reduce the potential of vector attraction (EPA, 1979). By contrast, digestion destroys or mineralizes the organic matter, reducing the mass sludge (not necessarily the volume). Digestion does not ensure the control of pathogens. Neither term involves removing metals or toxins. With the idea of promoting sludge reuse, US-EPA has created the term "biosolids" to refer to those sludges that have been stabilized and can be reused without causing any harm to soils, land fills or any other medium.

Nowadays, there are a great variety of processes and unitary operations that can be applied in sludge treatment (Figure 10.3). In the following sections complementary treatments are presented first (degritting, thickening, conditioning, and dewatering) followed by those that stabilize and digest sludges. Finally, storage considerations are discussed.

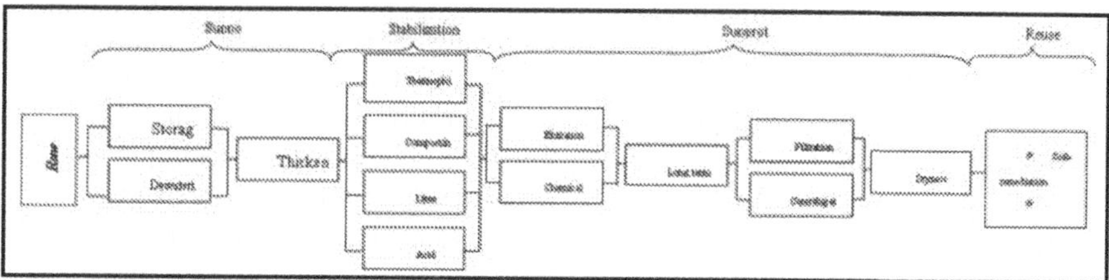

Figure 10.3 Processes considered applicable in developing countries within a sludge treatment scheme.

10.3.1 Degritting

Sludge degritting is a process whose application would be interesting in developing countries, as combined sewers convey a lot of sand and heavy material that reduce the active volume of reactors. Nevertheless, degritting is costly and difficult to operate and is not considered practical. The most effective method for removing sand is using cyclones, where a centrifugal force separates heavy inorganic matter from organic sludge, but is costly.

10.3.2 Thickening

Thickening is used to increase the solid content of sludge and reduce their volume through water elimination. For instance, if a sludge is thickened from 0.8% TS to 4% TS, not only more solids are present but also the volume is reduced by one fifth. Physical thickening is carried out by the procedures indicated in Table 10.8.

Table 10.8 Physical methods for sludge thickening (Metcalf & Eddy, 1991).

Method	Type of sludge	Frequency of use and achieved success
Gravity	Primary raw	Often used with excellent results
Gravity	Primary raw and activated Sludge purge	Often used, especially in small plants. Satisfactory results. Sludges TS concentration varies between 4 to 6%
Gravity	Activated sludge purge	Occasionally since TS content in sludge is poor (2-3%)
Dissolved air flotation	Activated Sludge purge	Limited use, similar results obtained to gravity thickeners
Dissolved air flotation	Activated Sludge purge	Limited use. Good results. TS concentration in sludges varies from 3.5 to 5%)
Basket centrifugation	Activated Sludge purge	Common use. Excellent results, Sludge concentrations vary from 8 to 10%.
Straight jacket centrifugation	Activated Sludge purge	Increasing. Good results. Sludge concentrations vary from 4 to 6%
Band filters	Activated Sludge purge	Increasing. Good results. Sludge concentrations vary from 3 to 6%
Rotary drums	Activated Sludge purge	Limited use. Excellent results. Sludge concentrations vary from 5 to 9%.

10.3.3 Conditioning

Some reagents are added to sludges in order to help suspended and colloidal particles to form floccules and eliminate capillary water in an easier way. Organic and inorganic products are used as well as some natural products, single or combined. The kind and type of reagents depends on the properties of the sludge and the type of equipment used for mixing and dehydrating. Thermal treatment also helps to condition sludge but is quite an expensive method.

Since most of the suspended solids in sludge have a negative electrical charge, cationic agents are used. Previously, ferric chloride and lime were used, but currently water soluble high molecular polymers are more common. Natural polymers are also used, like starches (from potato, corn, tapioca, arrowroot and wheat), polysaccharides (derived from galactose and extracted from leguminous), cellulose derived compounds, some polysaccharides produced by microorganisms, natural glues and gelatins from collagen (Schwoyer, 1981). At the moment, chitosan (which comes from animal skeletons) and bentonite are becoming the most popular natural polymers.

Dosification must be determined experimentally. Preliminary vessels, Specific Resistance to Filtration (SRF), and Capillary Suction Time (CST) tests are used for this purpose. First, the preliminary vessels test is applied to select the reagents that produce flocculation and then recommended doses are obtained from the SRF and CST tests.

In the preliminary vessels test, 100 ml of sludge is placed in one of the 250 ml plastic jars, and then 2 ml of a polymer solution, prepared following the manufacturer's indications, is added. The sludge is then drained 5 to 10 times from one jar to another in order to see whether flocculation exists or not. If flocculation does not happen, then more polymer must be added, repeating the procedure until a maximum dosification of 20 kg of polymer per dry ton of sludge is reached. If flocculation is obtained, the polymer is selected and the dose used is quoted for later tests.

In order to perform the SRF test, a Buchner funnel with a 9 cm diameter is used. This funnel is joined to a 100 ml graduated test tube to which negative pressure of 0.5 bars is applied. 100 ml of conditioned sludges are deposited over a *Whatman* No. 42 filter paper. The volume of the filtered water (V) is recorded at different intervals of time (t). Using this data, a *t/V vs.* V curve is formed and the slope (*b*) of the linear zone is determined. The SRF is subsequently calculated as follows (WPCF, 1988):

$$SRF = \frac{2bPA^2}{\mu c}$$

(3)

where

SRF:	Specific resistance to filtration, (m/kg),	
b:	Slope of the curve t/V vs. V, (s/m^6)	
t:	Filtration time (s),	
V:	Volume of filtrate obtained after time t (m^3)	
P:	Applied negative pressure (N/m^2),	
A:	Filter surface (m^2),	
μ:	Dynamic viscosity of filtrate (N·s/m^2),	
c:	Mass of solids by unit of volume filtrate (kg/m^3)	

When diverse doses of polymers are applied and SRF is measured, it is possible to select the recommended dose for conditioning and dewatering. The SRF test is highly recommended for its relative simplicity and reliable data. By contrast, the CST test can produce results that cause polymer overdose, which represents an unnecessary operating cost (Wu *et al.*, 1997).

In general, the increasing order of reagent demand for conditioning is: raw primary sludge, raw primary sludge blended with trickling filter sludge, primary sludge blended with activated sludge purge, anaerobically digested primary sludge, anaerobically digested primary sludge blended with activated sludge purge and aerobic digested sludge.

Mixing sludge with reagents is a fundamental part of the process and is easier if the coagulant is added in liquid form. While mixing must quickly disperse the reagent it does not have to break up the floccules that are formed. The mixture requires time to mature; however, it should not be left too long, as the reagent becomes less active with time. For appropriate handling, consult the supplier.

10.3.4 Dewatering

The main aim of dewatering is to eliminate as much water as possible to produce a non fluid material whose solid concentration is higher than 20% TS. Dewatering is carried out by mechanical and non-mechanical means. Mechanical means include dry filtration, centrifugation, filter press and filter band. These are recommended when there is insufficient space or adverse climatic conditions.

Non-mechanical methods are drainage and water evaporation. These are less complex processes and easier to operate than the mechanical ones. They consume less energy but require greater land extension and more manual labor, mainly to handle the sludge cake. As a result, they are ideal for developing countries. Drying beds and drying lagoons are some of the technologies used.

10.3.5 Stabilization

Stabilization methods are divided into biological and chemical. Another way to classify them is conventional and non-conventional, according to how frequently they are applied. Conventional processes include aerobic digestion, anaerobic digestion, composting, and alkaline stabilization. Non-conventional processes include irradiation and acid stabilization. Of them, and unlike the others, aerobic and anaerobic digestion use sludge without dehydrating. As a consequence, aerobic and anaerobic sludge must be dewatered prior to disposal. The application in the developing world of different stabilization processes appears in Table 10.9.

Table 10.9 Sludge stabilization processes and their international application (IA) (Adapted from Owen *et al*, 1983).

Processes	Method		IA
Lagoon			B
Anaerobic digestion	Mesophilic (25 – 35 °C)	One phase	A
		Two phases	C
		Two phases	D
	Thermophilic (45 – 55 °C)		C
	Without heating		B
Aerobic digestion	Autothermal (45 - 70 °C) With air or oxygen		C
Dual process	Aerobic autothermal digestion followed by anaerobic digestion		D
Composting with or without a bulking agent	Aerated static pile		B
	Turn around piles		C
	Closed composting systems		B
Alkaline stabilization	Quick Lime pre and post-stabilization		C
	Lime pre-stabilization		B
Addition of oxidizing agents	Reagents added are chlorine, peroxide acids and organic compounds.		C/D

A: Widely used in all the cities of Europe but not very often in America.
B: More common in Europe than in America
C: Limited use in Europe, and frequently in America.
D: Under experimentation and development.

10.3.5.1 Alkaline stabilization

Lime has been used for more than 2000 years to disinfect and deodorize excrements, latrines, manure and sludges. Nevertheless, only in 1970 was it recognized as a conventional stabilization method for sludge (EPA, 1991). Nowadays, lime stabilization is one of the processes most used after anaerobic and aerobic digestion. Its main advantages are its low investment and operation cost as well as the rapidity and facility of the operation.

Alkaline stabilization consists of adding an alkaline material to dewatered sludge, in sufficient amounts to raise pH above 12. This pH value has to be maintained for 72 hours for US EPA class A biosolids (see section 12.7) or 2 hours for class B.

Hydrated lime ($Ca(OH)_2$), quick lime (CaO) or alkaline ashes like those produced in the cement industry, are the products most used. Alkaline ashes have a greater ability to absorb odors and cost one third of what lime does; nevertheless, more than 2 to 3 times are required for the same stabilization results.

The main design criteria are: dose, pH, contact time and water content. The required dose to elevate and maintain pH depends on the chemical characteristic of the alkaline material and the chemical and physical characteristics of sludge (TS content and viscosity). No method has been developed to predict the specific required dose, which is why it must be determined in the laboratory. Literature recommends values between 0.1 to 0.6/kg TS and points out that demand increases as TS content in sludges increases to reach the same pH (Lue-Hing *et al.*, 1992).

There are two types of lime stabilization processes: pre-stabilization and post-stabilization. In the first one, lime is added as a conditioner prior to sludge dewatering, and sometimes aluminum or iron salts are applied for better results. Pre-stabilization can only produce US-EPA class B biosolids (see section 12.7). In lime post-stabilization, lime is applied to dewatered sludge and can produce class A or B US-EPA biosolids, according to the dose and contact time used. In addition, it has the advantage of avoiding the corrosion and abrasion of the equipment and installations and using a lower dose of lime. Post-stabilization with quick lime inactivates 6 to 8 Fecal coliforms log, 5 to 7 log of *Salmonella,* 0.5 to 2 log of helminth ova (Figure 10.4) and 4 to 5 logs of bactheriophages with doses between 20% and 40% dry weight. For this reason lime stabilization is considered appropriate for treating sludges with a high content of microorganisms (Jiménez *et al.*, 2000; Méndez *et al.*, 2002).

It has been demonstrated that by elevating pH to more than 12 units after 2 hours, high contents of bacteria are efficiently inactivated (Figure 10.5). This is why it is possible to use pH as a monitoring parameter for stabilization instead of making tedious microbiological determinations. Without a doubt, this facilitates and considerably reduces operating costs.

Lime stabilization is used for small plants with access to arable land or to landfills, where sludge must be stored before disposal; to replace for a short period of time processes that do not work; to absorb production peaks; to expand existing processes and at plants with variable loads. In particular, lime stabilization is useful in preventing vector attraction in septage. To treat septage, lime must be added until a pH of 12 for 30 minutes without adding more alkali, is reached. Lime stabilization is able to reach very low bacterial values, even below EPA standards, at a relatively low cost in a process easy to operate. This makes it a very valuable process for rural areas where people are used to applying lime to latrines.

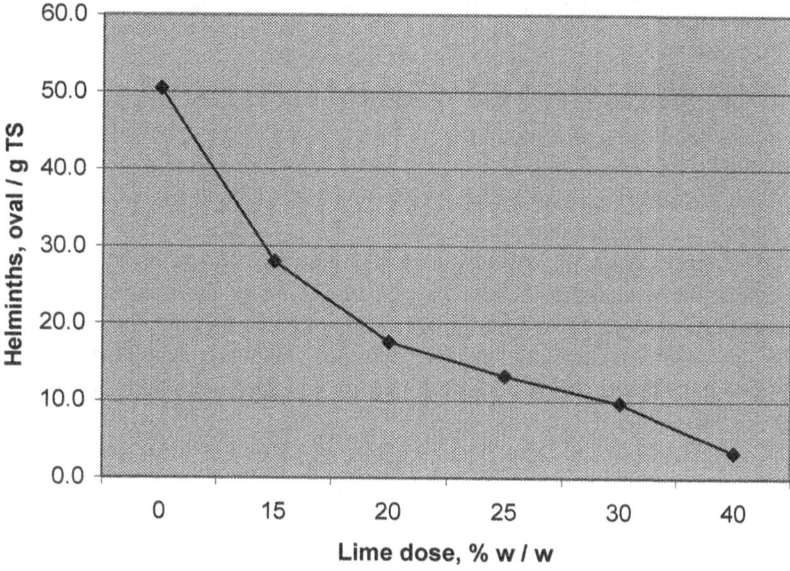

Figure 10.4 Helminth ova inactivation after 2 hours contact time. (Méndez *et al.*, 2002)

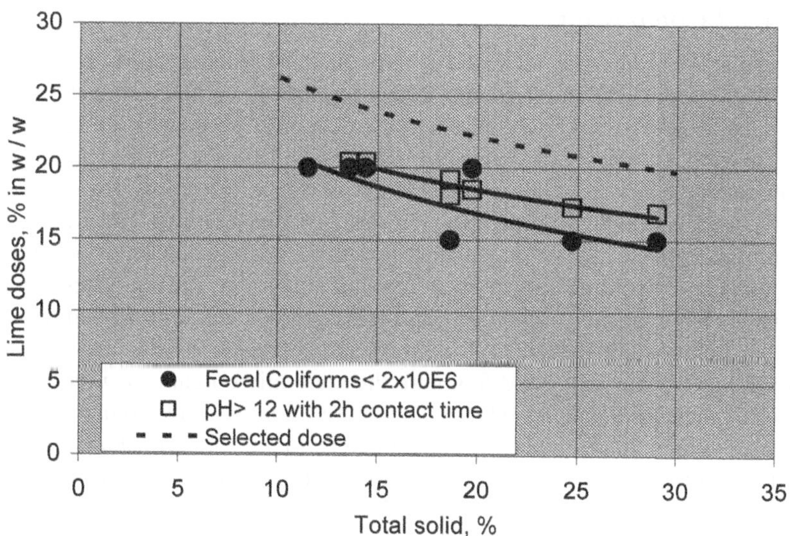

Figure 10.5 Experimental doses of lime to meet US EPA criteria in sludge with high microorganism content. (Jiménez *et al.*, 2000; Méndez *et al.*, 2002).

Even though companies offer diverse patented processes, given their simplicity, their use is not recommended; instead, design the process taking care to control ammonia emissions .

Biosolids produced from lime stabilization can be used as a partial fertilizer, for soil amendment, as remediation products or as cover material in landfills. In agriculture, the use of alkaline ashes from kilns constitutes a good source of nutrients, potassium, magnesium, sulfur and zinc. Limed biosolids can be used instead of traditional agents to recover acid soils if sludge contains aluminum sulfate coming from the wastewater treatment step. From an agricultural and environmental protection point of view, lime content is important in biosolids because of its capacity to immobilize heavy metals, thus protecting plants and aquifers from absorption and lixiviation, respectively. Fortunately, the application of limed sludges does not modify soil pH due to their buffer capacity and used rates.

With time (20 to 30 days) sludge pH tends to diminish in biosolids because of the CO_2 of the air and that produced by the metabolic activity of microorganisms present in sludges, which reacts with residual alkalinity and forms a weak acid soluble in water (H_2CO_3). This reactivates normal biological activity in sludges and soils.

10.3.5.2 *Composting*

Composting is a process in which organic matter is biodegraded (20 to 30% of the TVS) into carbon dioxide and water, producing a stable product similar to humus and odor free known as compost. Composting reduces sludge mass by 40% to 80%. Compost has several agronomic applications (even in gardens).

Composting consists of three phases: (a) sludge preparation, (b) composting itself, and (c) compost preparation. In the first phase, nutrients and a bulking agent are applied. A bulking agent is needed because optimal humidity for composting is 50 to 60 (50 to 10% TS) and sludge have more water than that. Usual bulking agents are wood shavings, straw, and organic solid wastes. During composting, microorganisms consume organic matter and reproduce. Bacteria, acthinomycetes and fungi are the main microorganisms involved. As a result of metabolic activity, sludge temperature increases by around 50° C (and even more than 70° C) inactivating pathogens. Final or intermediate products as well as the raw sludge can be used to inoculate. Composting takes from 2 to 4 weeks depending on temperature and the type of process. To produce USEPA class B biosolids, the average temperature must be 40° C for more than 5 days and during this period it must reach 55° C for 4 hours. Factors to consider during design are those presented in Table 10.10.

Table 10.10 Design considerations for aerobic composting of sludge. (Metcalf & Eddy, 1991).

Item	Comment
Type of sludge	Raw and digested sludge can be composted. Raw sludge has greater potential to produce odor nuisances than digested ones, especially in turnover piles. Raw sludge, in addition, has greater available energy and is degraded at a higher speed and with greater oxygen demand.
Amendment and bulking materials	Process and compost quality are affected by the amendment and bulking materials characteristics, especially humidity, particle size and available carbon.
Carbon/nitrogen ratio	C:N must be within a range of 25:1 to 35:1 in weight to assure biodegradation..
Volatile solids	TVS must be > 65%.
Air requirements	To obtain optimal results, more than 50% of oxygen must be present in the material to compost, especially in mechanical systems.
Humidity	Humidity must be < 60% in static and turn-over piles and for closed vessel < 65%.
pH	The pH of the mixture must be between 6 and 9.
Temperature	Optimal temperature is between 45°C and 55°C. Nevertheless, during the initial days it must be maintained at between 50°C and 55°C.
Mixing and turning	To avoid drying and solids agglomeration, material must be mixed and periodically turned.
Site location	The site must have enough surface available to compost; it must be near wastewater treatment plants, and have accessibility and adequate climatic conditions.

The process is relatively simple but requires space and a very long treatment time (2 to 4 weeks). The bulking agent requirement is also another drawback. Sludge usually has a high nutrient content (C:N < 20), potassium is low, pathogens are very high, the porous structure is poor and humidity content is high (> 60% produces anaerobic conditions during composting) thus the bulking agent is really needed (Dougherty, 1999). Depending on the initial quality of sludge, the following can be added (Dougherty, 1999):

- Not very old compost materials and not more than 10% in quantity to ensure appropriate microorganism content.

- Agricultural lime to correct calcium deficiencies in soils and to attenuate acidic conditions.
- Bones to add calcium and phosphorus
- Argillaceous soil or clays to favor the formation of argil humid compounds in sandy soils
- Gypsum to improve soil texture
- Rock phosphate to add slow available phosphorus.
- Sand or coarse granite powder (in small concentrations) to increase texture and favor drainage.
- Mineral dust to provide microelements and reduce nuisance odors, increase humus formation and improve drainage.

There are diverse composting methods: aerated static pile, turned piles and closed vessels are the most frequent.

Aerated static pile - This consists of a network of air conduction pipes over which a mixture of dewatered sludge with bulking material is distributed in 2-2.5 m height piles. Material is composted for 21 days to 28 days and is left a further 30 days to mature. Usually, to isolate piles, a compost layer is placed over the material.

Turn piles - This system is similar to the previous one, but aeration is done by mixing. The height of piles is 1 to 2 m, with a width of 2 to 4.5 m at the bottom. Piles must be turned around a minimum of 5 times while the temperature stays above 55°C. This operation is accompanied by the liberation of unpleasant odors. The time needed for composting varies between 21 days and 28 days.

Closed systems - In this case, closed vessels are used to diminish odor production and the process' duration. Relevant operating conditions are air flow, temperature and oxygen content. Closed systems are of two types: plug flow and dynamic beds (with mixing).

Composting cannot be performed near superficial and ground water bodies or populations in order to avoid pollution problems. Table 10.11 provides more details on the distances that must be respected.

Space for composting depends on the volume of sludge to be treated as well as the selected composting method and the pile and furrow forms (elliptical, in a triangle, trapezial, etc.). Facilities must consider a 25% to 200% additional area for compost maturity and storage (Dougherty, 1999).

Table 10.11 Commonly recommended distances for the location of composting sites (Dougherty, 1999).

Potentially affected area	Minimum range of separation, m
Distance at which the property of the composting site ends	15 (ideal 150)
Residential and commercial zones	61 (ideal 610)
Private wells or potable water sources	30 (ideal 300)
Wetlands or superficial water bodies (streams, lakes and ponds)	30
Discharge to superficial drainage	8
Groundwater level at the highest level	1
Bed rock	1

10.3.5.3 *Anaerobic digestion*

Anaerobic digestion is the biological transformation of 45% to 50% of the volatile solids into methane, carbon dioxide and water in the absence of dissolved or combined oxygen. Antagonistic microorganisms participate in this process: the acidogenic ones that produce acids, and the methanogenic ones that generate methane, but which are very sensitive to low pH. For that reason, to ensure the process' success, a rigorous control of alkalinity and pH above 6 must be performed.

The gas generated contains 65-70% of CH_4, 30-35% of CO_2, and small amounts of N_2, H_2, H_2S, steam and other gases. Methane is attractive to countries where energy resources are expensive and recovery competes with cost. Methane on the other hand must be collected and burned or retained and treated, as methane is a warming gas provoking the Greenhouse Effect four times powerful than CO_2

Anaerobic digestion produces a biologically stable sludge that can be used as a fertilizer in a great variety of soils. Drawbacks include a high detention time (20 days at 20 °C); the production of a supernatant with very high BOD content, the need for sludge mixing and heating as well as a post treatment process (dewatering, drying or incineration). Also, its high investment cost and operating complexity are other disadvantages of anaerobic digestion.

Heat is provided by the combustion of the methane produced or other combustible gases. Also, to simplify the operation, digestion can be followed up by measuring CO_2 production; when this falls below 25% or rises above 45%, the digester has problems.

There are two kinds of anaerobic digestion processes: the mesophilic and the thermophilic. The former operates at temperatures of between 27°C and 45°C with retention times of between 15 and 20 days. It can be carried out in a single mixed tank (one stage) in which digestion, thickening, and decantation of the supernatant are carried out simultaneously. Removal of pathogens in mesophilic digestion is very low, so only US EPA class B biosolids can be produced with a cellular retention time of 15 days at 35-55°C or 60 days at 20°C. It removes only some logs of fecal coliforms and less than 30% of helminth eggs, thus it is not recommended for developing countries.

Anaerobic thermophilic digestion operates at temperatures of between 45°C and 65°C. The rates of the biochemical reactions increase due to higher temperature, and consequently digestion is carried out faster than in the mesophilic version using much smaller reactor sizes. Anaerobic thermophilic digestion also improves the dewatering characteristic of sludge and increases pathogen inactivation via temperature. Nevertheless, it involves greater energy consumption, produces a bad quality supernatant (with a high content of dissolved solids), generates odors, and the process is unstable, which is why it has limited use (WPCF, 1988). As one might imagine, little research has been done on the removal of pathogens. Krugel *et al.* (1998) indicate, from a single measurement, that in a reactor operating at 55°C and 21 days retention time in Canada, *Salmonella* is not detected in sludge and coliforms are of the order of 10^2/g. In agreement with Oropeza *et al.* (2000), thermophilic digesters require a much longer period of time (months) than mesophilic ones to become stabilized. The reported removal rate of helminth eggs in sludge is 70% with high initial content.

In some countries, the treatment of sludge jointly with the organic fraction of municipal solids wastes has been studied to generate biogas, diminish the size of sanitary landfills and avoid the dewatering post-treatment of sludge. The final product can be used for application in soils as fertilizer and also as an end cover for sanitary landfills.

10.3.5.4 *Acid treatment*

Since 1984, patents have existed that use acids to stabilize sludges and inhibit or destroy pathogens with very good results (Lynch *et al.*, 1984). In addition, by means of this process, it is possible to extract metals like arsenic, cadmium, copper, manganese, nickel and zinc from sludges, and also to recover reagents such as aluminum . Another advantage of acid treatment is its efficiency in controlling odors (Fraser *et al.*, 1984). Compared with lime, acid stabilization has the advantage of oxidizing part of the organic matter and, thereby reducing the final sludge volume.

Besides the toxic effect caused by pH, acids can have another one that depends on the type of acid. In general, organic acids are more toxic than minerals since they interfere with cellular functions (El-Ziney *et al.*, 1997 and Barrios, 2003). Studies of acid stabilization have been reported using sulfuric, chloridric, peracetic, acetic and propionic acids (Roth and Keenan, 1971 ; Owen, 1984; and Kiff and Lewis-Jones, 1984). Of them, peracetic (APA) and acetic acids are the best disinfectants. In physiochemical sludges, 550 ppm of APA were sufficient to inactivate 5 - 6 logs of fecal coliforms, 4 - 5 logs of *Salmonella* and 2 log (95) of helminth ova, without bacteria regrowth in 10 min (Barrios, 2003).

Figure 10.6 shows three Ascaris eggs treated with acetic acid. The first clearly shows damage to the membrane; in the second, the nuclei is damaged by the acetic acid's penetration of the egg, preventing larvae development; the third is a normal egg for comparison.

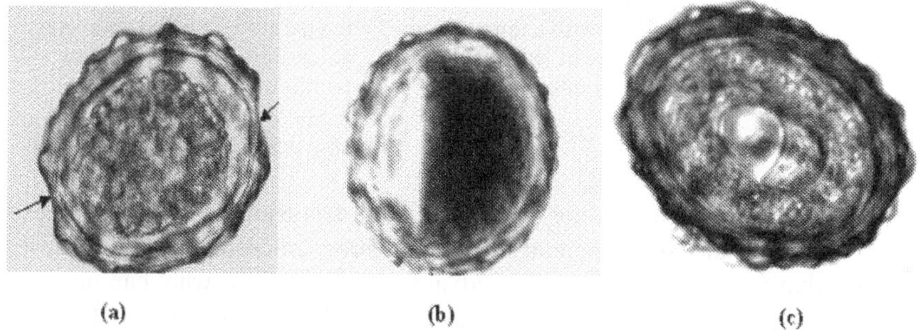

Figure 10.6 Ascaris eggs. (a) Membrane damaged by acetic acid (b) Nuclei damage by acetic acid; (c) Normal egg with larvae inside (Barrios, 2003).

10.3.6 Storage

Sludges must be stored because they are not continuously generated throughout the diverse stages of treatment but in batches (for example, when decanters of thickeners are purged). Short-term storage is performed in thickeners. Long-term storage can be performed in a long retention time process (e.g. anaerobic digesters) or in specific storage installations. Storage is particularly needed when lime or acid stabilization are used as well as mechanical dewatering and drying. The main risk during sludge storage is the generation of nuisance odors.

10.3.7 Considerations for process selection

Table 10.12 contains EPA-accepted technologies for the production of class A or B biosolids (section 12.7). It is important to note that the same process for producing class A or B biosolids depends basically on how high the temperature is.

Table 10.12 US- EPA technologies with potential to be applied in developing countries.

Technology	Type of biosolids that can be obtained	
	Class A	**Class B**
Aerobic digestion	Thermophilic, 10 days retention time at 55-60^{0}C*	40 days retention time at a 20^{0}C or 60 days at 15^{0}C*
Anaerobic digestion	Not recognized by the EPA.	15 days at 35^{0}C or 60 days at 20^{0}C.
Alkaline stabilization	PH > 12 during 72 hours with a temperature > 52^{0}C for at least 12 hours and a pH > 12. Dryness of 50% TS	pH of 12 during 2 hours
Composting	In aerated vessel at 55^{0}C for more than 3 days. In furrow with a temperature >=55^{0}C for at least 15 days and with a minimum of 5 turns*	In a closed vessel, aerated static pile or in furrows, the temperature must be on average at least 40^{0}C during 5 days; within this period the temperature must increase above 55^{0}C for at least 4 hours.

Table 10.13 summarizes the advantages and disadvantages of the processes based on cost and the potential for inactivating microorganisms. But it is obvious that many of the theoretical advantages differ from what can happen in reality, due to the operating procedures applied in different countries. For instance, in a study done in Mexico (Barrios *et al.*, 2001) in 14 plants in operation, 5 with alkaline stabilization, 3 with aerobic digestion and 6 with mesophilic anaerobic digestion, only the first ones, when well operated, produce US EPA class B biosolids. The main problem observed was the helminths' resistance and deficient operating conditions. Cardoso and Ramirez (2002) indicate that only lime treatment compared to anaerobic digestion can produce biosolids with < 2 fecal coliforms MPN/g from sludges with 5x10^6 MPN/g. Anaerobic digestion can only reduce the initial content to 1x10^6 MPN/g. Another observation from table 10.12 is that a process classified as Class A by EPA might not attain a helminth egg content less than 1 HO/gTS when initial content is very high as happens in sludge from developing countries.

Table 10.13 Advantages and drawbacks of sludge stabilization processes, with information from WEF and ASCE, 1992.

Process	Advantages	Disadvantages
Anaerobic digestion	<ul><li>Good TVS reduction (40 to 60%)</li><li>Operating cost can be low if methane is used to heat reactors</li><li>Wide application of the Mesophilic conditions</li><li>Biosolids obtained are appropriate for agricultural use</li><li>Medium pathogens inactivation</li><li>Sludge mass reduction</li><li>Energy requirements w if methane is used.</li><li>Thermophilic conditions can inactivate helminth ova</li><li>1-3 log removal in mesophilic condition and possibly up to 4 in thermophilic conditions.</li></ul>	<ul><li>High initial investment and requires skilled operators</li><li>Methanogenic microorganism grows slowly in acid digester and recovers very slowly after collapse</li><li>Supernatant has a high content of COD, BOD, TS and NH_3</li><li>Cleaning is difficult</li><li>Can generate nuisance odors</li><li>Potential of mineral incrustation and foam formation</li><li>Safe plans and risk studies are required due to the production of a flammable gas</li><li>Thermophilic version is not yet a well-known process.</li><li>Requires energy for mixing and heating.</li></ul>
Lime stabilization	<ul><li>Low cost and easy to operate</li><li>Good, like emergent method of stabilization</li><li>Very good pathogen control. Removes 6 to 8 log of coliforms, 5 to 7 of *Salmonella* and > 98% of helminth ova.</li><li>Biosolids produced can be use to desalinate and to amend acid soils.</li></ul>	<ul><li>Greater sludge production than any other treatment</li><li>Produce bad odor from ammonia during process.</li></ul>
Composting	<ul><li>Low cost and easy to operate</li><li>Reduces TVS by 20 to 30%</li><li>Produces biosolids with a high level of acceptance in agriculture and odor free</li><li>Good pathogen inactivation. Removes 5 to 7 log of fecal and coliforms, > 95% of helminth ova.</li></ul>	<ul><li>Demands a bulking material</li><li>Demands great area</li><li>It can contaminate water bodies and aquifers</li><li>Produces odor during treatment.</li></ul>
Acid treatment	<ul><li>Useful to remove heavy metals and recover reagents from sludge.</li><li>Produce odor free biosolids with potential use in agriculture, especially in saline and alkaline soils.</li><li>Oxidizes organic matter and reduce sludge volume (in a variable degree).</li><li>Good pathogen inactivation. Inactivates 5-6 log of fecal coliforms, 4-5 of *Salmonella* and more than 90% of helminth ova.</li><li>It has a residual disinfection effect.</li></ul>	<ul><li>Not recognized as a conventional process by US-EPA.</li><li>No full scale application.</li></ul>

10.4 SLUDGE REDUCTION

Sludge might be a new pollution source. To reduce this problem innovative technologies have been developed, such as:
- Bacterial predation method, which suggested that sludge in the presence of oligochaetes was found to be 0.15g MLSS/g chemical oxygen
- demand (COD) removed compared to 0.40g MLSS/g COD removed under normal operating conditions. 0.05-0.17g TSS/ g COD removed of sludge yields was achieved in Sweden by using 2 stage processes with a completely mixed first stage and no biomass retention to promote dispersed bacterial growth and a second stage with long sludge age and optimal condition for predation;
- Bioaugmentation method, by means of selecting bacterial strains or genetically altered strain introduced to the works, improved plant efficiencies have been obtained 16% sludge reductions on secondary sludge production and septic tanks can be greatly improved by this method;
- Metabolic uncoupling which employs complex metabolic pathways to control bacterial degradation processes involving catabolism that breaks down organic substrate and anabolism that harnesses released energy into new cell growth to stop at the catabolism stage inhibiting anabolism;
- The enzyme dosing method;
- Ultrasound method;
- Bio-lysis system;
- Extended aeration system

All above methods have been exploited from different aspects to reduce the sludge production a. Tilche *et al.* (1999) mention five alternatives for doing this in biological processes, these are:
- To use biochemical energy contained in sludges for conversion processes, such as denitrification and phosphorus removal.
- To use processes with low biomass growth.
- To apply systems of greater biomass age, like extended aeration, biofilm processes, biomembrane reactors, etc.
- To manipulate the nutritional chain in the activated sludge process to favor growth of bacterial predators.
- To enhance biological stabilization of sludges (pretreatment, thermophilic anaerobic digestion, etc).

Sludge reduction options for physiochemical processes are:
- To use methods that promote flocculation without adding reagents like ultrasound, mechanical agitation and ozonization.

- To use polymers with a high density charge and a high molecular weight instead of metallic salts or low efficient polymers.
- To combine small doses of metallic salts with anionic polymers
- To use flocculants or coagulants that can be used as bulking material during composting.

10.5 SEPTIC TANKS

The septic tank system is popularly used as a primary treatment system in a dormitory building, a group of dormitory buildings, or a small community in a decentralized wastewater management area. It is used to treat domestic wastewater and to digest the settled sludge. Sludges digestion time usually is very long, say half a year or even longer. Effluents from septic tanks are discharged into subsoils or diverted to the sewerage system. For its simplicity, septic tanks are very useful in rural areas and low income cities, and are a common practice in most cities in China. Even though sludge are biologically digested in septic tanks, and thus its volume is reduced, the reactors must be periodically cleaned. The cleanup period depends on the climate and influent composition. It is recommended to clean septic tanks 1 or 2 times per year, leaving 20% of mature sludge as inoculum for digestion. Design considerations for septic tanks are not considered in this chapter, however it is noted that to treat septages, lime must be added until a pH of 12 for 30 min without adding more alkali is reached.

10.6 BENEFICIAL USES OF BIOSOLIDS

There are several beneficial uses of biosolids, some of which involve soil application and others their transformation into something else, like bricks. Application in soils includes agricultural, forest and restoration uses. Table 10.14 shows the rates at which biosolids for these purposes are applied and compares them with confinement in landfills.

Table 10.14. Typical rates of biosolid application. EPA, 1997.

Final disposal method	Frequency of application or final disposal	Disposal rate (Ton/ha)	
		Range	Typical
Agricultural use	Annual	2 to 60	10
Forest	1, 3 or 5 times per year	8 to 200	40
Soil remediation or restoration	Once	6 to 400	100
Confinement in landfills	Annual	200 to 800	300

Biosolids can be applied in (a) Sites with no public contact, like agricultural grounds (land grass and crop soils), forests, restoration sites (minefields and construction sites), airports and highways, and in (b) Sites with public contact, such as, public parks, greenhouses, golf courses, prairies, family gardens and cemeteries.

The form of application depends on the physical characteristic of sludges, basically its water content. Liquid application (1% to 10% TS) permits the transportation of sludges by pumping and then aspersion, drainage into furrows, injection or pouring with a hose. Application of dry sludges (15% to 30% TS) requires ground transport and similar procedures to those used with solid fertilizers, manure or organic semisolid wastes. Dewatered sludges are applied at a rate superior to the liquids, and in both cases sludges must be plowed or disked using the appropriate equipment during the following 6 hours. When liquid sludge is injected, mixing can be avoided. Soil mixing has the aim of creating a barrier between solids and vectors. The ground eliminates water from the biosolids, limits its mobility and prevents odor release.

The rate of biosolid application must allow for the passage of vehicles that distribute them and maintain aerobic conditions in soils to avoid odor generation.

Biosolids can also be distributed in bags, bulks or other containers for soil application. These biosolids are sold and used in turfs, superficial layers of grass, golf courses, parks and gardens.

Any type of soil application (agricultural, forest or soil remediation) requires auxiliary facilities for storage (lagoons, sludge storing bays, etc.). The aim is to provide a cushioning capacity between sludge production and the intermittent application and to preserve sludge from extreme climatic conditions (rain, snow and wind). In order to avoid the pollution of superficial and ground water bodies during storage, sites with impermeable soils and hard layers underneath the soft land or rock should be selected. Lixiviates must also be controlled by placing drains around the storage sites. The area required for storage facilities depends on the amount of sludge generated and the application rate. Generally, it varies from 4 to 40 hectares and must have a flat topography.

Following, the main considerations for different biosolid applications are presented.

10.6.1 Agricultural application

Using biosolids in agriculture reduces the fertilizer consumption cost, but in fact, in poor countries more than representing a saving it represents the possibility of using fertilizers, increasing productivity and the possibility of controlling soil degradation by erosion. Erosion is equivalent to 5 to 10% of the agricultural production cost and causes the loss of 5 to 7 million of arable hectares per year (Rhoades, 1997; Young, 1998). The process consists of placing biosolids on the surface, incorporating them into the soil or injecting underneath it in order to improve soil characteristics increasing agricultural productivity. Agricultural application is used when fields are less than 50 km away from the sludge treatment site in order to be able to afford the biosolid transportation cost (OTV, 1997).

Some of the biosolid benefits of the agricultural application are soil improvement through the addition of organic matter and nutrients (Table 10.15) and the immobilization of some polluting agents. Nevertheless, the risk exists of contaminating plants and aquifers with pathogens and toxic substances and of producing disagreeable odors in the application sites. The pollution of aquifers with nitrates is the case to which most attention is paid and the amount of biosolids applied is frequently limited by this fact. Given the magnitude orders of nitrates present in sludges, when limiting application rates to avoid ground water pollution, the metal's risk is controlled. In order to avoid nitrogen pollution, applied rates must not exceed 100% of the demand for each crop. For instance, 200 kg/ha.year for grass, 100 kg/ha.year for corn, 60 for oats, barley and sorghum and 30 kg/ha.year for soya.

Table 10.15. N and P content in municipal sludge and comparison with traditional fertilizers. (Metcalf & Eddy, 1991 and EPA, 1994.)

Nutrient	Concentration g/kg (dry weight)			
	Activated sludge purge	Anaerobic digested sludge	Green fertilizers (sugar beet)	Manure (from cows)
Total Nitrogen	35	15 a 30	5	4
Total Phosphorus	70	15 a 30	0.5	1

One of the very favorable characteristics that biosolids present is the slow liberation of nutrients, which fits in very well with the crop's rate of absorption. In addition to N and P, other nutrients are available in sludge for plants, like potassium, sulfur (in sulfate form), calcium, iron, magnesium, manganese, copper, zinc, molybdenum and boron (Girovich, 1996).

Organic matter.- Organic matter improves the aggregate formation, which also improves air diffusion, water movement, infiltration rate and penetration of plant roots. A well-structured soil is more resistant to erosion. Given its negative charges, organic matter also increases the number of available sites for ion adsorption, which is why it retains nutrients and heavy metals. Organic matter represents a new energy source (organic carbon) for microorganisms (bacteria, protozoan, fungi and others) that participate in recycling and decomposing other organic compounds.

Pathogens - During the application of biosolids to soil it is important to consider pathogen content because they are extremely resistant to most of the sludge stabilization process. There is epidemiological evidence that relates health risks to exposure to unsanitary sludge. Of all the organisms involved, the main danger is *Salmonella spp* and helminths ova (Pike *et al.*, 1988).

Heavy metals - The EPA undertook exhaustive studies during the 1980's on the presence of heavy metals in sludge. The concentration, toxicity and toxic effects on humans and the environment were analyzed. From these studies, maximum permissible limits were laid down for arsenic, cadmium, chromium, lead, molybdenum, nickel, selenium and zinc, since they were fournd the only elements to display potential risks. Table 10.16 describes the source of these elements and the damage they can cause. When the heavy metals limits in sludge and the application rate allowed is used it is considered that virtually no risk exists.

Organic compounds - Pesticides, industrial solvents, tensioactives, colorants, plastifying agents, polynuclear aromatic hydrocarbons (PAHs) and many other complex organic molecules, generally not very soluble in water and with a high capacity to be adsorbed, tend to be accumulated in sludges. The main concern these compounds cause is not their absorption by plants but direct ingestion by animals, particularly cattle, although there is evidence that organic compounds can be absorbed by the surface of root crops, such as carrots.

A WHO work group on health risks related to the presence of chemical agents in sludge applied to soils, reached the conclusion that "the total absorption by humans of identified organic polluting agents coming from the application of sludge in crop soils is of little importance and probably does not cause adverse effects". Nevertheless, the ecotoxicological effects of the organic polluting agents in the soil-plant-water system and the nutritional chain continues to be investigated.

In order to use sludge in agriculture, not only their quality and characteristics should be considered but also the site conditions. Based on this information, the agronomic rate, soil requirements and needs, application methods and the schedule can be laid down. All these considerations can be established by legislation as well as by procedure manuals.

Table 10.16. Heavy metals sources in sludge and possible effects (Modified from Girovich, 1996).

Metal	Sources	Toxicity In high concentrations	Carcinogenic effect
Arsenic	Metal smelting and mineral coal combustion. Present in some pesticides.	Skin alteration with indirect effects on the nervous, gastrointestinal and hematopoyetic systems. Irritation of the respiratory system.	Carcinogen
Cadmium	Plating and mining	Toxic for plants Kidney damage. Hypertension and destruction of erythrocytes	Possible carcinogen
Chromium	Silver plated and platting. Tailings from mining	Toxic for plants and animals	Carcinogen
Copper	Industrial discharge Mining	Toxic for plants and very toxic for animals	No carcinogen
Lead	Industrial discharges and mining. Gasoline with lead	Kidney problems, alterations in the reproductive system, liver, brain and central nervous system	Possible carcinogen
Mercury	Combustion of mineral coal, pesticides, fungicides and pharmaceutical products	Damage to the nervous system	Not determined
Molybdenum	Industrial discharges	Toxic for animals	Not determined
Nickel	Mineral and industrial discharges	Highly toxic for plants, animals and humans	Possible carcinogen
Selenium	Mineral and industrial discharges	Highly toxic for plants and animals	Not determined
Zinc	Industrial discharges and plating	Highly toxic for plants and animals	Not determined

10.6.2 Remediation

Some sites need to be habilitated or remediated due to their originally bad quality, pollution or erosion problems. Biosolids can be used for these purposes in quantities that must be specifically defined for each site in terms of soil and environmental conditions. Most common problems to be addressed with biosolids are erosion and soil desalinization, problems often common in developing countries. For application, biosolids are mixed with soils (if they exist), and grasses or fodders are planted. It is emphasized that in these circumstances the primary target is still the final disposal of sludge and that vegetative growth is a secondary factor. The principle underlying application is similar to that of the agricultural application. Sites can receive one or more applications depending on soil conditions and, in general, at higher rates than those used for agriculture. The range varies a lot, from 7 to 585 dry ton/ha depending on site conditions (EPA, 1995 and 1997; Metcalf & Eddy, 1991; Girovich, 1996; Lue-Hing, 1992).

10.6.3 Desalinization of soils

A common problem in soils is the loss of agricultural productivity as a result of salinization, due to the increase in sodium content or SAR (Sodium Absorption Ratio). Biosolids can be used to control this problem if they contain Ca^{2+}, which can come from the sludge stabilization process itself when lime is used. Ca^{2+} ions from biosolids replace Na^+ ions in soils if the application is accompanied by washing and suitable drainage is provided, in accordance with the following reaction.

$$SAR = \frac{[Na^+]}{\sqrt{\frac{[Ca^{2+} + Mg^{2+}]}{2}}} \qquad (4)$$

Beltrán et al., 2002 used biosolids combined with hydro technique to correct salinity in soils with conductivity from 10 to 87 dS/m. With only one application and 10 months of washings, salinity was reduced to < 4 dS/m and sodicity was reduced to less than 15.

10.6.4 Forest use

The forest use permits the application at great rates of biosolids in areas where public access can be easily controlled with the advantage of creating green areas. As in the case of remediation, rates must be determined for each site depending on local conditions. Nevertheless, it is necessary to cautiously consider the costs, since forests are frequently far from the biosolids production sites and in places of very high elevation.

10.6.5 Non conventional uses of sludges

Sludge can be used for diverse purposes, like the preparation of animal feed, sources for different metals, brick manufacturing, along with cement, slag, Portland cement, artificial light weight aggregates, hydrocarbon production and for road construction materials. These techniques are still undergoing research and development and respond to local necessities. Also, they have been used to elaborate ceramic materials (Teratani et al., 2001) for metal revalorization (Meunier et al., 2001) and for biopesticide production (Tyagi et al., 2002). However, all these uses together with the energy revalorization of sludges are not considered economically feasible (Okuno, 2004 and Lee, 2003).

10.7   CONFINEMENT

Of the final disposal methods, given the cost and environmental concerns, incineration and dumping into the sea are not considered in this text. Thus, confinement options include monofills, ponds and lagoons, specific disposal sites, and landfills (Table 10.25).

10.7.1 Monofills

Monofills are sites where dewatered sludges (TS of 15%) are exclusively placed. Sludges are covered daily with a ground layer thicker than the sludge layer. As in the case of landfills, monofills can be units excavated in soils, ditches, piles or ramps that take advantages of ground accidents. Monofills have a very high disposal rate that distinguishes them from lagoons, ponds and specific application sites. A well-operated landfill can ultimately be used for a recreational site. Also, in monofills, there is not enough oxygen for aerobic decomposition, which is why anaerobic digestion occurs at a slow rate. For this reason biogas must be recovered and lixiviates must be treated.

In monofills, a distance must be respected between the sites where sludge is applied and the end of the facility, which depends on the arsenic, chromium and nickel content (Table 10.17). These limits are applied in sites without an impermeable bottom layer (such as clays or synthetic geomembranes) and a lixiviate collection system.

10.7.2 Ponds and lagoons

Ponds and lagoons are used to dispose of sludge with high water content. In these facilities, sludge is placed in cavities without a daily cover. The solid content generally varies from 2% to 5% TS. When sites are below ground level, they are called lagoons; if they have borders they are known as ponds. These sites are different from sludge treatment lagoons because they retain sludges for more than two years.

10.7.3 Specific disposal sites

These are sites in which sludges are injected below the surface of the ground or incorporated after aspersion or simply distributed over the soils with the single aim of carrying out the final disposal. These sites are usually near the wastewater treatment plants and receive sludges in greater rates than are used in agriculture.

Table 10.17 Comparisons between sludge confinement sites (US EPA, 1997[a])

Method	TS, %	Sludge[b]	Hydrogeology[c]	Ground slope
Monofill				
Narrow trench	15 a 28	With or without stabilization	Deep groundwater or rock bed	<20%
Wide trench	$\supseteq$ 20	With or without stabilization	Deep groundwater or rock bed	<10%
Area fill mound	$\supseteq$ 20	With or without stabilization	Shallow aquifers or rock beds	Convenient in soils with irregular topography
Layers	$\supseteq$ 20	With or without stabilization	Shallow aquifers or rock beds	Convenient in soils with medium slopes or flat grounds
Diked Containment	$\supseteq$ 20	With or without stabilization	Shallow aquifers or rock beds	Advisable in steep areas if dikes are constructed
PILES				
Piles	$\supseteq$ 28	Stabilized	Shallow aquifers or rock beds	<5%
METHOD	TS, %	SLUDGE[b]	Hydrogeology[c]	Ground slope
SUPERFICIAL DEPOSITS AND LAGOONS				
Surface impoundment and lagoons	$\supseteq$ 2	Stabilized	Shallow aquifers or rock beds	
SPECIFIC DISPOSAL SITES				
Dedicated surface disposal sites	$\supseteq$ 3	Stabilized	Shallow aquifers or rock beds	<3%
CODISPOSAL WITH MUNICIPAL WASTE IN SANITARY REFILLS				
Mixture of sludges and domestic wastes	$\supseteq$ 20[d]	With or without stabilization	Deep aquifers or rock beds	<30%
Mixtures of sludges and soils	$\supseteq$ 20[d]	Stabilized	Deep aquifers or rock beds	<5%

a: A comparison is made based on the design requirements and not in legislation needs.

b: To protect health and the environment the US EPA establishes three aspects that must be met for sludges: metal and pathogen content and vector attraction. The disposal of sludges together with solid wastes in monofills does not need prior stabilization but does require a daily cover.

c: Final disposal must not contaminate groundwater

d: Sludges to be disposed of in landfills must pass the filter painting test (FPT).

10.7.4 Municipal sanitary landfills

If the conditions of the site permit, most of the solid wastes produced by wastewater treatment plants (sand and sludge) can be sent to the closest sanitary landfill. Sludge must be drained prior to this to reduce volume and lixiviation. All the landfills in use have to be adequately designed and respect distances to the aquifer level and population areas as well as biogas and lixiviation control. Once the landfill is closed, it can be used for public parks or to host facilities that are not affected by potential collapses. Table 10.18 shows the main criteria in the design of landfills.

Sludges can be codisposed in landfill in two ways: (a) Mixed with solid wastes (b) Mixed with soil that serves as material for the daily cover. Table 10.19 shows the requirement in each case.

Table 10.18 Design considerations for landfills.

• Not affect endangered species.
• Not interrupt the flow of a flood with a 100-year return period.

• Be in sites free of
    ▪ Geological faults in a distance superior to 60 m
    ▪ Unstable soils. Landfills cannot be built near precipices, underground excavation, calcareous soils or mines.
    ▪ Seismic zones with a more than 10% probability of an earthquake with an horizontal acceleration of more than 0.1 times the gravity in 250 years.

• Wetlands or saturated soils cannot be used for landfills
• Facilities should collect run-off from a 24 hour storm with a 25-year return period.
• In areas with impermeable soils, lixiviates must be collected.
• In cover disposal areas, limits on methane discharges into the atmosphere must be respected.
• The selected site must not be used in the future for agriculture or animal pasturing.
• Public access must be limited

Table 10.19 Requirements for placing sludges in municipals landfills.

Type of mixture	Type of sludge	Total solids, %	Mixture ratio	Slope
Sludge and municipal solids wastes	With or without stabilization	≥ 20	4 to 7 tons of wastes/tons of wet sludge	< 30%
Sludge and soils for daily cover material	Stabilized	≥ 20	1 m^3 of soil/ m^3 of wet sludge	< 5%

<u>Sludges mixed with solid wastes</u>
Sludges with a minimum content of 20% TS are thrown over the solid wastes and mixed in the best possible way, with or without prior stabilization. The mixture is later compacted and covered with soil. In order to adequately handle the solid wastes/sludge mixture a ratio of 4 ton/ton of 20% TS sludges (dry weight) is recommended. The disposal rate for the solid wastes/sludge mixtures oscillates between 900 and 7000 m^3 of sludge/ha.

<u>Sludge use as daily and final cover in landfills.</u>
One of the more practical and very useful ways to dispose of sludge is their use as a daily or final cover in sanitary landfills. In effect, sanitary landfills require a great amount of material to ensure their proper operation and biosolids can be used for this. Municipal sludge with a TS content of at least 50% can replace soil or clay used in landfills. Biosolids act like a physical barrier that prevents vector attraction. In addition, they have low humidity and absorb excess water from solid wastes (Lue-Hing *et al.*, 1992). McBean *et al.* (1995) mention that for a solid wastes volume of 385,100 m^3, 142,870 m^3 of biosolids are needed (67% for daily cover, 28% for final cover and 5% for vegetal growth).
One advantage of using sludge as a cover is that operational problems like equipment slipping or becoming stuck in place are avoided. Another advantage is that their presence in the final cover promotes the growth of vegetation and reduces the need for fertilization. A drawback is that it requires greater equipment and more people for operation. Also, very disagreeable odors can be present since sludge are not buried, which is avoided with appropriate stabilization. Table 10.20 summarizes the main operation criteria for different sludge confinement systems.

Table 10.20 Design criteria for different sludge confinement options (US EPA, 1994-).

Method	TS, %	Ditch Width, m	Filling Material[a]			Cover Thickness		Soil Needs	Disposal Rate[b]	Equipment
			Use	Agent	Agent/Sludge Ratio	Daily	Final		m³/ha	
MONOFILL										
Narrow trench	15 - 20	C.7 - 1.1	No	-	-	-	0.7 a 1.1	No	2 260 a 10 600	Backhoe with loader
	20 - 28	1.1 - 3.5	No				1.1 4 1.4			Excavator trenching machine[c]
Wide trench	20 - <28	≅5	No	-	-	-	a 1.4	No	6 000 a 27 500	Track loader, dragline, scraper, track
	⊇28	≅5	No	-	-		a 1.8			dozer[d]
Area fill mound	⊇20	-	Yes	Soil	0.5 - 2:1	0.9	1 a 1.4	Yes	5650 a 26 500	Track loader, backhoe with loader, track dozer[e]
Area fill layer	⊇20	-	Yes	Soil	0.25 – 1:1	0.2 a 0.35	0.7 a 1.2	Yes	3700 a 17 000	Trackdozer, grader, track loader[e]
Diked containment	20 - <28	-	No	Soil	0.25 – 0.5:1	0.35 a 0.7	1 a 1.2	Yes	9 000 a 28 500	Dragline, track dozer, scraper[d]
	⊇28	-	No	Soil		0.7 a 1.1	1.2 a 1.8			
PILES										
Piles	⊇28	-	No	-	-	-		No	15 100 a 60 500	Spreader, bulldoze[f]
SURFACE IMPOUNDMENTS AND LAGOONS										
Surface impoundment and lagoons	⊇2	-	No	-	-	-	-.	No	9 000 a 28 500	Dragline, front-end loader[e]
DEDICATED SURFACE DISPOSAL										
Dedicated surface disposal sites and dedicated beneficial use sites	⊇3	-	No	-	-	-	-	No	95 a 3 800	Tank trick, subsurface injector, rotary sprayer, bulldozer[e]
LANDFILLS DISPOSAL										
Mixture of sludges with solids wastes	⊇20[f]	-	Yes	Municipal wastes	4 - 7:1 wet basis	0.2 a 0.34	0.7	No	940 a 7 950	Dragline, track loader
Mixture of sludges with soils	⊇20[f]	-	Yes	Soil	1:1	0.2 a 0.35	0.7	No	3 000	Tractor with disc, track loader

A: Volume basis unless anything else is specified

B: Rate established according to experience and engineering practice

C: Ground equipment

D. Ground equipment for sludges< 28% and solids wastes.

E: Sludge equipment

F: Sludges to be disposed in sanitary fills must fulfill the filter painting test

10.8   LEGISLATION

Due to the practical technical content of part 503 of the US-EPA's "Standards for the Use or Disposal of Residual Sludges" this standard is almost always used as a reference. The norm establishes the limits of polluting agents in the reuse and final disposal of sludge, procedures for beneficial uses and for the confinement, design and operating criteria for the treatment process, monitoring issues and record formats. One of the most practical advantages is that it also regulates operating conditions so they can be followed and inspected instead of measuring the fulfillment of each limiting parameter. Monitoring and analyses are much more complicated in sludge than in wastewater. Concerning biological parameters, the most important ones mentioned in part 503 is sludge classification into A or B. Class A considers a fecal coliform content < 1,000 MPN/g TS dry weight, *Salmonella spp* < 3 MPN/4g TS (or 0.75 MPN/g TS) and enteric viruses < 1 PFU/ 4 g TS. Viable helminth ova must be les 0.25/ g TS dry weight. Class B has < $2X10^4$ FC in MPN/g TS, and does not limit S*almonella* or helminth ova content, as long as in the United States they do not represent a great concern.

In the following sections the sludge legislations of different countries are described.

10.8.1 South Africa

Sludge criteria were established in 1997. They are very strict due to the need to protect the aquatic environment, seriously limiting agricultural application. Legislation classifies sludges into four types (Snyman *et al.*, 2000). Classes A, B and C are useful for improving soil conditions, whereas D does not have use restrictions. Each class corresponds to a certain level of treatment. For example, class A are sludge requiring no treatment, B are aerobically digested sludges while classes C and D are obtained by lime stabilization, composting, irradiation and pasteurization. The difference between classes C and D is the metal content.

10.8.2 Mexico

Previously, municipal sludge was considered in Mexico hazardous wastes, but the new proposed criteria changed this concept. They were published in 2001 as NOM-004-ECOL-2001. These criteria adopted the metal and fecal coliform limits of part 503 of the US-EPA standard, but modified those of *Salmonella* and helminths ova to reflect local conditions. Three sludge classes are considered (Table 10.21). An additional class C was defined based on the initial helminth ova content in sludge and values feasible and economical to obtain using conventional treatments. Class A is currently considered difficult to obtain due to the high initial pathogen content. In addition, legislation considered total ova instead of viable ova to facilitate monitoring. The standard does not consider viruses, as they cannot be measured by laboratories. It is assumed that in the future, they will be considered in legislation. Sludge are also classified by the heavy metal content in good or excellent, in the same way USEPA does. Class C Biosolids can be used in forest, for soil improvement or agricultural uses. Class B, is for urban uses without public contact curing application as well as for those uses for class C. And, Class A biosolids with very low metal content can be used for urban purposes with public contact as well as for the uses established for classes B and C.

Table 10.21 Maximum permissible limits for pathogens in sludges.

Class	Bacteriological Pollution Indicator	Pathogens	Parasites
	Fecal Coliforms MPN/g TS dry weight	*Salmonella spp.* MPN/g TS dry weight	Helminths ova/g TS Dry weight
A	< 1 000	< 3	< 1(a)
B	< 1 000	< 3	< 10
C	< 2 000 000	< 300	< 35

(a) Viable helminth ova

10.8.3 Chile

The elaboration of a new regulation began in January of 1999, with the participation of the Ministries of Health, Agriculture, Economy, the Sanitary Services Superintendence, Water Companies, Universities and Professional NGO. Before this legislation, sludges were dried until 40% TS and confined in landfills. For sludges from drinking plants, this new legislation considers their discharge into superficial water bodies prior authorization, while for wastewater treatment sludge it considers their reutilization for agricultural purposes. Chilean legislation contains the same biological criteria for classes A and B as the US-EPA but replaces the MS-2 bacteriophage virus indicator with the enteric virus, due to the facility in measuring them. However, there is evidence to suggest that the microbiological levels established are not attainable in 1 to 3 orders of magnitude by digestion or composting (Castillo *et al.*, 2002).

In Chile, own limits for heavy metals were established based on the purpose of application (agricultural or remediation; Table 10.22). These limits are three times more restrictive than the European criteria and six times US ones. Even so, sludge fulfills them from the beginning, because they contain a smaller amount of metals. (Castillo *et al.*, 2002).

Table 10.22 Maximum permitted limit of heavy metal content in sludges in Chile (mg/kg) (Mena 2002).

Heavy Metal	Kind of soil, mg/1g	
	Agricultural	Eroded
Arsenic	20	40
Cadmium	8	40
Copper	1000	1500
Mercury	4	20
Molybdenum	10	20
Nickel	80	420
Lead	300	400
Selenium	50	100
Zinc	2000	2800

Table 10.23 Maximum heavy metal limits in soils before sludge application (Mena 2002).

Heavy Metal	Total concentration in soils, mg/kg		
	North Macrozone		South Macrozone
	pH >6,5	pH >6,5	
Arsenic	20	12.5	10
Cadmium	2	1.25	2
Cupper	150	100	75
Mercury	1.5	1	1
Molybdenum	2	3	3
Nickel	112	50	30
Lead	75	50	50
Selenium	4	3	4
Zinc	175	120	175

Maximum application rates are established for (a) agricultural and forest soil or eroded land with great potential for immediate uses (15 ton/ha.year), (b) turfs, gardens and green areas (2 ton/ha.year), and (c) degraded soil without immediate potential uses (30 ton/ha.year). For the application of biosolids, maximum heavy metals limits are defined for two macro zones of applications (Table 10.23). The norm also establishes a condition for final confinement in monofills, sanitary landfill and mining sites. Prior to the first application, the pH, N, P and K content as well as heavy metal concentration in soils should be analyzed. For sampling, 25 samples have to be collected from the arable layer of an area measuring < 5 ha. Analyses must be repeated at least every two years.

10.8.4 China

China basically focuses its legislation on inorganic compound control and regulates the rate and duration of soil application in terms of soil texture and groundwater level (Wang, 1997). Heavy metal limits are stricter that those of the US-EPA. China does not establish limits for pathogens.

10.8.5 Brazil

Brazil has legislation for each state. In Paraná, for instance, the values for metals and the microbiological ones for USEPA Class A helminths and fecal coliforms were adopted.

To summarize, Table 10.24 contains limits for metals for the legislation presented (Brazil, Parana State) and compares them to those of the United States.

Table 10.24 Heavy metal limits in sludges for different countries (all values in mg/kg dry sludge)

	USA	Brazil	Chile		China		Mexico		South Africa
			Agric. Appl.	Erod. soils	Soil <6.5	Soil >6.5	Good	Excel-ent	
As	75		20	40	75	75	75	41	15
Cd	85	20	8	40	5	20	85	39	15.7
Cr		1000			600	1000	3000	1200	1750
Cu	4300	1000	1000	1500	250	500	4300	1500	50.5
Hg	57	16	4	20	5	15	57	17	10
Mo	75		10	20					25
Ni	420	300	80	420	100	200	420	420	200
Pb	840	750	300	400	300	1000	840	300	50.5
Se	100		50	100					15
Zn	7500	2500	2000	2800	500	1000	7500	2800	353.5

10.9 SAMPLING AND MONITORING ISSUES

Biosolids quality must be guaranteed by means of regular monitoring programs that include heavy metals, nutrients, microbiological indicators, organic matter, pH and toxic compounds, among others. Monitoring frequency and the number of parameters to be considered depends on the class of the desired biosolids, the disposal method, and the real analytical capacity of each country. It goes without saying that it is impossible to use developed countries as a model.

10.9.1 Sampling

The EPA indicates monitoring methods that require a large number of samples that are not viable in countries with limited resources. Therefore, it is suggested that practical and realistic solutions be used in order to have at least some data on which to base useful decisions. Table 10.25 contains a solution provided in Mexico where the number of samples is defined based on the amount of sludge produced or the capacity of the treatment plants. This solution considers economic feasibility.

Table 10.25 Monitoring frequency to demonstrate fulfillment of the Mexican norm.

Wastewater treatment plant capacity, L/s	Sludge production ton/year	Frequency per year
< 100	0 a < 290	Once per year
> 100 y < 250	> 290 a < 1,500	1 every three months
> 250 y < 500	> 1,500 a < 15,000	1 each 60 days
> 500	> 15,000	1 per month

For sampling sludge, it emphasizes their high variability, even in the same lot, which is why taking a representative sample becomes a problem. It is recommended that for liquid sludge samples be taken at the furthermost point of the pipes in which they are transported in order to take advantage of the conditions of maximum mixing. For sampling lagoons and ponds, samples must contain representative portions of the floating and the sediment material, as well as the entire covered area. In both cases, aliquots should be collected in a 10 L container, all the portions mixed and the required volume taken for analysis. For drained sludges (10 - 40% TS) sampling must be carried out when sludge are moving. In dry or solid sludge the grip point sampling technique is used. In addition, for metals, the sampling can be carried out after treatment, just before disposal or application.

Concerning the type of containers, these must be wide mouth. The form of preservation, the maximum storage time and the minimum sample volume are those shown in Table 10.26 for metals and Table 10.27 for biological analysis.

Table 10.26 Metals sampling requirements.

Type of sludge	Material of the container	Preservation[a]	Maximum storage time	Minimum volume of sample
Solids and semisolids	P, G	4°C	24 h	300 mL
Liquids for Hg analysis	P, G	HNO_3 until pH 2	28 days	500 mL
Liquids for all other metals	P, G	HNO_3 until pH 2	6 months	1000 mL

Table 10.27 Biological sampling requirements.

Material of the container	Preservation[a]	Maximum storage time	Sample minimum volume
G, P, B, AI In case of helminth ova use a glass bottle	Cool with a mixture of ice and water until samples reach a temperature < 10°C. Cool quickly to 4 °C	For bacteria, 6 h. 24 hours for bacteria and 1 month for helminth ova	1 - 4 L o 500 g wet weight

[a] Preservers must be added to the containers before taking the sample. The storage time begins when sample is placed in the container. It varies for each analytical method
P: Plastic (polyethylene, polypropylene and Teflon)
G: Glass
B: Presterile bags (for dewatered or liquid sludges)
AI: Stainless steel (without zinc or steel cover)

When sludge is to be applied in agricultural soils, N, P, K and SAR are also measured.

To express the biological content, the density of microorganisms is defined as the most probable number (MPN) or only the number per mass of total solids in dry weight basis. US-EPA expresses their criteria by 4 g TS because it is the weight of 100 ml of domestic sludge (EPA, 1992) but data is more useful per 1 g TS. Also, due to the high values that are used to measure microorganisms, the geometric average of seven simple samples taken at different points during the warmest month of the year is used to assess biological quality before disposal or application. Only values obtained at the end of stabilization are accepted as representative of those at the moment of application or disposal for helminth ova and virus. For bacteria (fecal coliforms and *Salmonella*) it is possible to use them as long as the application or disposal does not take more than 1 month.

Concerning the analytical methods, those presented in Table 10.28 are recommended.

Table 10.28 Analytical methods for sludges.

Parameter	Method
Fecal coliforms	Part 9221 or Part D, Standard Methods for the Examination of Water and wastewater, 18 edition, American Public Health Association, Washington DC, 1992
Helminth ova	Yanko, W.A. Occurrence of Pathogens in Distribution Marketing Municipal Sludges, EPA/600/1-87/014, 1987. PB 88-154273/AS, National Technical Information Service, Springfield VA; (800)553-6847
Inorganics	Test Methods for Evaluating Solid Waste, Physical/chemical Methods, EPA Publication SW-846, 3rd edition (1986) PB 87-120291, National Technical Information Service, Springfield, VA 3rd edition Don. No. 955-001-00000-1, Superintendent of Documents Government Printing Office, Washington D.C.
Salmonella	Part 9260 Standard Methods for the Examination of Water and wastewater, 18 edition, American Public Health Association, Washington DC, 1992; or Kenneth, B.A. and H.P. Clark, Detection and Enumeration of Salmonella and Pseudomonas aeruginosa, J. Water Pollution Control federation, 46(9):2163-2171, 1974
Paint Filter Liquid Test	Method 9095 "Paint Filter Liquid Test", Test Methods for Evaluating Solid Wastes, Physical/Chemical Methods. EPA SW 846, 3rd edition (1986)
Oxygen consumption specific rate	Part 2710B Standard Methods for the Examination of Water and wastewater, 18 edition, American Public Health Association, Washington DC, 1992
Specific resistance to filtration	WPCF, 1988. Sludge conditioning Manual of Practice FD-14
Total, volatile and fixed solids	Part 2540 G. Standard Methods for the Examination of Water and wastewater, 18 edition, American Public Health Association, Washington DC, 1992
Enteric viruses	D 4994-89 Standard Practice for Recovery of Viruses from Wastewater sludges, Annual Book of ASTM Standards: Section 11. Water and Environment Technology, ASTM, Philadelphia, PA, 1992

10.9.2 Vector attraction

Vectors are animals that constitute a means of transmitting a disease. Flies, mosquitoes, birds and rodents are attracted by sludge and can act as disease spreaders. Vector attraction has to do with the organic content, odor and degree of sludge hygiene. In order to reduce vector attraction, sludge must be efficiently treated by any digestion or stabilization process, or suitably incorporated into the ground. The latter is obtained when sludge is injected below the surface soil or is mixed within a few hours after application. Stabilized sludge with TS content of 75%-90% do not attract vectors.

10.10 COSTS

Treatment and disposal costs depend on the plant size, the type of technology used to treat sludge, distances to the application, or confinement sites and other factors influenced by local conditions. The cost of sludge management can represent up to 60% of the cost of operation and maintenance of the wastewater treatment plant, although sludge volume is less than 1% that of treated water.

Sullivan and Oerke (1997) did a study that compares, under same conditions, the costs of alkaline stabilization, composting and thermal treatment to produce US EPA class A Biosolids. The investment and the operating costs were determined for facilities with capacities of between 40,000 and 230,000 m^3/d. Lime stabilization was the process with the lowest unitary cost. Composting was 16% to 32% greater than the lime stabilization process and heating 58% to 78% greater.

10.11 PRACTICAL EXPERIENCES

10.11.1 Mexico

Of the 187 m^3/s of municipal wastewater that Mexico produces, only 22% receives treatment. For this reason, sludge production is low, but it is considered that this situation can be taken advantage of to develop and compare treatment and disposal options before massive sludge generation begins. Sludge production is estimated to be 640,000 ton dry weight when 140 cities of the country, with more than 50,000 inhabitants, are treating wastewater. According to legislation, sludge must be stabilized and not necessarily digested prior to their application in soils or their confinement in landfills as solid waste. 63% of the Mexican territory (i.e. 2 millions km^2) has soils with moderate to severe erosion so it is considered that the application of sludges can help remedy this situation and legislation in fact promotes it.

At the moment, 304 ton/day of sludge is generated. It is stabilized by anaerobic digestion (51%), aerobic digestion (22%); or lime stabilization (25%). Only the latter manages to significantly reduce pathogens in sludge. The final disposal of sludge is landfill (76%), sewerage systems (2%) and soil application (22%). In Mexico it is considered that the only viable method for sludge management in treatment plants smaller than 60 litres (2500 inhabitants) are drying beds (Cardoso and Ramirez, 2002).

Because few reports on sludge quality exist, Barrios *et al.*, (2001) did a national study to measure metal content. It was demonstrated that concentration in raw sludges is smaller that EPA criteria (Table 10.29). Other studies have confirmed this (Cardoso and Ramirez, 2002).

Demonstrative project of sludge application.- Several demonstrative application projects have been performed all over the country. Such is the case of Ciudad Juárez, Monterrey, Chihuahua, Toluca and Morelos. In Ciudad Juarez, the treatment of 3.5 m^3/s (1.2 million inhabitants) produced between 50 to 70 dry tons of sludge. These are lime stabilized and are intended to be applied in sodic soils to control salinity. Garciapiña *et al.*, 2000, have demonstrated that lime stabilization reduced 4.4 log of fecal coliforms and 96% of helminth ova and that limed sludges when applied in regional sodic soils at a rate of 340 ton/ha (50 times the agronomic rate by nitrogen) reduces sodicity by 85%. In Monterrey, there are 26 wastewater treatment plants for 3 million people. Accumulation of heavy metals in diverse crops due to biosolid application has been studied, without observing any effect. This city has the first registered brand of biosolids in the country, called Nutririego (Martínez *et al.*, 2002).

Table 10.29 Heavy metal concentration in Mexican sludge (Barrios *et al.*, 2001).

Parameter	Concentrations			US EPA Limits	
	Minimum	Maximum	Mean	Maximum	Monthly average
Arsenic	5	62	29	75	41
Cadmium	3	11	7	85	39
Copper	190	378	287	4300	1500
Mercury	0	44	8	57	17
Molybdenum	0	26	714	75	
Nickel	26	208	102	420	420
Lead	58	112	75	840	300
Selenium	1.3	57	24	100	100
Zinc	415	1,080	766	7500	2800

In Chihuahua, studies of metal accumulation in alfalfa, oats, maize for cattle and cotton have been performed. With rates from 45 to 400 ton/ha, no effect has been found in soils or crops (Uribe *et al.*, 2002).

10.11.2 Brazil

Generally, sludge in Brazil is returned to water bodies or deposited in land near to cities without control due to a lack of convenient disposal options. The Agricultural Interdisciplinary Program from SANEPAR (Water Company of Parana State) promotes sludge reutilization in agriculture (Andreoli, 1999). Wanke *et al.*, (2002) has a managing plan for communities with 20 000 inhabitants that consists of composting and lime addition (30 to 50% w/w) to produce USEPA Class A Biosolids, drying, transport and soil application at a cost of 1.8 USD/inhab that is considered economically feasible.

Texeira *et al.*, (2002) and Comparini and Sobrinho (2002) showed that metal content in Brazil is also well below the EPA criteria for non-treated sludges. Sludge biological content for Sao Paulo city is presented in Table 10.30 . It is observed that storage improves quality. Storage in open tanks and cover ponds at room temperature (mean of 23.5°C, ranging from 4.4 to 42.7) reduces *E. coli* from 10^5 to 10^3 MPN/gTS in 168 days, bacteriophages F from 10^4 PFU/gTS to 10^2 in 30 days and completely inactivates helminth ova by 252 days. In 44 to 70 days *E. coli* was reduced to 10^3 MPN/g dry basis and bacteriophages F were completely inactivated, but the agronomic quality of sludge was decreased (Table 10.30).

Luca *et al.*, 2002 have applied oxidation with Iron VI to sludge. They managed to control odors and produce a final product similar to lime stabilized sludges. Andreoli *et al.*, (2001) studied sludge disinfection of anaerobic digested sludges in drying beds heated with biogas. They obtained very good helminth egg inactivation when sludge wereperiodically turned and heat waswell distributed.

Table 10.30 Pathogen inactivation due to storage in Sao Paulo sludge (Comparini and Sobrinho, 2002).

Organism	Initial value	70d	250d	357d
Bacteriophages PFU/gTS	10^5	10^2	10^2	10^2
E. Coli MPN/gTS	10^5	10^3-10^5	10^2-10^3	10^2-10^3
Total Coliforms MPN/g TS	10^6 - 10^7	10^4-10^6	10^4-10^5	10^5
Helminth ova/g TS	60 - 67	48 - 50		24 - 37
Viable Helminth ova/gTS	11 - 21	13 - 16	0	0

10.11.3 Argentina

In Argentina, wastewater and sludge reuse is considered important in the Patagonia, Andean, and extra Andean zones, since these regions have soils with a poor P and N content. Nevertheless, The National Hazardous Waste law prevents it.

10.11.4 Chile

Wastewater treatment is 20% of the total volume and is performed in stabilization ponds and submarine emitters, Mena (2002). For a future treatment coverage of 90%, a sludge production of 250 000 ton/year is estimated. In Chile, sludge irradiation is the only disinfection method that can fulfills their proposed criteria (Mena, 2002), since it reduces 4 to 5 log units of bacteria. The microbiological quality of Chilean sludges digested is presented in Table 31; given their characteristics, it is considered that a final composting stage is needed prior to soil application.

10.11.5 China

In this country, a dramatic increase in wastewater and sludge production is observed due to population growth, industrial and agricultural development. It is estimated that of total wastewater generated, only 7% (44 m^3/s) receives treatment and produces 400 000 tons of sludge (dry weight) per year, Wang (1997). Only one third of the wastewater treatment plants have facilities for stabilizing sludge and less than 10% of the plants use a complete treatment and disposal process. There have been no in-depth published studies which contribute data about the sludges' characteristics. At present, from an environmental and economic point of view, land applications are considered to be the best option. Agriculture application has been practiced to some degree for more than 30 years and has become more common in the last 20 years, especially in big cities like Beijing, Tianjin and Shangai. However, sludge is still discharged directly into superficial water bodies. Some revalorization projects have been performed: biogas digestion and brick construction. Nevertheless, the more attractive option from the economic point of view is an agricultural application.

Table 10.31 Microbiological quality of Chilean sludge after biological digestion (Castillo *et al.*, 2002).

Microorganisms	Aerobic sludge	Anaerobic sludge
Fecal coliforms MPM/g	2.3 X 10^9	1.5 X 10^6
Escherichia coli MPN/g	8.7 X 10^8	5.0 X 10^5
Salmonella MPN/4g	Negative	1.2 X 10^4
Phages MS-2 PFU/4g	Not determined	1.8 X 10^3
Humidity %	85	84
VTS %	63.5	68
C %	35.3	37.8
N %	0.8	1.5

Sludge reduction-Present situation of sludge production in China - In China, the wastewater treatment rate was only 13.8% in 1998. Lots of effort has been made to improve wastewater treatment, however, only a little attention has been paid to sludge treatment, and sludge reduction in particular. Therefore, even although there were 398 wastewater treatment plants in commission until 1998, only half of these plants have sludge treatment facilities and only 30% of these sludge facilities operated normally, with 70% of sludge being discharged directly into the water bodies nearby without any treatment, thus causing serious pollution.

For example, in Daqing and Chengfengzhuang wastewater treatment plants, 30,000m^3/d wastewater is treated each day, with nearly 110 metric tons of excess sludge produced each day, which is discharged directly into the water body nearby, without any treatment. As a result, the water body is polluted again. Even although, a wastewater treatment plant has been in operation since 1993 (8 years), the water environment has not been improved significantly. Another example is in Shenzhen, Guangdong, the most advanced city in China with high economic development. A modern wastewater treatment plant was commissioned in 1997 with sludge treatment facilities, such as sludge dewatering and digestion. However, all these sludge treatment facilities have never operated normally, since it was put into operation. It was only operated during a special period when it was inspected by EPB officials. In Gaobeidian wastewater treatment plant, the sludge treatment facilities including sludge digestion and dewatering are operated normally, however, during dewatering, some chemicals are put into the sludge, which make the sludge quality not suitable for agriculture. Thus it was rejected by farmers to use on their farmland. These examples reflect different aspects currently existing in China, as well as showing the potential problems threatening the water environment and regenerated pollution sources with wastewater treatment. As also indicated, if we want to solve the water pollution thoroughly, we need to pay much more attention to sludge treatment and we need to find an economic way to solve this problem by means of sludge reduction.

10.11.6 Accra, Ghana

Accra has 1.5 million inhabitants and 4 wastewater treatment plants. One of these has anaerobic sludge treatment. The others consist of stabilization ponds and are overloaded because they receive the septages. Sludges are extracted from ponds, accumulated, dried and manually composted. Compost is an odor-free product with good physical properties and is well accepted by local farmers. There are no sludge regulations, although enteric diseases are very common.

High values of fecal coliforms and an extraordinarily high number of helminth ova have been detected in sludge (Hall, 2000).

10.11.7 Alexandria, Egypt

Alexandria is Egypt's second largest city with 4 million inhabitants and 40% of the country's industry. The city has two primary treatment plants from which raw sludges are obtained. Sludge is mechanically dewatered and transferred to a remote site for composting. Heavy metal contents in sludge is below EPA limits, in spite of the large amount of industries present. During a 12-month evaluation of pathogenic organisms, very high values were detected. The Egyptian microbiological standard for sludge is similar to EPA's. Therefore, there is a tremendous need for appropriate technologies to lower the contents. Actual standards are considered difficult to meet. Composting is considered a low cost, convenient process in Egypt (Hall, 2000).

10.11.8 Europe

Until now, the main disposal method for sludge in Europe has been dumping in sanitary landfills, soil application, sea dumping (mainly in the United Kingdom) and incineration. Table 10.32 displays a mixture of European countries with very different economic conditions. In order to ensure suitable handling of sludges in Europe, a European Standardization Committee has been established. At present, some physical and chemical parameters have been standardized. The main challenge in Europe is to apply common standards to different countries. It is estimated that more than 1 million Euros is needed annually to handle sludges in Europe (Wang, 1997).

Table 10.32 Final sludge disposal methods in some European countries (Lue-Hing *et al.*, 1992).

Country	Final destination (%)			
	Agriculture	Landfills	Incineration	Sea dumping
Germany	25	65	10	0
Spain	61	10	0	29
France	27	53	20	0
Greece	10	90	0	0
England	51	16	5	28
Portugal	80	12	0	8

10.12 ACTIVITIES NEEDED

Much still needs to be done in developing countries, including:

- Improve technology efficiencies to reduce high levels of very resistant pathogens.
- Promote sector qualification for sludge management as a part of sanitation programs.
- More data on the chemical and biological characteristics of sludges in developing countries.
- Appropriate laws for each country which promote reuse instead of confinement, as the fertilizing potential of sludge can be lost and problems arise from competition with solid wastes for space in landfills. Legislation with rational economic and technical bases which can be implemented gradually, rather than the sophisticated legislation of many developed countries is recommended. In other words, use of the "sufficient challenge" instead of "zero risk" principle.
- Another problem for regulators is the evident absence of skilled laboratories to perform routine sludge analyses. Cost is another issue, as helminth ova detection represents 200 USD per analyses and requires 4 hours to determine viability.

REFERENCES

Andreoli, C., Von Sperling, M., and Fernandes, F. (2001) *Lodo de esgotos: tratamiento e disposiçao final.* Belo Horizonte: Departamento de Engenharia Sanitária e Ambiental-UFMG;Ed. Compañía de Saneamento do Paraná. Brasil. (in Portuguese).

Andreoli, C., Pegorini, E., Lara, A.., Ferreira, A., Bonnet, B., and Fernández, F. (1999) Distribution plans and legal permission requirements for the biosolids land application programs in Parana State, Brazil. *Proceedings of the IAWQ Specialised Conference on Disposal and Utilisation of Sewage Sludge: Treatment Methods and Application Modalities.* 13-15 october , pp.71-78. IAWQ and European Commission Directorate General XIII, Athens, Greece

Barrios, J., Jiménez, B., Rodríguez, A., González, A. and Maya, C. (2001) Application of peracetic acid to physicochemical sludge to reduce its microbial content. *6th European and Organic Residuals Conference, Workshop and Exhibition.* Wakefield, UK. Nov. 11-14. *6th European Biosolids and Organic Residual Conference.* Lowe P., and Hudson J.A. 1(2): 1-7, CIWEM/Aqua Environ Consultancy Service Wakefield UK

Barrios, J.A., Rodríguez, A., Gonzalez, A., Jiménez, B. and Maya, C. (2001) Quality of sludge generated in wastewater treatment plants in Mexico: meeting the proposed regulation. www.iwaponline.com/wio/2002/06/wio200206010.htm

Barrios-Pérez, José (2003) Sludges Acid Stabilization. PhD-Thesis in Environmental Engineering, DEPFI-UNAM. D.F., Mexico (in Spanish).

Beltrán R., Lucho C., and Dendooven L. (2002) Evaluación de la recuperación de un suelo salino sódico después de la incorporación de biosólidos y de un tratamiento hidrotécnico.

Proceedings XXVIII Interamerican Congress of Sanitary and Environmental Engineering, 27-31 October, FEMISCA, AIDIS, CWWA, Cancún, Mexico (in Spanish) These proceedings are on a CD; the identification of each paper is made by the author's name therefore; there are no volume or page numbers.

Castillo, G., Mena, M. and Alcota, C. (2002) Experiencias sobre compostaje de lodos de digestion aerobica y anaerobica. *Proceedings XXVIII Interamerican Congress of Sanitary and Environmental Engineering,* 27-31 october, FEMISCA, AIDIS, CWWA Cancún, Mexico (in Spanish)

Cardoso, L. and Ramírez E. (2002) Aplicación en un suelo agrícola de lodos residuales tratados. *Proceedings XXVIII Interamerican Congress of Sanitary and Environmental Engineering,* 27-31 october, FEMISCA, AIDIS, CWWA, Cancún, Mexico (in Spanish)

Carrington, E. and Harman, S. (1984) The effect of anaerobic digestion temperature and retention period on the survival of Salmonella and Ascaris ova. In *Sewage sludge stabilization and disinfection,* (ed. A. Bruce), pp. 369-380, Ellis Horwood Limited, UK.

Comparini, J. and Sobrinho, P. (2002) Decaimento de patógenos em biosolidos submetidos á secagem em estufa agrícola. *Proceedings XXVIII Interamerican Congress of Sanitary and Environmental Engineering,* 27-31 october, FEMISCA, AIDIS, CWWA, Cancún, Mexico (in Spanish)

Crewe, W. (1984) Transmission of Taenia saginata in Britain. *Annals of tropical Medicine and Parasitology* **78**, 249-251.

Dougherty, M. (1999) *Field Guide to Farm Composting.* Natural Resources, Agricultural and Engineering Service. New York, USA.

El-Ziney, M., De Meyer, H. and Debevere, J. (1997) Growth and survival kinetics of Yersinia enterocolitica IP 383 O:9 as affected by equimolar concentrations of undissociated short-chain organic acids. *International Journal of Food Microbiology* **34**, 233-247.

EPA (1979) *Process Design Manual. Sludge Treatment and Disposal.* EPA 625/1-82-014. Washington, D.C., USA.

EPA (1991) *Preliminary risk assessment for parasites in municipal sewage sludge applied to land.* Office of Research and Development. EPA/600/6-91/001. Washington, D.C., USA.

EPA (1992) *Control of Pathogens and Vector Atraction in Sewage Sludge* EPA/625/R-92-004. Washington, D.C., USA.

EPA (1994a) *A Plain English Guide to the EPA, Part 503 Biosolids Rule.* EPA/832/R-93-003. Washington, D.C., USA.

EPA (1994) *Biosolids Recycling: beneficial technology for a better environment, Audience concerned citizens, industry.* EPA/832R94009, http://www.epa.gov/owm/mtb/biosolids/biorecyc.htm

EPA (1994) *Land application of sewage sludge A guide for land appliers on the requirements of the Federal standards, 40 CFR Part 503, office of enforcement and compliance assurance.* EPA/831-B-93-US-EPA Washington, D.C.

EPA (1995) *Process design manual, Land application of sewage sludge and domestic septage.* Office of Research and development, EPA/625/K-95/001 EPA Washington, DC , USA.

EPA (1997) Process Design Manual: Surface Disposal of Sewage Sludge and Domestic Septage. EPA/625/K-95/002 EPA Washington, D.C.

EPA (1999) *Biosolids generation, use, and disposal in the United States.* EPA530-R-99-009.US EPA, Washington D.C:

Fraser, J., Godfree, A. and Jones, F. (1984) Use of peracetic acid in operational sewage sludge disposal to pasture. *Water, Science and Technology* **17**, 451-466.

Garciapiña, T., Jiménez, B., Barrios, J.A. and Garibay, A. (2000) Application of Limed Biosolids to Improve Saline-Sodic Soils from Nothern Mexico. *5th European Biosolids and Organic Residuals Conference, 20-22 November* Lowe P., and Hudson J.A. **1**(3): 1-4, CIWEM/Aqua Environ Consultancy Service Wakefield UK

Gaspard, P., Wiart, J. and Schwartbrod, J. (1997) Parasitological contamination of urban sludge used for agricultural purposes. *Waste Management and Research* **15**, 429-436.

Girovich, M. (1996) *Biosolids treatment and management process for beneficial use,* Marcel Dekker, Inc. N.Y., USA.

Hall, J. (2000) Sludge management in developing countries. *Proceedings of the Joint CIWEM Aqua Enviro Consultancy Services 5th European Biosolids and Organic Residuals Conference.* Aqua Enviro Consultancy Services, (eds. P. Lowe and J. Hudson) Seminar 1, paper 3. Wakefield, United Kingdom.

Hays, B. (1977) Potential for parasitic disease transmission with land application of sewage plant effluents and sludges. *Water Research* **11**, 583-595.

Jiménez-Cisneros and Chávez Mejía, A. (1997) Treatment of Mexico City Wastewater for Irrigation Purposes. *Environmental Technology* **18**, 721-730.

Jiménez, B., Maya, C., Sánchez, E., Romero, A. and Lira, L. (2002) Comparison of the quantity and quality of the microbiological content of sludge in countries with low and high content of pathogens. *Wat. Sci. Tech.* **46**(10), 17-24.

Jiménez,B., Barrios, J. and Maya, C. (2000) Class B Biosolids Production from Wastewater Sludge with High Pathogenic Content Generated in an Advanced Primary Treatment. *Wat. Sci. Tech.* **42** (9), 103-110.

Jimenez, B., Barrios, J., Mendez, J. and Díaz, J. (2004) Sustainable sludge management in developing countries. *Wat. Sci. Tech 40 (10):251-258*

Kiff, R. and Lewis-Jones, R. (1984) Factors that govern the survival of selected parasites in sewage sludges. In *Sewage sludge stabilization and disinfection.* (ed. Bruce, A.M.),pp. 426-439, Ellis Horwood Limited, UK.

Krugel, S., Nemeth, L., Peddie, C. (1998) Extending thermophilic anaerobic digestion for producing class A biosolids at the greater Vancouver Regional Districs Annscis Island Wastewater Treatment Plant. Wat. Sci. Tech. **38** (8-9) 1998

Lee and Tay (2003) Energy recovery in sludge management process. *Proceedings of Biosolids 2003. Wastewater Sludge as a Resource, 23-25 June. Trondheim, Norway.* [EDITOR?], pp. 299-308, International Water Association, London, UK.

Leppe, A., López, A. and Nelson, P. (2002) Lodos provenientes de plantas de aguas servidas: potencialidades y restricciones; temores y realidades. *Proceedings XXVIII Interamerican Congress of Sanitary and Environmental Engineering,* 27-31 october, FEMISCA, AIDIS, CWWA, Cancún, Mexico (in Spanish)

Luca, A. (2002) Tratamiento de lodos de esgostos con Ferrato XXX *Proceedings XXVIII Interamerican Congress of Sanitary and Environmental Engineering,* 27-31 october, FEMISCA, AIDIS, CWWA, Cancún, Mexico (in Spanish)

Lue-Hing, C., Zenz, D.R. and Kuchenrither, R. (1992) Municipal sewage sludge management: processing, utilization and disposal. Technomic Publishing Company, USA.

Lynch, J.M., Pfaffin, J.R., Pecker, C., Cárdenas, R., Cunningham, S., Bozzone, R.T. and Borg, S. (1984) *Method for the treatment of wastewater sludge.* Patente No. 4'500,428, E.U.

Martín del Campo, S., Vaca, P. Lugo, FJ., Gómez, B. Estelle,r A. and Garrido, S. (2002) Aplicación de lodos residuales municipales en cultivo de haba (viciafaba l.) En suelos agrícolas del valle de Toluca. *Proceedings XXVIII Interamerican Congress of Sanitary and*

Environmental Engineering, 27-31 october, FEMISCA, AIDIS, CWWA, Cancún, Mexico (in Spanish)

Martínez, J., Hinojosa, J. and Romero, L. (2002) Biosolids in México -Success Story. *16th Annual Residuals and Biosolids Management Conference. 3-6 March, WEF, Austin, Texas, USA.*

McBean, E., Rovers, F. and Farquhar, G., (1995) *Solid Waste Landfill engineering and design.* Prentice Hall, USA.

Mena, M. (2002) Avances en el marco legal para el uso y disposición de biosólidos en chile: efectos. *Proceedings XXVIII Interamerican Congress of Sanitary and Environmental Engineering,* 27-31 october, FEMISCA, AIDIS, CWWA, Cancún, Mexico (in Spanish)

Méndez, J.M., Jiménez, B. and Barrios, J.A. (2002) Improved Alkaline Stabilization of Municipal Wastewater Sludge. Wat. Sci.Tech. **46**(10), 139-146.

Metcalf & Eddy Inc (1991) *Wastewater Engineering, Treatment, Disposal, and Reuse.* 3rd edition, revised by G. Tchobanoglous and F.L. Burton. McGraw-Hill International Editions, USA.

Meunier, N., Blais, J.F., Lounès, M., Tyagi, R.D. and Sasseville, J.L. (2001) Different Options For Metal Recovery After Sludge Decontamination at the Montreal Urban Community Wastewater Treatment Plant. *Specialized Conference on Sludge Management: Regulation, Treatment, Utilization and Disposal, International Water Association. Acapulco, Mexico. 25-27, October.* Femisca, D.F. Mexico: 304-311

Michel, I. and Rooksby, F. (2000) A continuous discharge-batch pasteurization process for the treatment of sewage sludges. *Proceedings of the Joint CIWEM Aqua Enviro Consultancy Services 5th European Biosolids and Organic Residuals Conference.* (eds. P. Lowe and J. Hudson) Seminar 6, paper 46, Aqua Enviro Consultancy Services, Wakefield, UK.

Murray, C. and Lopez, A. (1996) *Global Health Statistics.* Harvard University Press. USA.

Oropeza M., Castro P., Ortega S. and Chabirol N. (2000) Digestión anaerobia mesofílica y termofílica de lodos biológicos de desecho y lodos de tratamiento primario avanzado. Puesta en marcha del proceso. *XII Congreso Nacional 2000 Ciencia y Conciencia "Compromiso Nacional con el Medio Ambiente Morelia, México" Agust FEMISCA, D.F. Mexico* **(I):** *789-803.* (in spanish).

Okuno N., Ishikawa Y., Shimizu A. and Yoshida, M. (2004) Utilization of sludge in building material. *Wat. Sci. Tech.* **49**(10): 225–232

OTV (1997) *Traiter et Valoriser les Boues.* Infinités Collections, Saint Maurice, France (in French).

Owen, R. (1984) The effectiveness of chemical disinfection on parasites in sludge. In Sewage sludge stabilization and disinfection. (ed. Bruce, A.M.), pp. 426-439, Ellis Horwood Limited, UK.

Pike, E., Carrington, E. and Harman, S. (1988) Destruction of Salmonellas, enteroviruses and ova of parasites in wastewater sludge by pasteurization and anaerobic digestion. Wat. Sci. Tech. **20**(11/12), 337-343.

Reimers, R., Oleszkiewickz, J.A., Sheperd, S.L., Bakeer, R.M, and Fitzmorris, K.B. (2001) Advances in alkaline stabilization/disinfection of agricultural and municipal biosolids. *Proceedings of the IWA Specialised Conference on Sludge Management Entering the 3rd Millennium: Industrial, Combined, Water and Wastewater Residues.* 25-28, march pp. 104-110 [IWA, National Taiwan University, Chinese Institute of Environmental Engineers Taipei, Taiwan.

Roth, L.A. y Keenan, D. (1971). Acid injury of *Escherichia coli. Canadian Journal of Microbiology.* Vol. 17, No. 8. pp. 1005-1008.

Rhoades, J.D. (1997) Sustainability of irrigation: an overview of salinity problems and control strategies. p. 1–42. *In* Footprints of Humanity: Reflections on Fifty Years of Water Resource Developments. Proc. Canadian Water Resources Assoc. (CWRA) Conf., 50th, Lethbridge, AB. 3–6 June 1997. CWRA, Cambridge, ON. Canada

Sidhu, J., Gibbs, R., Ho, G. and Unkovich, I. (2001) The role of indigenous microorganisms in suppression of Salmonella regrowth in composted biosolids. Water Research **35**(4), 913-920.

Schuh R., Philipp, W. and Strauch, D. (1985) Influence of sewage sludge with and without lime treatment on the development of Ascaris suum eggs. In *Inactivation of micro-organisms in sewage sludge by stabilization processes,* (eds. P. Strauch, A.H. Havelaar and P. L´Hermite), pp. 100-113, Elsevier Applied Science, Hohenheim, The Netherlands.

Schwoyer W. (1981) *Polyelectrolysis for water and Wastewater Treatment.* CRC Press, Florida, USA.

Stien J.L. (1989) Oeufs d´helminthes et environnement: le modèle oeufs d´Ascaris. Thèse, Université de Metz, France.

Smith and Vasiloudis, 1989 en Snyman, H.G., Terblanche, J.S. and van der Westhuizen, J.L.J. (2000) Management of land disposal and agricultural reuse of sewage sludge within the framework of the current South African guidelines. *Wat.Sci.Tech.* 42(9), 13-20.

Snyman, H.G., Terblanche, J.S. and van der Westhuizen, J.L.J. (2000) Management of land disposal and agricultural reuse of sewage sludge within the framework of the current South African guidelines. *Wat.Sci.Tech.* **42**(9) 13-20.

Sullivan D.G. and Oerke D.W. (1997) Which Class A stabilization process is the most economical: Lime stabilization, Composting or Thermal Drying. In *Proceedings of the 10 th Annual Residuals and Biosolids Management Conference: 10 years of progress and a look to the future. Water Environment Federation,* **17**, 5-42.

Teixeira, S., de Souza, S., de Souza, N, Job, A., Gómes, H. and Neto, J. (2002) Caracterização de resíduo de estações de tratamento de àgua (eta) e de esgoto (ete) e o estudo da viabilidade de seu uso pela indústria cerâmica Brasil *Proceedings XXVIII Interamerican Congress of Sanitary and Environmental Engineering,* 27-31 october, FEMISCA, AIDIS, CWWA, Cancún, Mexico (in Spanish)

Tyagi, R.D., Sikati Foko, V., Barnabe, S., Vidyarthi, A.S. and Valèro, J.R. (2001) Simultaneous Production of Biopesticide and Alkaline Proteases by Bacillus Thuringiensis Using Sewage Sludge as a Raw Material. *Specialized Conference on Sludge Management: Regulation, Treatment, Utilization and Disposal, International Water Association. Acapulco, Mexico. 25-27, October,* pp. 312-319, FEMISCA, D.F. MEXICO

Tilche, A., Bortone, G. and Dohányos, M. (1999) Alternative waste water treatment processes to reduce sewage sludge production. In *Workshop on problems around sludge. Heinrich Langenkamp.* (eds. Luca Marmo), pp. 184-194, European Commission Joint Research Centre, EUR 19657 EN.

Thomaz-Soccol, V., Paulino, R.C., and Castro, E.A. (2000) Metodologia para análise parasitológica em lodo de esgoto. In *Manual de métodos para análises microbiológicas e parasitológicas em reciclagem agrícola de lodo de esgoto.* (eds. Andreoli C. and Bonnet R. Ed.), pp. 28-41, Sanepar, Curitiba, Brazil. (In Portuguese)

Teratani T., Okuno, N. and Kouno, K. (2001) *New Technology to Manufacture Fine Spherical Ceramic From Sewage Sludge. Specialized Conference on Sludge Management:*

Regulation, Treatment, Utilization and Disposal, International Water Association. *Acapulco, Mexico. 25-27, October.* pp. 296-303. [FEMISCA D.F. Mexico

Uribe H., Orozco G., Chávez N. and Espino S. (2002) Factibilidad económica del uso de biosólidos en el cultivo de maiz forrajero. Proceedings *XXVIII Interamerican Congress of Sanitary and Environmental Engineering,* 27-31 october, FEMISCA, AIDIS, CWWA, Cancún, Mexico (in Spanish)

Watanabe, H., Tomokazu, K., Ochi, S. and Ozaki, M. (1997) Inactivation of pathogenic bacteria under mesophilic and thermophilic conditions. Wat. Sci.Tech. **36**(6-7), 25-32.

Wang, M. (1997) Land application of sewage sludge in China. *Sci Total Environ.* **197**(1-3):149-60

Wanke, R., da Silva, G., Sant'Ana, T. and Goncalves, R. Soluções integradas para gerenciamento de lodos de pequenas estações de tratamento de esgoto sanitário na região sudeste do Brasil. *Proceedings XXVIII Interamerican Congress of Sanitary and Environmental Engineering,* 27-31 october, FEMISCA, AIDIS, CWWA, Cancún, Mexico (in Spanish)

WHO (1998) *WHO strategy on sanitation for high-risk communities.* Press Release WHO/18. January 29th.

WHO/UNICEF (2000) *Global Water Supply and Sanitation Assessment Report, Joint Monitoring Programme for Water Supply and Sanitation.* World Health Organization. Geneva.

WPCF (1988) *Sludge Conditioning, Manual of Practice FD-14.* Water Pollution Control Federation, United States of America.

Wu, C.C. Huang, C. and Lee, D.J. (1997) Effects of polymer dosage on alum sludge dewatering characteristics and physical properties. Colloids and Surfaces J. of Colloids and Intreface Science, **122**, 89-96.

Water Environmental Federation (WEF) and American Society of Civil Engineers (ASCE) (1992) *Design of Municipal Wastewater Treatment Plants, Volume II.* USA.

Young, A. (1998) *Land Resources: Now and for the Future.* Cambridge University Press. UK.

11

Management of Decentralised Sewerage Systems

Haniffa Abdul Hamid and Zaini Ujang

11.1 INTRODUCTION

Decentralised and small sewerage systems are widely used in both developing and developed countries. Currently, more than sixty million people in the United States live in homes where individual decentralised systems are used for wastewater management. The USEPA estimated that about 40% of new homes built are served with on-site small systems (Crites and Tchobanoglous 1998).

In the early 1970s, with the passage of the Clean Water Act, it was planned that sewerage systems would be centralised or regionalised in terms of facilities to almost all residents of the United States. 30 years later, however, it is recognised that complete sewered systems in the country may never be possible or desirable, due to both geographic and economic constraints (Crites and Tchobanoglous 1998). Hence, it is clear that effective decentralised and small wastewater management systems are needed for the protection of public health and the environment, including for developed countries.

Typical examples of various types of decentralised and small wastewater management system used worldwide are shown in Table 11.1.

Table 11.1 Typical types of decentralised wastewater management systems.

Types of systems	Comments
Individual premise	Vary from a simple system with a septic tank and a gravity fed soil absorption area to a system comprised of a septic tank with an internal or external wastewater treatment unit and distributed soil absorption system.
Cluster system	Catering 4 to 12 or more homes or premises, which are grouped together to form a cluster system to minimise cost of wastewater management.
Residential developments	Isolated housing developments (shop houses could be part of them) can be grouped together to achieve wastewater management objectives. In Malaysia for example, these vary from 50 to 5,000 units.
Integrated system (commercial, residential, institutional and recreational premises)	Wastewater from commercial buildings, apartment, institutional, industrial parks (only sewage components) and recreational facilities can be managed with complete recycle systems
Satellite treatment plant	Satellite wastewater treatment plants are integrated with centralised systems, usually for sludge management

In addition, the types of wastewater management facilities are closely associated to the types of technologies used for decentralised and small system. It covers the collection, transfer, degree of treatment and reuse options, which are important considerations particularly in water-poor areas. In general, the types of technologies for wastewater management systems are presented in Table 11.2.

Table 11.2 Typical technologies used for decentralised wastewater management systems.

Collection and Transport Of Wastewater	Wastewater Treatment and/or Containment	Wastewater Reuse Options
Conventionanal gravity sewers	Pre-treatment Lint Filters on washing machines	Cooling water
Small-diameter sewers variable - slope gravity	Primary treatment Septic tank (with effluent filter vault) Imhoff tanks	Constructed wetlands Wildlife habits Wetland mitigation
Pressure sewers (with non grinder pumps)	Disk and spin filters	Crop irrigation Groundwater recharge
Pressure sewers (with grinder pumps)	Secondary treatment Aerobic / Anaerobic unitsb	Land application for reuse Subsurface Drip
Vacuum Sewers	Aerobic treatment unitsb Anaerobic treatment unitsb	Shallow trench Shallow sand filled trench
Combinations of the above	Constructed wetlands Intermittent and recirculating packed-bed filtersMembrane bioreactors	Surface application Drip Spray
	Tertiary treatment Carbon adsorption Filtration Membrane filtration Reverse osmosis	Landscape and tree watering Streamflow augmentation Toilet flushing, clothes washing, car washing, etc.
	Home distillation unit	Combinations of the above
	Disinfection Chlorine UV radiation Ozone	
	Recycle treatment systems (e.g. toilet flushing and landscape watering)	

[a] Adopted from Tchobanoglous (1995) and Crites & Tchobanoglous (1998)
[b] Both suspended and attached growth processes are employed either single system or in various combinations

Decentralised and small systems are appropriate for many types of communities and conditions. Normally, cost-effectiveness is the primary consideration for selecting these systems. In general, the objectives of implementing such systems are summarised as follows:

Protects Public Health and the Environment Properly managed decentralised and small wastewater systems can provide the appropriate treatment necessary to protect public health and comply with the water quality standards. Decentralised systems can be sited, designed, installed and operated to meet various Federal and State required effluent standards. Effective advanced treatment units are available for additional nutrient removal and disinfection requirements. These systems can help to promote better watershed management by avoiding the potentially large transfers of water from one watershed to another that typically occur with centralised systems.

Appropriate for Low Density Communities In small communities with low population densities, the most cost-effective option is often a decentralised and small system. This system will reduce capital cost of wastewater collection and treatment by retaining water and solids near the point of waste generation, especially when a reuse scheme is available. This is more obvious in developing countries.

Appropriate for Varying Site Conditions Decentralised and small systems are suitable for a variety of site conditions, including shallow water tables or bedrock, low-permeability soils, and small lot sizes.

Other Objectives and Benefits Decentralised systems are suitable for ecologically sensitive areas where advanced treatment, such as nutrient removal or disinfection is necessary.

Since centralised systems require collection of wastewater for an entire community at substantial cost, decentralised systems, when properly installed, operated and maintained, can achieve significant reduction in capital and operating costs, while recharging local acquifers and providing other water reuse opportunities close to points of wastewater generation.

Despite the many advantages of small and decentralised sewerage systems, particularly its affordability and suitability within the framework of a sustainable sanitation approach, there are many aspects to be improved on management issues and technical advancement to ensure effective and efficient services. The shortcomings have been reported in recent publications (for example, Wilderer 2001 and Ødegaard 2001). This chapter examines the appropriate institutional framework and management issues for small and decentralised system to illustrate the importance of proper planning, strategy and monitoring systems. Special emphasis will be given to the experiences of the sewerage industry in Malaysia.

11.2   MAJOR ISSUES

In actual practice, developing countries, including fast-industrialising developing countries such as China, Malaysia, Thailand, Vietnam and South

Africa have adopted and are familiar with many types of small and decentralised wastewater systems. These range from treatment technologies, collection and catchment systems, to disposal and reuse schemes. However there are six major issues pertaining to management and operation of small and decentralised system, particularly in developing countries, as follows:

- **Treatment technologies are usually inadequate to meet the effluent standards.** The typical understanding of on-site or small sewerage system will be either primitive or low-tech options, which will not be able to meet high effluent standards mainly due to sludge loss (Ødegaard 2001). For example, effluent from septic tanks hardly comply with the Malaysian Environmental Quality (Sewage and Industrial Effluent) Regulations 1979, i.e. BOD of 20 mg/l and 50 mg/l for Standards A and B, respectively. Therefore the level of treatment technologies has to be refurbished or upgraded after several years of operation, and this could be very costly. The question will be on the funding ability for individuals or small business entities to invest in effective but expensive technologies.

- **Poorly operated and maintained**. The decentralised approach allows many on-site and small treatment plants to be built and operated. Due to lack of manpower and financial resources among owners or Local Authorities, many plants are poorly operated and maintained. For example in the sewerage catchment in the city of Johor Bahru, Malaysia, there are more than 50 sewage treatment plants and more than 5000 septic tanks. Most of the plants are owned and maintained by individuals or business entities with low level of expertise in wastewater engineering practice.

- **Difficult to supervise and controlled by authorities.** This is due to the fact that most of the on-site facilities and sewage plants are owned by individual or private business entities. And the volume is too numerous to be properly supervised and controlled by authorities. For residential areas, particularly in slump urban areas, septic tanks or pit latrines are located at the backyards of the houses, with no access road. Even if desludging services are available, they will be difficult to carry out.

- **Difficult to plan and upgrade**. It is a common situation in developing countries that small sewerage facilities, such as communal septic tanks and Imhoff tanks, are used for many years without improvement or upgrading by individual owners. In many cases it will be difficult for authorities to plan and invest to refurbish the facilities, or to connect to the public sewer.

- **Institutional support and management framework**. In many countries, environmental-related projects such as sewerage infrastructures are planned and regulated by the Department of

Environment, or the Environmental Protection Agency. However the Department is normally in the position to execute infrastructure projects or capital works because of its nature as regulatory agency. In Malaysia, the Sewerage Services Department, under the Ministry of Housing and Local Government is responsible for planning, managing and regulating the sewerage sector. On the other hand the Department of Environment is responsible for regulate industrial and hazardous waste management.

- **Piecemeal type of development.** This current practice poses a major drawback for the more efficient basin-wide sewerage system development in many countries. Not only does the number of small package sewerage treatment plants increase over the years, but the different type and mode of operations of these plants need further investment in equipment and trained personnel by the operators. Such substantial investment will need to be recovered through the tariff and appropriate charging schemes.

It is also important to note that the concept of separation at source for domestic wastewater is not new to developing countries. However, this is not intentionally designed for in rural areas but rather an improvement in housing and communal facilities, particularly in rural and urban slump areas. In the past few decades storm water drainage systems have been constructed for flood mitigation. Now the systems are also used for collection and transport of greywater. In such situations, domestic wastewater is separated into two separate pipelines, as shown in Figure 11.1.

The separation at source could be an option for public health protection: *Blackwater* from toilets is treated in individual septic tanks or small sewage plants, and the *greywater* or sometimes called sullage, is discharged to a hydraulically well-designed stormwater drain, or into a sullage soakaway. Communal septic tanks are also commonly used to treat blackwater from a cluster of houses up to about 300 PE (Mara 1996).

However, separation at source can also contribute to water pollution. In this situation stormwater drains contain high concentrations of washing chemicals, oil and grease and residuals of food. Most of the stormwater drains in urban areas in tropical countries are connected to tributaries of major rivers. As a result it is a common phenomenon that tropical rivers are mainly polluted due to untreated greywater. In the city of Johor Bahru, Malaysia, for example, the concentration of COD of greywater at various points prior to discharge to the Segget River was recorded on average to be 1400 mg/l.

In other words, separation at source in the present context in developing countries is not an ideal situation. A lot more improvement is urgently required, especially in the collection system and pre-treatment stage. In certain countries, including recently in Malaysia, greywater and stormwater are collected from households and their surrounding areas. Then it will be channelled to sand filtration basin, which is an open impoundment. The basin contains filter media, normally sand and gravel, and an underdrain system. In addition, constructed wetlands could also be used. It is an effective pre-treatment method for conventional organic pollutants with partial nutrient removal.

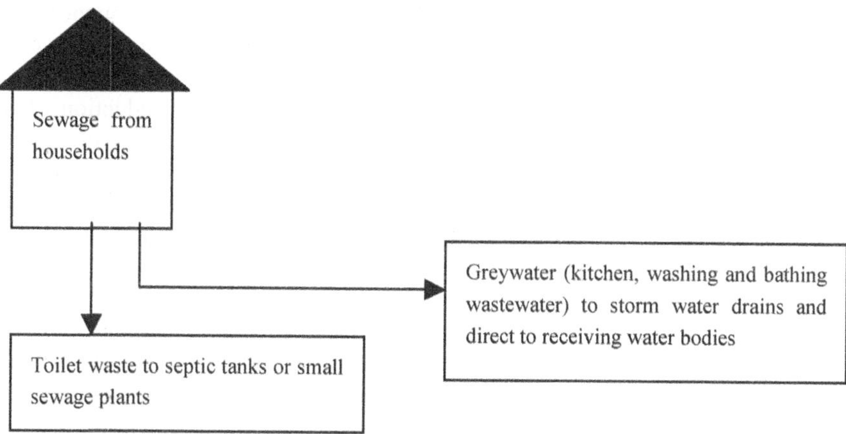

Figure 11.1 Typical separation of domestic wastewater at source in residential areas in developing countries.

11.3   PLANNING AND STRATEGY FOR SEWERAGE CATCHMENT

11.3.1 Institutional framework

The institutional framework is one of the most important aspects in decentralised and small sewerage management. It covers policy statements, regulatory framework, management and planning and financial mechanisms. In many developing countries the demarcation line between the implementing and regulatory agencies has been poorly defined. As a result there are many agencies planning the same projects or activities to provide appropriate sewerage facilities and services for the same target groups. The overlapping of work and authority is a common situation in most developing countries, partly due to the fact that the related agencies are competing to get development funds both from local and foreign sources.

In a few fast-industrialising developing countries, there is a coordination of sewerage policies, project management and regulatory compliance. Generally the coordination defines the boundary of responsibilities, tasking, scope of work, management team, annual operating budget, development budget and regulatory enforcements. In Malaysia, for example, the institutional framework and responsibilities are illustrated in Table 11.3.

The Ministry of Health has been instrumental in introducing septic tanks to rural areas under its public health and engineering divisions since 1960s. The function of the Department of Environment is mainly to set and regulate the effluent quality, both from sewage and industrial wastewater treatment plants. The major portion of the empowerment and responsibilities in the sewerage sector lies on Department of Sewerage Services that regulate the sewerage industry as a whole since its formation in 1994, after the introduction of the Sewerage Services Act 1993. Local authorities, in addition, are responsible for approving new development projects including sewage treatment facilities, as well as maintaining the urban services including solid waste and wastewater management in their operational areas. Certain local authorities have a bigger role, such as Pasir Gudang Local Authority, including managing industrial waste facilities and providing integrated solid waste management.

In addition the Standards & Industrial Research Institute of Malaysia (SIRIM Berhad) provides the code of practice on sewerage services, particularly on the design of sewer and treatment plants. Indah Water Konsortium is the private consortium, awarded a 28-year contract to operate, maintain and build sewerage infrastructures in Malaysia. In addition, the Malaysian Water Association provides training together with local universities and professional networking among practitioners, researchers and policy makers to enhance human resource development.

In general, an effective institutional framework is critical for the success of decentralised sewerage systems, particularly in developing countries where political stability and economic situations are not always supportive. The framework defines the level of system reliability and funding mechanism. Each agency plays their role accordingly and supports others to serve the public. The organisational framework of Sewerage Services in Malaysia is presented in Figure 11.2.

Table 11.3 Institutional framework and responsibilities within the sewerage sector in Malaysia.

Agencies Functions/Policies	Ministry of Health	Department of Environment	Department of Sewerage Services	Local Authorities	Standards & Industrial Research Institute of Malaysia	Indah Water Konsortium	Malaysian Water Association
Environmental Quality (Sewage and Industrial Effluents) 1979		■					
Introduction of pour-flush latrines for rural areas & "conventional centralised sewerage", 1960s	■						
Policy on new housing development to provide local sewage treatment system, 1980			■	■			
Approval of housing and commercial developments				■			
Malaysian Standard 1228:Code of Practice for Design & Installation of Sewerage Systems, 1991					■		
Establishment of the Sewerage Services Act 1993			■				
Privatisation of O&M of sewerage services in Local Authority operational areas 1993			■			■	
Publication of Guidelines for Developers: Vol. 1–V, 1996			■				■
Policy on sewerage capital contribution, i.e. 1.65% of the property value, 1996			■				
Regulation: all septic tanks to be desludged every two years, 2002			■			■	
Human resource development		■	■			■	■

11.3.2 Planning

Normally the size and coverage of small and decentralised municipal wastewater services are growing proportionally to the housing facilities provided in residential development areas. In many major cities in developing countries, private developers build residential development areas with appropriate sanitation facilities, at least with septic tanks. However, rapid urbanisation creates urban squatters who are living in unplanned residential or slump areas with inappropriate living conditions.

At present 60 percent of the world's population lives in urban areas. The United Nations World Water Development Report (2003) indicates that at present 1.1 billion people lack access to improved water supply and 2.4 billion to improved sanitation and they are mainly living in urban areas. Asia shows the highest number of people unserved by either appropriate water supply or sanitation as shown in Table 11.4. In terms of proportionality, this group is larger in Africa due to the difference in population size between the two continents.

In general, the statistics show the need for a policy shift in lower-income countries towards better household water-quality management, as well as the continued expansion of water supply and sanitation coverage. It is also important to upgrade the existing levels of services and facilities in order to protect public health and environmental degradation. Planning for sewerage infrastructures require extensive data on the following items as listed in Table 11.5. It needs a detailed description of the catchment, including boundaries and topography, land use and existing and future conditions and facilities.

Table 11.4 Water supply and sanitation distribution of unserved population (United Nations World Water Development Report 2003).

Continents	Water supply	Sanitation
Asia	65%	80%
Africa	27%	13%
Latin America and Caribbean	6%	5%
Europe	2%	2%

Figure 11.2 (opposite) The organisational framework of Sewerage Services in Malaysia.

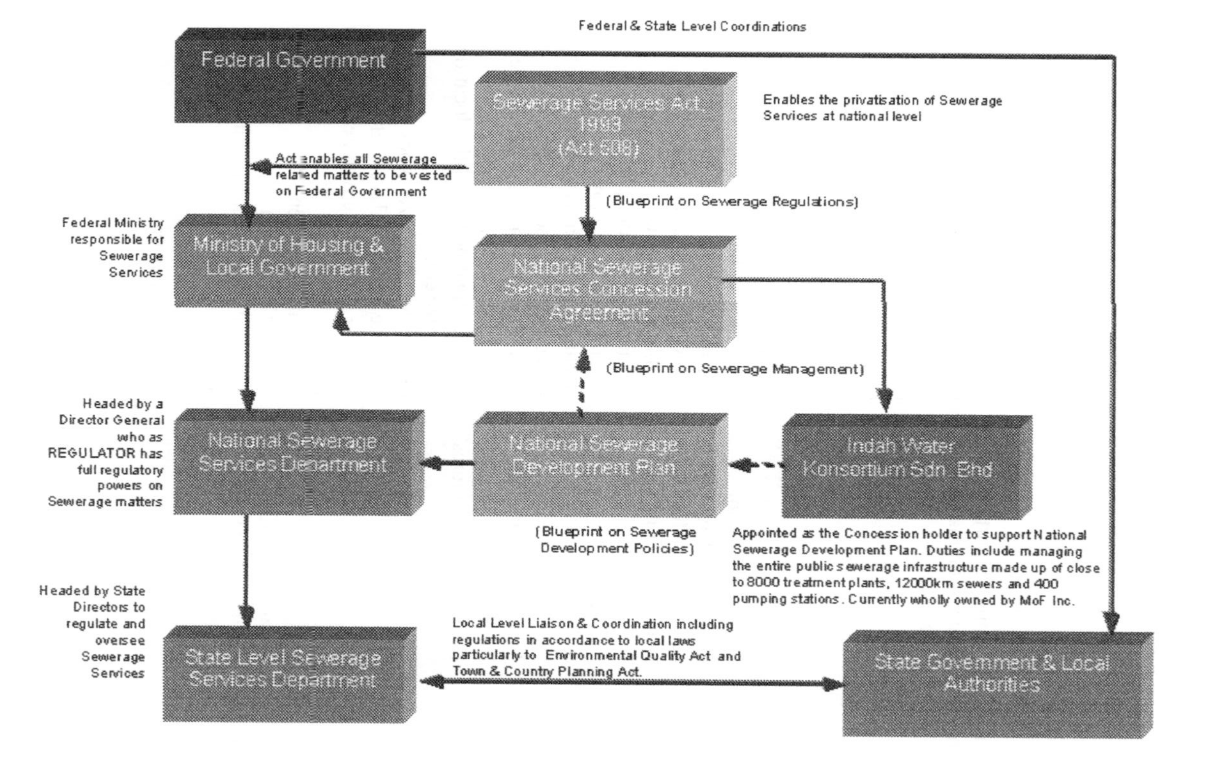
Federal & State Level Coordinations
Federal Government
Sewerage Services Act, 1993 (Act 508)
Enables the privatisation of Sewerage Services at national level
Act enables all Sewerage related matters to be vested on Federal Government
Federal Ministry responsible for Sewerage Services
Ministry of Housing & Local Government
National Sewerage Services Concession Agreement
(Blueprint on Sewerage Regulations)
(Blueprint on Sewerage Management)
Headed by a Director General who as REGULATOR has full regulatory powers on Sewerage matters
National Sewerage Services Department
National Sewerage Development Plan
Indah Water Konsortium Sdn. Bhd
(Blueprint on Sewerage Development Policies)
Appointed as the Concession holder to support National Sewerage Development Plan. Duties include managing the entire public sewerage infrastructure made up of close to 8000 treatment plants, 12000km sewers and 400 pumping stations. Currently wholly owned by MoF Inc.
Headed by State Directors to regulate and oversee Sewerage Services
State Level Sewerage Services Department
Local Level Liaison & Coordination including regulations in accordance to local laws particularly to Environmental Quality Act and Town & Country Planning Act.
State Government & Local Authorities

Table 11.5 Information and data required for planning of a sewerage catchment.

Items	Objectives	Details
Catchment descriptions	To provide a brief background data	-Local area description -Boundaries of local authorities -Topography overview -Historical landuse summary -Current landuse status -Externalities – airport,major developments? -Geology -Water intake points, if appropriate
GIS for the Catchment	To provide further data for decision making processes	-Topography – drainage, roads, pipelines, railways -Boundaries – major developments, area of local authorities, rationalisation coverage -Natural sub-catchment break-up -Landuse zones -Water intake points – present and future
Issues	To describe major issues that lead to the need to upgrade the sewerage facilities	-Rezoning of land -Change in government policy -New programmes or policies -Public complaints -Change to previous planning assumptions (e.g. growth rates) -Change to standards of service -Specific developer requests
Existing conditions	To provide sufficient information on the existing infrastructures	-Sewerage sub-catchment and coverage -Sewage pumping stations -Sewage treatment facilities,locations, types, size etc. -Sewer network, size, types, materials -Sewage treatment levels, efficiency -On-site systems – coverage, types, size etc. -Greywater treatment?
Future conditions	To provide information on the planned developments	-Population growth -Residential, commercial, institutional and industrial -Premises– staging, timing, funding -Sewerage coverage and strategy

11.3.3 Planning of sewerage catchment

Catchment Planning (CP) is critical for the long term efficiency of sewerage infrastructures. The objective of CP is to minimise the whole life costs for a sewerage system. This is achieved by considering the various ways in which a catchment can be provided with sewerage services, listing the options available, calculating the present worth of both the capital and operational costs over the life of assets for the various options and finally selecting the option with least present worth. Catchment planning has many benefits as follows:

- Planned sewerage works will be discussed with Federal, State and Local Government to ensure that the proposed works integrate with services provided by other utility companies.
- Land for sewage treatment plants, sludge treatment facilities and network pump stations will be identified well in advance of the need for the works thereby permitting various interest groups to comment on the suitability of the sites and the application or acquisition of the land to be made
- Works constructed by developers will be integrated with those constructed by the Government ensuring only the most acceptable and cost efficient works are installed
- Works identified in a catchment plan will be placed on Capital Works program and prioritised based on the greatest need

There are many stages involved in developing an objective sewerage catchment plan. Specifically, the stages could be simplified as follows:

Establish Catchment Boundaries A sewerage catchment can be reviewed as an area over which raw sewerage can be conveniently conveyed, predominantly by gravity, and at reasonable costs, to a treatment plant located within its confines. A demarcated sewerage catchment area is usually segregated into a number of drainage cells or individual sub-catchments dictated mainly by physico-graphic profiles. Nonetheless existence of natural and artificial barriers which impact sewerage flows by gravity can influence the number of sub-catchment calls. The factors defining catchment limits include topography and administrative borders.

Appraisal of Existing Sewerage System An essential task to be carried out in any Sewerage Catchment Planning exercise in an assessment of the different forms of sewerage systems existing within the demarcated catchment area. A consequential task is to carry out a comprehensive inventory of existing sewerage assets within the catchment's boundaries, and an assessment of their structural state, hydraulic capacity, functional capability and performance efficiency. Assets under construction, or to be constructed in the near future, need also to be included in the inventory. The assessment should generate information which can be utilised to evaluate their further usefulness, i.e. whether they can be rehabilitated and upgraded, and whether there is a need to improve their effectiveness and performance efficiencies to cater for additional loads and to conform to stricter effluent discharge standards. The data obtained, and the deductions arising from this review, will assist in identifying and evaluating alternative sewerage upgrading and expansion strategies.

Sewage Flow and Sludge Generation Estimate Estimation of sewage flow and associated pollutant loading and sludge generation rate are important considerations as they dictate the sizing of collection and conveyance systems, and the determination of treatment capacities. The sewage flows estimation covers the various flow regimes of Average Dry Weather, Average Wet Weather and Peak Wet Weather flows. Estimation of future Population Equivalent (PED) counts within the catchment will then provide the pollutant load projections of specific sector contributions from residential, commercial and industrial sources.

Assess Existing Sewerage Deficiencies Constraints that exist within the current sewerage infrastructures to receive future sewage loads must be identified. Among the major factors that would be considered to propose alternatives in solving existing constraints are:

- Barrier to sewage conveyances
- Site availability for pumping stations and treatment plants
- Sludge management
- Environmental factors

Optional Sewerage Strategies The number of different sewerage strategic options available for any catchment are large where several possible sites for treatment plants within a catchment may be possible. Elimination of relatively impractical and cost prohibitive options would be needed usually by the following steps:

- Identify issues and constraints that will apply
- Identify available options
- Assess the impact of issues and constraints on the option

Financial Analysis Comparison of costs is often the factor by which options can quantitatively be assessed against each other. A sewerage scheme will require to be constructed and this entails capital expenditure. In addition recurrent operating costs will have to be incurred to maintain or replace equipment in order to achieve operational objectives. The combination of these components of the cost needs to be accounted for in any analysis. The NPV and other financial tools such as the rate of return are commonly used to gauge the financial merits of options.

Selection of Preferred Option The selection of the sewerage strategies would be finalized after the assessment made encompassing the technical, financial and environmental analysis of each option, subjecting them to a composite ranking exercise with weighting factors to emphasise the important criteria. Upon selection, a programme for a staged installation of facilities under the Catchment Plan should be prepared.

Components of Sewerage Systems They are a basic infrastructure component of built-up areas that support relatively high population densities. They include all physical facilities involved with the collection of sewage at source, conveyance to a treatment facility, its treatment to conform to effluent discharge standards, and finally either its disposal to the environment, or its reuse as irrigation waters or for other accepted purposes. The physical assets associated with a sewage system would include:

- Property connections which channel sewage from individual building(s) to the nearest public reticulation sewer, or directly to an on-site treatment plant (minimum 150mm in diameter).
- Reticulation sewers of relatively small diameters (225 mm to 300 mm diameter) which channel sewage collected from groups of individual properties to branch sewers (300 mm to 450mm diameter).
- Branch sewers (9300mm to 450mm diameter) which receive sewage from reticulation sewers and thereafter transfer the flow to small sewage treatment plants, or to main sewer (450mm to 900mm diameter).
- Main sewers (450mm to 900mm diameter) which receive sewer from branch sewer and thereafter transfer flows to small sewage treatment plant or trunk sewers (greater than 900mm diameter).
- Trunk sewer (greater than 900mm diameter) from the spine of a large catchment and transfers sewage to a large sewage treatment plant.
- Sewage treatment plants (STP) which converts the sewage into a form which can be safely discharged to the environment, or reused for various compatible purposes.
- Sludge treatment and disposal facilities which cater for low solids content sludge that are generated by STPs. Sewage sludge management facilities can be part of an overall STP, or installed as a separate entity at centralized sludge processing centers. Facilities are provided to:
- thicken the sludge in order to reduce its volume so that it can be more economically processed thereafter.

- stabilise sludge by reducing its degradable organic content and pathogenic properties.
- dewater the stabilized sludge to further increase its solids contents (25%) such that it becomes a "spadeable". And more easily handled and transported to approved disposal centers.
- Systems to dispose or reuse the processed sludge.
- On-site (decentralised) treatment facilites.

Separate Sewer System Sewerage Catchment Planning should focus only on implementation of separate sewer system that cater for raw sewage discharges exclusively. They should not be planned to accommodate storm water runoffs (e.g. rain water collected from building roofs, courtyards and street pavements). However, allowances should be incorporated in sewerage planning to accommodate storm waters that may unintentionally or inadvertently be introduced into the sewerage system (e.g. through uncovered manholes, leaking pipe joints and manhole structures, and illegal roof drainage connections).

Cater for Growth Areas In sewerage catchment planning special emphasis should be made to cater for Growth Areas. Growth Areas are defined as new developments occurring within a catchment.

Environmental Implications Sewerage Planning should take due cognizance of protecting the aesthetic, recreational and biological values of water bodies. In addition, sewerage planning for catchments which contribute to the water resources potential of public water supply systems should be prioritised to ensure adequate protection of raw water intakes.

11.4   CAPITAL COST

One of the reasons for adopting small and decentralised sewerage systems is lower capital investment cost (compared to centralised systems) that can be implemented in stages based on the need, affordability and urgency. Various parties, mainly the Governments and private sector (with various type of incentives) can share the capital cost and spend it invarious stages. In general, capital cost for a decentralised sewerage system is closely associated with the ownership and types of the served facilities or premises. The served facilities for developing countries can be divided into four major categories, as follows:

- **Individual-household system.** This is mainly pour-flush, ventilated improved latrines and individual septic tanks.

- **Communal-based on-site system**. The typical system used in many urban and peri-urban areas is individual septic tank. In some areas communal septic tanks are connected to individual pour-flush latrines.
- **Household-centred low-cost sewerage systems**. This is a system that is commonly used for a proper residential setup. The system includes toilets, sewer network and low-cost small sewage treatment plants such as waste stabilisation ponds and constructed wetlands. In some areas the system also accommodates greywater.
- **Decentralised and small mechanical sewerage systems**. This is a system that is commonly used for a proper built environment urban setup. The system includes toilets, sewer network and small sewage treatment plants. Normally the system accommodates greywater.

The four categories as above are further described in Table 11.6 and related to the responsibility to provide the capital costs. House or premise owners normally pay capital cost for their sewerage facilities, which includes toilet, connecting sewer, main sewer and sewage treatment plants. The unsewered on-site treatment system will be cheaper and affordable. This applies to many housing schemes in urban areas in most developing countries. However this option is still debatable from a cost-effective perspective.

In Malaysia, the Government has decided on a policy for housing developers to provide local sewage treatment plants since 1980. This framework was further improved by the introduction of Sewerage Capital Contribution (SCC) for new developments in 1996. The principle is to share the responsibility of sewerage infrastructure cost among developers when centralised sewage treatment plants are available.

However, the housing developers can only charge their customers who bought the middle and high cost premises. Low-cost premises are cross-subsidised by the SCC scheme. While the government is responsible for ensuring provision of adequate sewerage facilities, developers also play their roles to partly bear the cost of building the sewerage infrastructure for their new developments. Developers (including government agencies) continue to be responsible for bearing the cost of the sewerage infrastructure for new developments, which include the following components:

- sewer reticulation and pumping stations
- a share of the cost of trunk sewer and pumping stations
- a share of the cost of centralised sewage treatment plants catering for sewage and sludge treatment
- full cost of temporary sewage treatment plants

The funding of the sewerage infrastructure by the developers constitutes the SCC. Based on the amount to be spent on capital works including operation and maintenance of new sewerage infrastructure of approximately RM24 billion over the next 28 years concession period, a new sewerage tariff at least 70% higher than the existing rates would be required to fund the additional works.

The Government of Malaysia has seriously reviewed this and finally, the Federal Cabinet approved a sewerage capital contribution rate of up to 1.65% of the property value to be levied on all properties, except low cost houses. This SCC has been made part of the approval conditions of all new development projects by the Government. The SCC will be paid into the Government Trust Account, which will be used for the purpose of upgrading of sewerage infrastructure by the Government. More details about SCC are discussed with two calculation examples in this chapter.

The pressing problem on sewerage planning is on logistic issues, especially on individual premises in urban slump and remote sparsely populated areas. It will be very costly and legally unacceptable to connect to public sewers. For urban slump, there are many options available that mainly focus on the household-centred environmental sanitation concept. For remote areas, individual on-site systems such as septic tanks, composting toilets or *johkasou* are encouraged in many developing countries, including in environmental-proactive developed countries, such as Norway, Denmark, Sweden and Germany. In Malaysia for example, two-chamber septic tanks are commonly used in rural and remote areas. The design should comply with the *DSS & MWA Guidelines for Developers* (Vol. V on Design of Septic Tanks). In addition, Johkasou is a Japanese technology specifically designed for rural areas in Japan, as well as has been part of the foreign aid package in the sewerage sector to many developing countries (Yang *et al.* 2001).

The capital cost for constructing septic tank systems (consists of pour-flush and the septic chambers) is not expensive and considered as affordable for most of the people in many developing countries. Composting toilet are also cheap but requires proper plumbing connections in order to reduce odour and collection problems. However, *johkasou* is still expensive for developing countries standards, since it is a package plant consisting of many mechanical and electrical parts.

Table 11.6 Categories of small and decentralised wastewater facilities in relation to the capital cost.

Systems	PE	Target populations	Types of wastewater	Technologies options	Responsibility for capital cost
Individual on site systems	1-10	-Rural areas -Urban poor	-Blackwater	-Septic tanks -Johkasou -Vacuum toilets -Composting toilets	-Premise owners
Communal on-site system	10-100	-Urban poor -Government-sponsored houses	-Blackwater	-Communal septic tanks -Imhoff tanks -Membrane johkasou -Package plants	-Premise owners / developers -Subsidy by government
Household-centred low-cost sewerage systems	100-5,000	-Urban poor -Middle class urban	-Blackwater -Greywater	-Sewer networks -Waste stabilisation ponds -Wetlands, land treatment -Sea outfall	-Premise owners (provided by developers) -Subsidy by government
Decentralised and small mechanical sewerage systems	5,000-50,000	-Low cost flats -Middle class houses -Resort areas -Business areas -New townships	Blackwater -Greywater	-Sewer networks -Package plants: activated sludge systems -Package plants: biofilm systems -Package plants: hybrid system	-Premise owners (provided by developers) -Subsidy by government for rationalisation scheme

Example 11.1 In a sewerage feasibility study in Taman Mutiara Rini, Malaysia, the population and growth trends suggest that wastewater flow rate will double over the next 10 years. The cost of expanding the existing sewerage infrastructures will be compared with a phased program of expansion. Immediate development would cost RM1,000,000 with annual maintenance costs of RM500,000/year. A phased programme would involve a capital cost of RM600,000 and an estimated expenditure of RM700,000 in 5 years. Annual maintenance cost is estimated at RM100,000/year for the first 5 years and at RM50,000/year following. Assuming a perpetual period of service for each system and an interest rate of 7%, evaluate the two alternatives on a capitalized cost basis.

Solution: Capitalized cost may be considered the same as a present worth analysis when an infinite period time is considered.

For the immediate expansion program:

Initial Cost = RM 1,000,000

$$\text{Maintenance Cost} = \frac{RM\,500,000}{0.07} = RM\ 7,142,860$$

Capitalised Cost = RM 1,000,000 + RM 7,142,860
                 = RM 8,142,860

For the phased expansion program:

Initial Cost = RM 600,000

$$\text{Present worth factor (pwf)} = \frac{1}{(1+i)^n}$$

where i = interest rate and n = number of interest periods

5-year cost = RM700,000 (pwf)
             = RM700,000 (0.7130)
             = RM499,100

Maintenance cost is calculated as RM 50,000/year for an infinite period of time plus the present worth value of RM 50,000/year (RM 100,000-50,000) for an initial 5-year period.

$$\text{Maintenance Cost} = \frac{RM\,50,000}{0.07} + (RM\ 100,000 - RM\ 50,000)\ (\text{pwf})$$
$$= RM\ 749,936$$

Capitalised Cost = RM 600,000 + RM 499,100 + RM 749,936
$$= RM\ 1,849,036$$
The phased expansion program has the lower capitalised cost.

Example 11.2 Two types of sewer pipe are being evaluated for a construction project. Pipe A will cost RM 100/m and have an estimated service life of 60 years. Annual cost for pumping and maintenance is estimated at RM 250,000 per year. Pipe B has an estimated service life of 40 years and an annual pumping and maintenance cost of RM 200,000 / year. Approximately 20,000m of pipe is to be installed. Calculate how much more can be paid for Pipe B than Pipe A. Assume the interest rate of 7%.

Solution: Problem will be solved on an annual cost basis and the break-even value of Pipe B calculated.

For Pipe A,

Initial Cost = RM 100/m x 20,000m = RM 2,000,000

Annual Operating Cost = RM 250,000

Capital recovery factor (crf) @ 7%, 60 years = $\dfrac{i[1+i]^n}{[1+i]^n-1}$ = 0.0712

For Pipe B,

Annual operating cost = RM 200,000/year

Capital Recovery Factor @ 7%, 40 years = 0.075

Annual Cost Pipe A = Annual Cost Pipe B

RM 2,000,000 (0.0712) + RM 250,000 = B(0.075) + RM 200,000

Cost Pipe B = RM2,565,300

The cost of pipe B should not exceed RM128.27.

11.5   OPERATIONS AND MAINTENANCE COSTS

Operation and maintenance costs in sewerage services cover the following activities:

- operating and maintaining sewage and sludge treatment plants
- maintaining sewer network
- clearing blockages in sewer line
- refurbishing and upgrading sewage treatment plants
- desludging of sewage sludge from septic tanks
- monitoring effluent quality and public awareness programmes.

In many developing and developed countries the responsibility for operation and maintenance of sewerage services, particularly sewage treatment facilities, lies on Local Authorities. The consumers pay the service under council or municipal tax.

However the quality of such services has always been at minimum level due to the fact that Local Authorities are also responsible for providing other urban facilities which can be considered higher priority, such as solid waste management, road construction and maintenance, flood mitigation and public recreational facilities. In many cases the portion of investment on operation and maintenance of the sewerage system is limited.

This is the main motivation behind privatisation of the sewerage sector in many countries, including Malaysia. In other terms, the privatisation of sewerage sector allows the Federal Government to take over from Local Authorities the responsibility to develop, manage and operate the sewerage facilities and services. The national sewerage company, Indah Water Konsortium, has been appointed to implement the policy. The monthly fee for sewerage services in Malaysia is based on the type of houses and customers, set by the Department of Sewerage Services.

The billing structure of sewerage services, as practised in Malaysia and many other countries, can be classified into four (4) categories as follows:-

- **Domestic Customers** are customers who occupy or own domestic premises, low cost dwelling house, new village dwelling house, village dwelling house and government quarters. All customers in this category will pay fixed sewerage charges based on type of premises. The tariffs for this category range from RM2 to RM8 per month.
- **Commercial Customers** are customers who occupy or own any commercial premises used wholly or partly for trade, business, provisions of services or facilities or any other activity whether for profit or otherwise.

Commercial customers will pay fixed charges based on the annual value of their property, plus a charge for additional cubic volume of water consumption, which exceeds 100 cubic meter per month.

- **Industrial Customers** include all customers who own industrial premises in which the principal activity carried out involves the making, altering, blending, ornamenting, finishing or otherwise treating or adapting any article or substance with a view to its use, sale, transport, delivery or disposal and includes the assembly of parts and ship repairing but does not include any activity normally associated with retail or wholesale trade. Industrial customers will be charged based on the domestic portion of the wastewater i.e. by total number of employees in the factory.

- **Government Premises** (excluding government quarters) are any premises owned or occupied by the Federal Government, the Government of a State, a local authority, a statutory body established by Federal or State Law, a court or tribunal. The Government Premises will pay a fixed charge (RM25 for Septic Tank and RM40 for Connected Services), plus a charge of 65 sen per additional cubic meter of water consumption exceeding 100 cubic meter per month.

NOTE: "Annual Value" of a property is defined as the estimated yearly rental, which can be derived from the said property and is not the same as the property value.

The **types of services** provided in Malaysia are categorised into two (2) main categories: (a) connected sewerage services and (b) septic tanks services.

- **Connected sewerage service** customers are those with their sewage outlet linked directly to a sewage treatment plant via an underground network of sewer pipes. The typical connected services include a chamber with a metal cover, which is known as inspection chamber. The inspection chamber is to access the private connection pipe in case of blockages. This system does not require scheduled desludging. The sewer pipes are eventually connected to a sewage treatment plant where sewage undergoes treatment.

- **The septic tank services** are scheduled desludging or removal of sludge from septic tanks which are carried out once every two (2) years or less depending on the capacity and design of the septic tanks. Desludging of septic tanks is a mandatory service and is regulated and monitored by Department of Sewerage Services. Desludging of septic tanks is very important to prevent untreated sewage and sludge being released into our rivers and / or water sources. This is because the maximum volume of sludge that a septic tank can store is approximately one third of its total volume.

However in many developing countries, particularly in less developed countries, communities take part to operate and maintain individual on-site systems and low-cost communal sewage treatment facilities. This is due to the fact that the institutional and financial frameworks are not sufficient to maintain the facilities, and poor living conditions among served population. Individual on-site sewerage systems, such as pit latrines and septic tanks need good operation and maintenance at the household level. Mara (1996) proposed a household-level operation and maintenance (HLOM) concept in which household members (especially women) may need special training in HLOM techniques.

Community-level operation and maintenance is also appropriate for low-cost communal sewage treatment facilities such as waste stabilisation ponds or constructed wetlands. However many communities require specialised training to maintain the systems. Thisinvolvement has been normally coupled with economic benefits such as fishing stocks at the final ponds.

11.6   DATABASE AND ZONING

11.6.1 Types of sewage plants

Malaysia currently has almost 8,000 sewage treatment plants and approximately 12,500 kilometer length of sewers, mostly situated in urban areas, serving more than 12 million population. Specifically, approximately 55% are communal septic tank type, 15% is Imhoff tank type, 10% is waste stabilisation ponds/aerated lagoons type, and only about 20% are of mechanised treatment plant type which provide secondary treatment. Besides these plants, Malaysia also has about 1.2 million individual septic tanks or other types of sanitation systems such as pour flush latrine etc, serving more than 6 million population. The distribution and types of sewage treatment plants in Malaysia are shown in Table 11.7.

Table 11.7 Distribution of treatment facilities by types and states (May 2003).

States	Sewer (km)	Plant types					Total
		CST	IT	WSP/ AL	MP	NPS	
Johor	877.59	450	49	77	155	50	781
Kedah	367.14	415	24	31	154	19	640
Melaka	428.21	300	34	16	248	10	608
Negeri Sembilan	731.69	438	84	83	188	36	829
Pahang	235.01	175	39	24	128	8	374
Perak	1041.28	574	111	139	293	53	1170
Perlis	11.07	116	1	0	11	0	28
Pulau Pinang	1740.43	94	131	21	310	32	588
Selangor	4787.35	911	195	167	745	137	2155
Trengganu	54.96	149	21	4	22	6	202
WP Kuala Lumpur	1534.95	23	89	47	73	52	284
WP Llabuan	24.07	5	3	2	12	6	28
WP Putrajaya	65.95	0	0	0	1	8	9
TOTAL	11898.70	3547	781	611	2340	417	7696

Note: CST = communal septic tanks, IT = Imhoff tanks, WSP/AL = waste stabilisation ponds / aerated lagoons, MP = mechanised plants, NPS = number of pump stations

Table 11.8 Treatment facilities by type and size (May 2003).

PE RANGE	CST	IT	WSP/ AL	MP	Total No. Plants	% Plants	Total PE (Estimate)	% Total PE
< 150	3022	67	1	51	3141	43	232758	2.1
151 - 1,000	490	560	107	1116	2273	31	1048648	9.3
1,001 - 5,000	33	150	360	883	1426	20	3220892	28.7
5,001 - 10,000	1	4	85	172	262	4	1849496	16.5
>10,000	1	0	58	118	177	2	4869951	43.4
TOTAL	3547	781	611	2340	7279	100	11221745	100.0

Note: Not included –   One Marine Outfall for PE - 432,235
                                          NPS 417 for PE - 1,780,148

Table 11.8 shows the distribution of sewage treatment plants by type and size. It should be noted that about 80% of existing treatment plants are less than

1,000 population equivalent (PE) but serve only 13% of the population. The balance 20% of the treatment plant serves 87% of population. This should be an important factor for consideration in planning future improvement strategies i.e. decentralisation impacts.

11.6.2 Typical mechanised plants

There are two main types of mechanised plants widely used for sewage treatment as discussed in Chapter 4, i.e. (a) biofilm system and (b) activated sludge system. In general biofilm system in Malaysia consists of a bio-reactor, clarifier and sludge digester. The biofilter system, for example, consists of biofilter tower for oxidation of organic materials by attached bacteria, clarifier for sedimentation of sloughed-off biomass and sludge digester for stabilisation of biomass.

Typical problems which may be encountered by the system are odour from the biofilter tower due to lack of oxygen for microorganism oxidation for organic materials, difficult to maintain clarifier due to its placement underneath the biofilter tower and discharging of screenings to the digester may result in choking of sludge digester and desludging may be difficult. Other biofilm systems used in Malaysia are rotating biological contactor and trickling filter.

Activated sludge systems, in general consist of primary sedimentation where settlable solids are removed, followed by aeration tank and clarifier for oxidation of organic materials by bacteria and sedimentation of biomass from the liquid stream respectively. The excess biomass or sludge from the system is normally thickened and discharged to a sludge digester for stabilisation before being dewatered at sand drying beds which are also provided on-site. The treatment systems in this group includes conventional activated sludge, extended aeration, oxidation ditch and package plants.

11.6.3 Other sanitation facilities

The rural and other remote areas such as villages and squatter areas are still using individual on-site sewage treatment systems, such as pit latrines, bucket latrines and overhanging latrines. The systems can only provide primary sedimentation and are unable to comply to the regulated effluent standards. These systems directly pollute our water sources.

11.6.4 Current sludge management

All sewage treatment systems are capable of producing massive quantities of sludge, which is an active organic matter, which can rapidly turn septic if not further treated. Untreated sludge is a significant environmental and public health hazard. Treated and stabilised sludge is safe for reuse as soil conditioner or as top cover for landfill sites.

It is estimated that Malaysia generates 3.2 million cubic meters of sludge annually. However, appropriate facilities for sludge treatment and disposal is limited. At present, existing sewage treatment plants with excess capacity are being used to treat septage from septic tanks. It is estimated that by year 2020, Malaysia will be producing approximately 7.2 million cubic meter of domestic sludge per year. More new sludge treatment and disposal facilities will be required to manage this large volume of sludges. Figure 11.3 gives an idea of the volume of sludge produced in Malaysia over the next 20 years.

Decentralisation must address sludge issues in a very strategic and practical way. Steps taken by Malaysia to resolve these issues could be a 'model' for others in need.

11.6.5 Sewerage guidelines and standards

Prior to 1993, the Environmental Quality Act (EQA) 1974 and the Code of Practice for Design and Installation of Sewerage Systems 1991 (Malaysian Standard 1228) were the main guideline standards for sewerage development in Malaysia. The EQA provides standards for effluent discharge into the environment as discussed in Chapter 3, whilst the MS1228 which was introduced in 1991 to provide system designs.

The introduction of the Sewerage Services Act (1993) paves the way for the development of more standards and regulations for use by the local sewerage industry.

The Environmental Quality Act 1974 specifies 2 standards for sewage effluent as follows: (a) Standard A for discharges upstream of raw water intake (b) Standard B for discharges downstream of raw water intake. These standards are interpreted as absolute standards, which are not to be exceeded at any point and enforcement would be based upon grab samples only.

Consent or wastewater discharge standards are established for the sole purpose of regulating the disposal of pollutants into a receiving water course to enhance and protect its identified economic resource potentials, by supporting viable aquatic ecosystems, and by protecting public health and welfare. The stringent standards have been influenced by the fact sewage discharges are accountable for the largest amounts of biodegradable organic matter, suspended solids and ammoniacal nitrogen imposed on the nation's waterways.

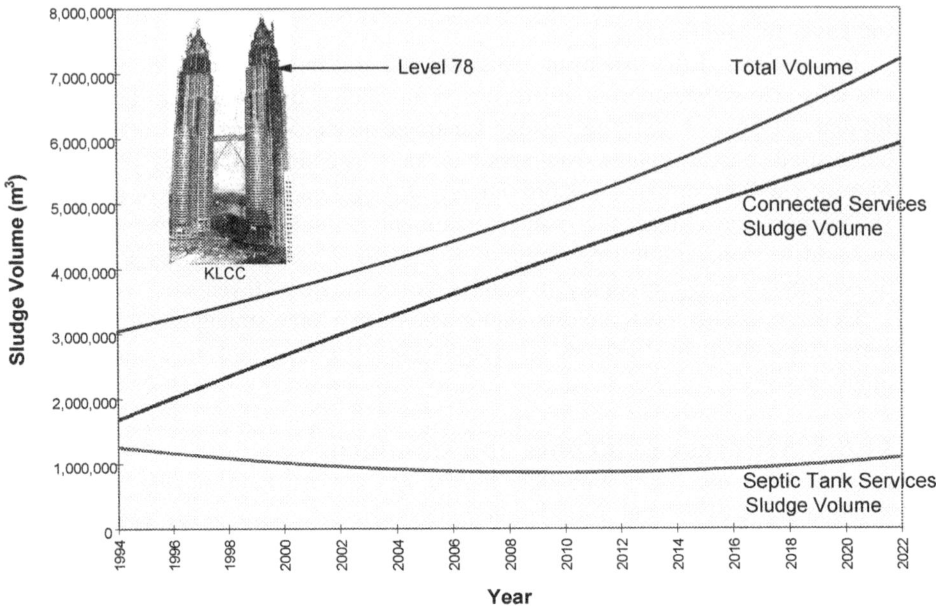

Figure 11.3 Estimated sludge quantities in Malaysia.

The current Standard A and Standard B are absolute limits (not average limits as commonly adopted) when designing treatment plants. New sewage treatment plants are designed to produce average effluent qualities below those of standard figures to guarantee almost 100% compliance. For example to achieve Standard A effluent of 20mg/L BOD_5 consistently, the designer must design the plant for a final effluent BOD_5 of 10 mg/L. Most of the existing sewage treatment plants are not in compliance with the standards due to their original designs that are unable to provide sufficient treatment for removal of BOD and SS.

11.6.6 Application of innovative technology management

The current scenario of sewerage infrastructure in most developing countries, including Malaysia require continuous improvement by adopting sustainable, innovative technology and proactive policy. Amongst the areas for innovative technology to be considered can be summarised as follows:

- Improvement of septic tanks and Imhoff tanks to meet more stringent regulatory effluent standards

- Understanding and applications of microbial culture and behaviours in hot climate environment
- Sustainable sludge recycling and wasting techniques
- Optimisation of aeration requirements and techniques
- Nutrient removal and recycling technologies
- Upgrading of existing plants using better treatment technologies
- Plant retrofit and optimisation of operational performances
- Utilisation and process optimisation of the appropriate microbes and nutrients
- Desludging and dewatering techniques
- Introduction of hybrid systems
- In-sewer processes and treatment capability, including denitrification
- Water recycling and waste reduction technologies
- Compact and reliable small-footprint-high-performance sewage treatment plant for environmentally-sensitive areas
- Integration of advanced control and automation technologies and operations

11.7 INSPECTION AND EFFLUENT QUALITY MONITORING

There is a need to regularly inspect the facilities and monitor the effluent quality from sewage and sludge treatment plants. Inspection of individual septic tanks, pit latrines or pour-flush latrines is normally conducted by Health Inspectors in most developing countries. This is conducted due to the fact that sanitation is under the jurisdiction of the health agency. In many developed countries, particularly in Japan and Scandinavia, effluent from *johkasou* and individual septic tanks are also regularly inspected and monitored by contractors to Local Authorities. This is to ensure that the treatment processes are meeting certain standards and comply the regulations set by the authority. Inspection and monitoring are also conducted to ensure the system reliability and quality control.

The initial inspection for individual septic tanks or other individual on-site systems is normally conducted to document the systems: types and system components, dimension and capacity, installation and commissioning dates, contractors and O&M requirements. Samples of effluent are normally collected to record the baseline quality.

The inspection of packaged plants for small communities is normally conducted before the commissioning stage by the sewerage regulator. In Malaysia, the inspection is conducted by the Department of Sewerage Services.

The developer or contractor will be informed, if there are deficiencies and defects on the structural, process or electrical systems, and appropriate remedial actions are required before a Certificate of Final Inspection can be issued. A standard checklist is normally used for inspection.

It is also important to note that monitoring is less extensive (and less expensive) than the assessment exercise. By definition, monitoring means collection of information at set locations and at regular intervals in order to provide the data, which may be used to define current conditions, establish trends and compliance level. On the order hand, assessment means the overall process of evaluation of the physical, chemical and biological nature of water in relation to natural quality, human effects and intended uses, particularly uses which may affect human health and the health of the aquatic system itself (Chapman 1996).

In order to minimise resources, particularly in effluent quality monitoring in developing countries, it is important to strategise the exercise. This is to optimise the short and long-term outputs to define the status and trends. Since monitoring is based on an extensive data collection, therefore specific objectives are vital in reducing the financial and analytical expenses. In general, objectives of monitoring, survey, assessments and surveillance exercises can be summarised as shown Table 11.9.

In many packaged plants used for small and decentralised wastewater systems, the effluent quality monitoring is coupled with an automatic control system. However most of the control systems maintain several parameters such as DO, SRT or clarifier sludge depth at fixed values which are not the parameters normally used in effluent quality standards. In Malaysia, for example, the typical parameters for design and monitoring are BOD, COD, SS and oil and grease. These parameters normally correspond directly to other important parameters, as shown in Table 11.9.

It is impossible to monitor all operational and effluent parameters on a regular basis for every small sewage treatment plants, and individual on-site treatment facilities. Therefore statistical methods, such as an experimental design, are required to be implemented to monitor the performance-related parameters to achieve specific objectives. From the experience in Malaysia, the most appropriate monitoring objective is *trend monitoring*, specifically to observe the long-term evolution of effluent quality in terms of concentrations and loads.

It is appropriate, cost-effective and legal to measure the parameters of small sewage treatment plants using composite sampling on a weekly basis. The results indicate the trend of the system performance as well as pollution loading to the receiving aquatic systems.

Table 11.9 Typical objectives of water quality assessment operations that are suitable for effluent monitoring exercise (adapted from Chapman 1996).

Type of operations	Major objectives or focus of water quality assessment
Common Operations	
Multipurpose monitoring	Space and time distribution of water quality in general
Trend monitoring	Long-term evolution of pollution (concentrations and loads)
Basic survey	Identification and location of major survey problems
Operational surveillance	Water quality for specific uses and related water quality descriptors (variables)
Specific operations	
Background monitoring	Background levels for studying natural processes; used as reference point for pollution and impact assessments
Preliminary surveys	Inventory of pollutants and their space and time variability prior to monitoring programme design
Emergency surveys	Rapid inventory and analysis of pollutants, rapid situation assessment following a catastrophic event
Impact surveys	Sampling limited in time and space, generally focusing on few variables, near pollution sources
Modelling surveys	Intensive water quality assessment limited in time and space and choice of variables, for example, eutrophication models or oxygen balance models
Early warning surveillance	At critical water use locations such as major drinking water intakes or fisheries; continuous and sensitive measurements

11.8 UPGRADING STRATEGY

Upgrading of existing sewage treatment plants or individual on-site treatment systems may become necessary for a variety of reasons. The main reason for small sewage treatment plant upgrading is growing population in many urban areas which has resulted in a significant increase of generated amount of wastewater. Increase in population served, or the desire to serve additional areas, may result in the need to increase and optimize the capacity of an existing treatment facility.

Obsolete equipment and technology may need upgrading to replace existing equipment that no longer functions as intended or to allow installation of brand new equipment, more efficient and cost-effective. Upgrading sewage treatment plants could reduce the operating costs because of the latter.

Many countries including Malaysia, have through legislation imposed more stringent effluent standards to meet the effluent quality standard and as a result of that, upgrading of sewage treatment plants is needed. In the larger context, Table 13.13 explains various reasons for sewage treatment plants upgrading, with the objectives to be achieved.

On-site systems also require upgrading and improvement. It is mainly due to increase in PE served or obsolete equipment. Septic tanks that have never been desludged require extensive overhaul. Often the chambers need to be reconstructed or new septic tanks constructed.

Table 11.13 Upgrading and objectives of sewage treatment plants.

Reasons	Objectives
Serve new areas or increase PE	-To increase and optimize plant capacity
More stringent effluent standards	-For example, Standards A or B in Malaysia as shown in Table 3.1 -Nutrient removal -Pathogen removal for effluent reuse -New effluent quality standards
Poor design criteria	-To optimize plant performance, capacity and reliability -To improve desludging exercise -To remedy the poorly designed compartments or facilities
Obsolete equipment/technology	-To increase and optimize treatment performance and reliability -To adopt a new improved treatment system
Odor problems	-To reduce H_2S production -To adopt and optimize non-odour associated system

On-site systems also require upgrading and improvement. It is mainly due to increase in PE served or obsolete equipment. Septic tanks that have never been desludged require extensive overhaul. Often the chambers need to be reconstructed or new septic tanks constructed.

11.9 RATIONALISATION STRATEGY

11.9.1 Background

The process of improving the condition and performance of the existing sewage treatment facilities in Malaysia has been accelerated since 1990s. In conjunction with the activities needed to achieve these improvements, it is also required to increase the coverage of the sewerage systems by increasing the number of households connected to sewerage network systems. This increased coverage will require connection of some individual and communal septic tanks, which at present discharge directly to the environment.

Malaysia has a refurbishment program continuously undertaken. In order to enable the improvement in performance to meet the specified effluent quality standards and increase the number of household connected to systems, it is necessary to rationalise some of the existing sewerage systems.

It is anticipated that a significant reduction from the present 8000 sewage treatment plants will need to occur as part of the refurbishment program. It is proposed to use the rationalisation method to make available a portion of the funds, which have been allocated for refurbishment, for upgrading of selected sewage treatment plants and abandoning others. The rationalisation concept forms part of the refurbishment strategy to maximise the technical and financial benefits.

These upgrades will provide concurrent increase in quality and capacity, thus allowing abandonment of a large number of plants. The sewage that is currently treated at plants identified for abandonment will then be transferred by new sewers to the central plants, which will have undergone refurbishment, upgrade and in some cases additional amplification.

Under the normal refurbishment program, there are five key steps, which are identified as work categories of refurbishment as follows:

Category 1	-	Process Upgrade
Category 2	-	Safety and Security
Category 3	-	Mechancal Retifications
Category 4	-	Electrical Retifications
Category 5	-	Aesthetics and Environment

11.9.2 Introduction

The general philosophy adopted for this initial methodology has been that the majority of sewage treatment plants (whether to be abandoned or upgraded) will be refurbishment for Categories 2 and 5. This will entail safety and security plus cleanliness and aesthetic type works to ensure minimising the potential for impact on the community. Assumptions have been made regarding the following issues:

- the cost of laying a normal 300 mm gravity main
- the cost of refurbishment categories 1,3 and 4 for Imhoff tanks
- the cost of upgrade, including refurbishment of Imhoff tanks to extended aeration facilities, thus providing a 2.5 times capacity increase.

It was assumed that an abandonment and transfer was financially viable if the money that could be saved by not carrying out the Categories 1, 3 and 4 refurbishment, at a site to be abandoned, could be applied to the cost of laying the transfer piping to the proposed centralised plants and the refurbishment and upgrade of the centralised plants.

In summary the philosophy was that the capital invested in the rationalised system was not to exceed that which was to be spent on refurbishment of the existing decentralised plants. The benefits of the rationalisation are as follows:

- Reduced capital outlay
- Reduced operational costs
- Decrease in land usage
- Reduced odour /noise sources
- Increase effluent quality
- Increase in connection coverage

11.9.3 Description of rationalisation method

The method to determine the initial system rationalisation method is generally summarised in seven steps, as follows:

STEP 1 Define Potential Grouping for Rationalisation of STPs

Choose a logical sewer catchment area defined by topographical barriers such as: Main roads, rivers, canals, ridgelines, etc. Then within the sewer catchment area, formulate possible groupings of treatment plants that are within approximately 250m to 300m of the nearest plant within the group. Thus a grouping of 3 plants, which were along a linear character could each be up to 300m from the next in line, i.e. plant 1 could be up to 600m from plant 3. When selecting groupings and transfer pipe routes between sewage treatment plants, it is essential to consider the potential to connect to the sewer main. If circumstances prevent immediate connection of nearby communal and individual septic tanks that are identified, then junctions could be installed in the main. These junctions should be capped and marked for future connection. The following points must be taken into account when considering the selecting of possible transfer sewer routes:

- The sewer must be able to be laid in a minor road or other accessible route, which has minimum disruption to the community.
- The sewer must be, in most cases in a group of sewage treatment plants, able to be laid in such a manner as to transfer flow by gravity to the new central location.

- The receiving point must be either the inlet to the new centralised plant or the receiving pump well of the existing feed pump station that services the proposed centralised plant.
- The capacity of interconnecting sewers and the receiving (centralised plant) shall, if possible, be adequate to accept additional (previously unconnected) communal and individual septic tanks flow.

STEP 2 Choose Central STP

Once possible groupings are identified a review of field survey data on each plant must be carried out. This review must consider the following factors:
- Size
- Condition
- Elevation
- Topography between sewage treatment plants
- Rated capacity
- Type of sewage treatment plant

From this review, judgement must be used to select the best choice for a centralised sewage treatment plant. The ideal selection for a centralised plant would be:
- The largest sewage treatment plant in the group
- The sewage treatment plant in best condition
- Farthest from the adjacent houses
- Easy access by vehicle for maintenance
- Largest site area
- Able to accept gravity supply from other sewage treatment plants in the grouping
- Able to give greatest capacity increase in upgraded state i.e. a waste stabilisation pond plant would be preferable to an Imhoff tank or mechanical plant

STEP 3 Calculate Upgrade Capacity of Centralised Plant

If the potential centralised sewage treatment plant is a site of an Imhoff tank, calculate the upgraded (new) capacity by multiplying the original rated capacity by 2.5. If the potential central sewage treatment plant is a waste stabilisation pond, multiply the original rated capacity of the facility by 6. If the potential central plant is an aerated lagoon multiply the original rated capacity by 3.5. If the potential central plant is a mechanical plant a more detailed review of the potential for upgrade and capacity increase would need to be calculated on the basis of the size of the unit operations (system components) at the plant in question, i.e. a case by case review would be needed.

STEP 4 Confirm Rationalised Grouping of Sewage Treatment Plants

Once the potential upgraded capacity of the proposed centralised plant has been calculated a check of the total load, proposed to be delivered from those plants, which are to be abandoned, must be made. The sum total of the original rated PE capacity of all the plants in the group must not exceed the calculated capacity of the upgraded centralised plant. The load of any individual and communal septic tanks, which will also be added to the system, must also be added to the total of the rated PEs of the plants in the group.

STEP 5 Adjusting Group Size

If the total of the PEs connected to the group plus any potential new individual and communal septic tanks connections exceeds the new upgraded centralised plant capacity, then a reduction in the proposed group size may be required. The plants that are either pumped or have the longest transfer lines should be removed from the group first. Adjacent groups of sewage treatment plants should be modified to manage maximum sewage capture at least cost. Considerations could be given to amplifying (beyond the upgrade) the central plant rather than removing plants from the group.

STEP 6 Confirm Financial Viability

Once the practical groupings have been finalised in Step 5 a check of the financial viability of the proposed rationalisation must be done. This check consists of comparing, the estimated cost of carrying out Category 1, 3 and 4 refurbishment on all plants in each group, with the sum of the costs of refurbishment and upgrade of the proposed central plant plus the cost to lay all transfer sewers. If the cost of the rationalised system exceeds the total cost of refurbishing all the individual plants, then different grouping must be tested to achieve the least cost combination. If a particular plant is unable to be attached to any group within the financial constraint described above then the cost of upgrading the single plant to meet the effluent standard must be checked against the cost of linking that plant to another group (that has adequate capacity, to accept that plant). If the cost to link to the group is greater than the cost to upgrade, the plant must be upgraded and become a single plant group.

STEP 7 Submit Recommendations

When the final groupings are formed and routes selected within a specific sewer catchment area, recommendation is made nominating the preferred configuration for the grouping of plants. The submission will be in the form of a

'Business Case' showing all the recommended groupings. Maps and plans showing names and locations of all plants and proposed transfer pipe routes will also be provided. Approval of the specific rationalisation plan is required before proceeding to detailed planning design prior to calling tenders and following the rationalisation acquisition process.

11.10 FUTURE CHALLENGES FOR MALAYSIA

There are many challenges ahead for the smooth implementation of sewerage projects in developing countries. Some of the challenges include:

Future Treatment Technology Based on experiences in developed countries, developments in sewerage systems have led to stringent effluent quality. It is expected that a stage will be reached whereby stringent effluent quality will be imposed in the future. It is a challenge for developing countries to upgrade the existing plants considering the number of plants to be upgraded as most current systems may not be easily upgraded or retrofitted with new treatment technologies to meet future standards.

Lack of Local Expertise Many countries have experienced obstacles in terms of local expertise in sewerage system designs. Most local consultants are not fully competent in handling the design works and presently they are quite dependent on foreign consultants, which delays the delivery of the projects. The same can be said of the local contractors and equipment suppliers. The other problem, which may be encountered in the future, is lack of operational and maintenance skills of operators and technicians. At the moment, there are not many mechanised plants in operation, and most of the existing plants are similar compared to the future plants to be built.

Insufficient of Regulated Sewerage Guidelines and Standards Currently, the sewerage industry in many parts of the world is in dire need of more comprehensive regulated standards. In Malaysia for example, apart from the EQA, MS1228 and the Guidelines for Developers (SSD and MWA), there are no other regulated guidelines and standards to enforce the private sectors to design proper sewerage facilities. The Government will be introducing more prescriptive standards and guidelines over the next few years in order to ensure that standards in the local sewerage industry will be comparable to recognised standards worldwide.

Technological Development Although the imported technology has been well developed and can function under environmental and tropical conditions, its long-term optimal performance can only be optimised with local adaptations. To date the 'localisation' of the imported technology has not been carried out systematically.

Local Industrial Capability Though Malaysia is not ready yet to develop its own range of sewerage products and equipments, it is more than capable of fabricating a variety of products and equipment under foreign licences. Perhaps it will soon be able to perform reverse engineering and eventually turn out some equipment of its own to suit similar requirements and conditions found locally.

Operation and Maintenance Lack of manpower and experience can no longer be accepted as an excuse for not carrying out the responsibilities associated with operations and maintenance of sewerage systems. Private contractors capable of performing the tasks are now available locally. What is required now are adequate funds, proper planning, training and effective contract management.

Research and Development At present, most of the sewerage technologies in Malaysia have been "imported" from foreign countries. While nobody will deny the fact that such technology works at optimum level under the set of design assumptions and criteria currently used in this country, the situation is not without financial implications. Imposing unnecessary conservative design criteria will result in wasting of financial resources of the precious and limited funds channelled to this sector. In addition, the extra expenditure might not necessary result in better sewerage systems. Instead it may well cause numerous operations and maintenance problems and consequently higher recurrent cost. Better understanding of the performance and behaviour of various systems under the local conditions will also enable the professional to select the most appropriate sewerage system for a specific project. Again it will result not only in better system performance but also in much superior operating economies.

Bill Collections and Public Acceptance The success of implementing a modern and efficient sewerage system for the country finally depends on the public's participation in playing their roles especially with regards to paying their bills promptly. While the Government of Malaysia may be able to provide its people with the most up-to-date facilities, it also needs commensurate improvements in the quality of its social infrastructure. Unless effective measures are taken with those who install and maintain the facilities with the spirit of professional ethics, there will always be bottlenecks in all services in years to come.

Malaysia has now arrived at the stage where development of the last of its major urban infrastructures i.e. sewerage system, has been made possible by the country's economic progress. In order to achieve the key objective of providing proper sewage collection, treatment, operations, maintenance and disposal for all communities i.e. to sustain adequate public health protection, enhance environmental quality and preserve our water resources, the realisation of the following factors should pave the way forward towards success:

- All sewage including blackwater and greywater shall be treated prior to disposal
- All existing sanitary facilities such as pit and overhang latrines shall be replaced with sanitary systems
- All individual septic tanks shall be regularly or periodically desludged on a scheduled routine basis.
- Catchment strategies shall be developed to integrate all new and old sewerage systems
- All existing plants shall be refurbished to meet the original intended designs.
- Planned capital projects shall be implemented in order of priority.
- Operations and maintenance of all systems must be enhanced.
- Introduce standard concepts in design, construction and also procurement of equipment.
- Ensure quality at least cost of the sewerage asset - whole life cost concepts.
- Introduce integrated training and accreditation programmes including research and development.
- Public education and a vigorous communication campaign is important.
- Make available manpower and financial resources for all the above purposes.

REFERENCES

Chapman, D. (1996) Water Quality Assessments: A Guide to the Use of Biota, Sediments and Water in Environmental Monitoring, 2nd edn. F&FN Spon, London, UK.

Sewerage Services Department (1998) Sewerage Services Report 1994-97. Ministry of Housing and Local Council, Malaysia, Kuala Lumpur.

SSD & MWA (1999) Guidelines for Developers Vol.1: Sewerage Policy for New Developments. 2nd edn. Sewerage Services Department and Malaysian Water Association, Kuala Lumpur.

SSD & MWA (1999) Guidelines for Developers Vol.V: Septic Tanks. 2nd edn. Sewerage Services Department and Malaysian Water Association, Kuala Lumpur.

Mara, D. (1996) Low-Cost Urban Sanitation, John Wiley & Sons, New York, USA.

Ødegaard, H. (2001) Compact on-site treatment methods for communities – Norwegian experiences. In Decentralised Sanitation and Reuse (eds. P. Lens, G. Zeeman and G. Lettinga), IWA Publishing, London, UK.

United Nations World Water Development Report (2003) Water for People, Water for Life. The United Nations and Berghan Books, United Kingdom.

Ujang, Z. and Buckley, C.A. (2002) Water and wastewater in developing countries: Present reality and strategy for the future. Wat.Sci.Tech. **46**, 1-9.

Ujang, Z. (2004) Hazardous waste management in Malaysia. UNESCO Encylopedia of Life Support System for Hazardous Waste Management in Developing Countries. UNESCO, Paris (in press).

Wilderer, P.A. (2001) Decentralised versus centralised wastewater management. In Decentralised Sanitation and Reuse (eds. P. Lens, Zeeman G. and G. Lettinga), IWA Publishing, London, UK.

Crites, R.W. and Tchobanoglous, G (1998) Small and Decentralised Wastewater Management Systems. McGraw-Hill, Singapore.

West, Sarah (2001) Innovative Decentralised Sewage, Internet Conference on Ecocity Development, 2003.

Yang, X.M., Yahashi, T., Kuniyasu, K. and Ohmori, H. (2001) On-site systems for domestic wastewater treatment (johkasous) in Japan. In Decentralised Sanitation and Reuse (eds. P. Lens, G. Zeeman and G. Lettinga), IWA Publishing, London, UK.

Index

agricultural application 266
potable water standards 229, 234
World Water Development Report 2003
  (United Nations) 302
WSP *see* waste stabilization ponds

yellow water *see* urine
Yemen 171
Y_{obs}, activated sludge process *87*

zoning, Malaysia 316–21

IWA Publishing's authorised EU representative for General Product Safety Regulations is Diane D'Arras, 15 rue Duret, 75116 Paris, France, e-mail: safety@iwap.co.uk.

Printed and bound by CPI Group (UK) Ltd, Croydon, CR0 4YY
06/07/2026
02157563-0004